BIOLOGICAL AND MEDICAL PHYSICS, BIOMEDICAL ENGINEERING

BIOLOGICAL AND MEDICAL PHYSICS, BIOMEDICAL ENGINEERING

The fields of biological and medical physics and biomedical engineering are broad, multidisciplinary and dynamic. They lie at the crossroads of frontier research in physics, biology, chemistry, and medicine. The Biological and Medical Physics, Biomedical Engineering Series is intended to be comprehensive, covering a broad range of topics important to the study of the physical, chemical and biological sciences. Its goal is to provide scientists and engineers with textbooks, monographs, and reference works to address the growing need for information.

Books in the series emphasize established and emergent areas of science including molecular, membrane, and mathematical biophysics; photosynthetic energy harvesting and conversion; information processing; physical principles of genetics; sensory communications; automata networks, neural networks, and cellular automata. Equally important will be coverage of applied aspects of biological and medical physics and biomedical engineering such as molecular electronic components and devices, biosensors, medicine, imaging, physical principles of renewable energy production, advanced prostheses, and environmental control and engineering.

U.R. Müller D.V. Nicolau (Eds.)

Microarray Technology and Its Applications

With 123 Figures
Including 16 Color Plates

Springer

Uwe R. Müller, Ph.D.
V.P., Applied Science
Nanosphere, Inc.
4088 Commercial Avenue
Northbrook, IL 60062
USA
e-mail: umuller@nanosphere.us

Prof. Dan V. Nicolau
Swinburne University of Technology
533-545 Burwood Rd.
Hawthorn, Victoria 3122
Australia
e-mail: dnicolau@swin.edu.au

ISSN 1618-7210

ISBN 978-3-642-06182-0 e-ISBN 978-3-540-26578-8

Springer is a part of Springer Science+Business Media

springeronline.com

Printed in Germany

Cover concept by eStudio Calamar Steinen

Cover production: *design & production* GmbH, Heidelberg

Preface

It has been stated that our knowledge doubles every 20 years, but that may be an understatement when considering the Life Sciences. A series of discoveries and inventions have propelled our knowledge from the recognition that DNA is the genetic material to a basic molecular understanding of ourselves and the living world around us in less than 50 years. Crucial to this rapid progress was the discovery of the double-helical structure of DNA, which laid the foundation for all hybridization based technologies. The discoveries of restriction enzymes, ligases, polymerases, combined with key innovations in DNA synthesis and sequencing ushered in the era of biotechnology as a new science with profound sociological and economic implications that are likely to have a dominating influence on the development of our society during this century. Given the process by which science builds on prior knowledge, it is perhaps unfair to single out a few inventions and credit them with having contributed most to this avalanche of knowledge. Yet, there are surely some that will be recognized as having had a more profound impact than others, not just in the furthering of our scientific knowledge, but by leveraging commercial applications that provide a tangible return to our society.

The now famous Polymerase Chain Reaction, or PCR, is surely one of those, as it has uniquely catalyzed molecular biology during the past 20 years, and continues to have a significant impact on all areas that involve nucleic acids, ranging from molecular pathology to forensics. Ten years ago microarray technology emerged as a new and powerful tool to study nucleic acid sequences in a highly multiplexed manner, and has since found equally exciting and useful applications in the study of proteins, metabolites, toxins, viruses, whole cells and even tissues. Although still relatively early in its evolution, microarray technology has already superseded PCR technology not only in the breadth of applications, but also in the speed with which this evolution has taken place. Note that the literature dealing with microarrays has increased dramatically from its humble beginnings in the mid-nineties to reach more than 2000 articles and almost 300 reviews in 2004 alone (Fig 1). Although a saturation point may have been reached - not surprisingly given that there is still a limit to the number of laboratories that have access to this technology-its impact is truly remarkable, especially when compared, for example, to the emerging and much touted field of Nanotechnology.

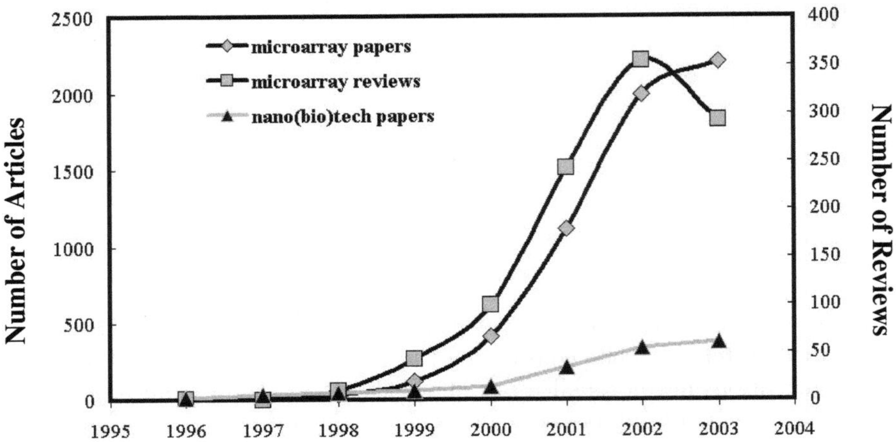

Fig. 1. Comparative evolution of publications regarding microarrays and nanobiotechnology

Amidst the pace of such rapid knowledge expansion, there is a challenge in trying to compose a book that does not face obsolescence by the time of its first publication. Alas, the breadth of this field is driving the growing knowledge base into many new directions, generating the need for different books at different levels and each with a different and unique focus.

As early participants in the development of microarray technology the editors have learned to appreciate the need for contributions from many different areas in the basic sciences and engineering that were crucial to its birth and continued healthy growth. In turn we have observed how the involvement in this particular scientific endeavour has affected many careers, turning physicist into oncologists, physicians into bioinformaticians, and chemists and biologists into optical engineers. Provided the diverse nature of backgrounds that are required to further propel this field, we thought it appropriate to aggregate this book around three aspects of microarray technology: *fundamentals*, designed to provide a scientific base; *fabrication*, which describes the current state of the art and compares 'old' and new ways of building microarrays; and *applications*, that are aimed to highlight only the amazing variety and options provided by these techniques. As an aid to the practitioner we have also asked the authors to provide a detailed method section wherever appropriate.

Part 1, General Microarray Technologies, opens with an overview on microarray formats. Chapters 2 and 3 cover the fundamentals of the physico-chemical aspects of immobilizing biomolecules on different substrates, while

Chaps. 4 and 5 describe the principal techniques used for array manufacture. Chapter 6 explores the limits of miniaturization with nanoarrays, and Chap. 7 illuminates various aspects of microfluidics for automation. Finally, Chaps. 8 and 9 deal with the principles of labelling and detection methodologies. The next parts are concerned with application of these fundamental techniques toward the development and use of specific types of microarrays. Part 2 describes DNA based microarrays in 4 chapters, covering SNP detection, high sensitivity expression profiling, comparative genomic hybridization, and the analysis of regulatory circuits. Part 3 contains 3 chapters that deal with microarrays for protein and small molecule detection, describing array technology for antibodies, aptamers, and lipid bound proteins, respectively. The final part comprises 4 chapters that introduce the most esoteric arrays, those that contain high information content in each feature (whole cells or tissues), and the capability of performing biological reactions, such as transfections. How the combination of these types of arrays generates new insights into the molecular basis of normal and malignant cell function is summarized in the last chapter.

It appears that given the dynamics of microarray technology any book would be a 'work in progress'. Rather than fighting this, the editors and the authors of this book embrace this concept: chances are that this book will grow in time in line with the new developments in microarray technology.

June, 2004

Uwe Müller
Dan Nicolau

Acknowledgements

The initial idea for this book emerged during a serendipitous meeting between the Editors and a representative from Springer Verlag during a Conference on Microarrays, Fundamentals, Fabrication and Applications that was chaired and organized by the Editors as part of the International Society for Optical Engineering (SPIE) Meeting in January 2001 in San Jose, CA. In fact, several Chaps. of this book were authored by people present at that Conference. The Editors wish to thank the organizers of SPIE, and in particular Marilyn Gorsuch and Annie Gerstl, who helped with the organisation of these Conferences in the last four years. Thanks also to the Conference co-Chairs, Ramesh Raghavachari and David Dunn. The Editors also wish to thank Peter Livingston and Gerardin Solana for the tedious work of converting the manuscripts into a camera-ready format.

Many contributors have specific acknowledgements.

The authors of Chap. 1 are grateful to Stephen Felder, Ph.D. and Richard Kris, Ph.D. of NeoGen, LLC. (Tucson, AZ), the inventors of the multiplexed nuclease protection assay, for proof-of-principle work on the mRNA assay and for the software for reagent design, image analysis and data interpretation.

The authors of Chap. 4 would like to thank Innovadyne Technologies for use of Fig. 4.5 and Peter Hoyt for helpful discussions. The research presented here was sponsored by the Laboratory Directed Research and Development Program of Oak Ridge National Laboratory (ORNL), managed by UT-Battelle, LLC for the U. S. Department of Energy under Contract No. DE-AC05-00OR22725 and by NIH Grant R01 HL62681-02. The manuscript has been authored by a contractor of the U.S.Government under contract DE-AC05-00OR22725. Accordingly, the U.S. Government retains a nonexclusive, royalty-free license to publish or reproduce the published form of this contribution, or allow others to do so, for U.S. Government purposes.

One of the authors of Chap. 6 (DVN) wishes to thank Dan V. Nicolau Jr. for discussions regarding the computational applications of nanoarrays.

The authors of Chap. 7 would like to thank Joe Bonanno and Dale Ganser (formerly Motorola Labs) for help in device fabrication, and Gary Olsen and

Pankaj Singhal (Motorola Life Sciences) for useful discussions on hybridization kinetics. This work has been sponsored in part by NIST ATP contract #1999011104A and DARPA contract #MDA972-01-3-0001.

Some of the authors of Chap. 8, i.e. JJS and SSM, acknowledge the NIH for support. CAM acknowledges the AFOSR, DARPA, and the NSF for support of this work.

The authors of Chap. 9 are very grateful to Gabriele Günther for excellent technical assistance, to Dr. K. Böhm, Jena, for kindly providing us with microtubules and kinesin samples, and to Dr. Wolf, PicoRapid GmbH Bremen, for help in spotting protein samples by an automatic arrayer.

The authors of Chap. 13 thank Dr. Tae Hoon Kim and Miss Sara Van Calcar for critical reading of the manuscript. We are also grateful to Drs. Hieu Cam, Yasuhiko Takahashi, Brian Dynlacht, Richard Young, and Mr. Tom Volkert for their help during the development of the technology described in this chapter. B.R. is supported by the Ludwig Institute for Cancer Research and a Sidney Kimmel Foundation for Cancer Research Scholar Award.

The authors of Chap. 14 wish to acknowledge the great support by Dr. Ronald Frank.

Finally, the authors of Chap. 20 thank Juha Kononen, Guido Sauter, Holger Moch, Lukas Bubendorf, Galen Hostetter, Ghadi Salem, John Kakareka and Tom Pohida for their contribution to the tissue microarray development, and Robert Cornelison, Don Weaver, Abdel Elkhahloon, and Natalie Goldberger and for their contributions to the cell microarrays.

Contents

List of Contributors

Heinrich F. Arlinghaus
Physikalisches Institut der
Universität Münster Wilhelm-
Klemm-Str. 10 D-48149
Münster, Germany
arlinghaus@uni-muenster.de

Paul Bao
Nanosphere, Inc.
4088 Commercial Avenue
Northbrook, IL 60062, USA
pbao@nanosphere.us

Mark Basik
Translational Genomics Research
Institute (TGen)
20 First Field Road
Gaithesburg, MA 20878, USA
mbassik@tgen.org

Ihab W. Botros
High Throughput Genomics, Inc.
6296 East Grant Road
Tucson, AZ 85712, USA
ibotros@htgenomics.com

Konrad Büssow
Max Planck Institute of Molecular
Genetics
Ihnestrasse 73
D-14195 Berlin, Germany
buessow@molgen.mpg.de

Natasha Caplen
Medical Genetics Branch
National Human Genome Research
Institute
National Institutes of Health
Building 10, Room 10C103
10 Center Drive
Bethesda, MD 20892 USA
ncaplen@mail.nih.gov

Alvaro Carrillo
Rensselaer Polytechnic Institute
Howard P. Isermann Department of
Chemical Engineering
Ricketts Building, 110 8th Street
Troy, NY 12180, USA
carria@rpi.edu

Linnette Demers
NanoInk, Inc.
1335 W. Randolph Street
Chicago, IL 60607, USA
ldemers@nanoink.net

Mitchel J. Doktycz
Life Sciences Division and Condensed
Matter Sciences Division
Oak Ridge National Laboratory
P.O. Box 2008
Oak Ridge, TN 37831-6123, USA
doktyczmj@ornl.gov

Andreas Eing
Nanofilm Technologie
Anna-Vadenhoeck-Ring 5
D-37081 Göttingen, Germany
aeing@nanofilm.de

Ye Fang
Corning, Inc.
Biochemical Sciences, Science and Technology Division
Corning, NY 14870, USA
fangy@corning.com

Luisa Filipponi
Swinburne University of Technology
Industrial Research Institute Swinburne
533-545 Burwood Road
Hawthorn, VIC 3122, Australia
lfilipponi@swin.edu.au

Anthony G. Frutos
Corning, Inc.
Biochemical Sciences, Science and Technology Division
Corning, NY 14870, USA
frutosag@corning.com

Mona Fredriksson
Uppsala University Hospital
Dept Medical Sciences
S-751 85 Uppsala, Sweden
Mona.Fredriksson@medsci.uu.se

Viswanadham Garimella
Nanosphere, Inc.
4088 Commercial Avenue
Northbrook, IL 60062, USA
vgarimella@nanosphere.us

Kunal V. Gujraty
Rensselaer Polytechnic Institute
Howard P. Isermann Department of Chemical Engineering
Ricketts Building, 110 8th Street
Troy, NY 12180, USA
gujratyk@rpi.edu

David S. Ginger
Department of Chemistry
University of Washington
Box 351700
Seattle, WA 98195-1700, USA
ginger@chem.washington.edu

Jörn Glökler
Max Planck Institute of Molecular Genetics
Ihnestrasse 73
D-14195 Berlin, Germany
gloekler@molgen.mpg.de

Chad Greef
SomaLogic, Inc.
1745 38th Street
Boulder, CO 80301, USA
chad.greef@somalogic.com

Karl-Otto Greulich
Institute for Molecular Biotechnology
Department of Single Cell and Single Molecule Techniques
Beutenbergstrasse 11
D-07745 Jena, Germany
kog@imb-jena.de

Piotr Grodzinski
Microfluidics Laboratory, PSRL, Motorola Labs
7700 S. River Parkway
Tempe, AZ 85284, USA
Current address:
Bioscience Division, MS J586
Los Alamos National Laboratory
Los Alamos, NM 87545, USA
piotrg@lanl.gov

Kristi L. Hanson
Swinburne University of Technology
Industrial Research Institute Swinburne
533-545 Burwood Road Hawthorn,
VIC 3122, Australia
khanson@swin.edu.au

Yulong Hong
Corning, Inc.
Biochemical Sciences, Science and Technology Division
Corning, NY 14870, USA
hongy@corning.com

Claus Hultschig
Max Planck Institute of Molecular Genetics
Ihnestrasse 73
D-14195 Berlin, Germany
hultschi@molgen.mpg.de

Anne Kallioniemi
University of Tampere
Laboratory of Cancer Genetics
Institute of Medical Technology
P.O. Box 607
FIN-33014 University of Tampere, Finland
anne.kallioniemi@uta.fi

Olli Kallioniemi
Medical Biotechnology Group
VTT Technical Research Centre of Finland
University of Turku
P.O. Box 106, 20521 Turku, Finland
Olli.Kallioniemi@vtt.fi

Ravi S. Kane
Rensselaer Polytechnic Institute
Howard P. Isermann Department of Chemical Engineering
Ricketts Building, 110 8th Street
Troy, NY 12180, USA
kaner@rpi.edu

Ravi Kapur
Anudeza Group
292 Morton Street
Stoughton, MA 02072, USA
rkapur@anudezagroup.com

Joydeep Lahiri
Corning, Inc.
Biochemical Sciences, Science and Technology Division
Corning, NY 14870, USA
lahirij@corning.com

Ralf Lenigk
Microfluidics Laboratory, PSRL, Motorola Labs
7700 S. River Parkway
Tempe, AZ 85284, USA
Current address:
Applied NanoBioscience Center
P.O. Box 874004
Arizona State University
Tempe, AZ 85287, USA
Ralf.Lenigk@asu.edu

Zirong Li
Ludwig Institute for Cancer Research
UCSD La Jolla Medical School Campus
9500 Gilman Drive
La Jolla, CA 92093-0653, USA
z3li@ucsd.edu

Peter Lichter
Molekulare Genetik
Deutsches Krebsforschungszentrum
D-69120 Heidelberg, Germany
p.lichter@dkfz.de

Ulrika Liljedahl
Uppsala University Hospital
Dept Medical Sciences
S-751 85 Uppsala, Sweden
Ulrika.Liljedahl@medsci.uu.se

Robin H. Liu
Microfluidics Laboratory, PSRL, Motorola Labs
7700 S. River Parkway
Tempe, AZ 85284, USA
Current address:

Applied NanoBioscience Center
P.O. Box 874004
Arizona State University
Tempe, AZ 85287, USA
Robin.Liu@asu.edu

Yingjie Liu
Microfluidics Laboratory, PSRL, Motorola Labs
7700 S. River Parkway
Tempe, AZ 85284, USA
Current address: Applied NanoBioscience Center P.O. Box 874004 Arizona State University Tempe, AZ 85287, USA
Jason_liu@asu.edu

Sudhakar S. Marla
Nanosphere, Inc.
4088 Commercial Avenue
Northbrook, IL 60062, USA
smarla@nanosphere.us

Ralph R. Martel
High Throughput Genomics, Inc.
6296 East Grant Road
Tucson, AZ 85712, USA
rmartel@htgenomics.com

Mirlacher Martina
University of Basel
Institute of Pathology
Schoenbeinstrasse 40
4031 Basel, Switzerland
Martina.Mirlacher@unibas.ch

Chad A. Mirkin
Northwestern University
Institute for Nanotechnology
2145 Sheridan Road
Evanston, IL 60208, USA
mirkin@northwestern.edu

Spyro Mousses
Translational Genomics Research Institute (TGen)
20 First Field Road
Gaithesburg, MA 20878, USA
smousses@tgen.org

Uwe R. Müller
Nanosphere, Inc.
4088 Commercial Avenue
Northbrook, IL 60062, USA
umuller@nanosphere.us

Michelle Nessling
Molekulare Genetik
Deutsches Krebsforschungszentrum
D-69120 Heidelberg, Germany
m.nessling@dkfz.de

Dan V. Nicolau
Swinburne University of Technology
Industrial Research Institute Swinburne
533-545 Burwood Road
Hawthorn, VIC 3122, Australia
dnicolau@swin.edu.au

Roeland Papen
Picoliter inc.
231 S Whisman Road,
Mountain View CA 94041
roeland.papen@picoliterinc.com

Bernhard Radlwimmer
Molekulare Genetik
Deutsches Krebsforschungszentrum
D-69120 Heidelberg, Germany
b.radlwimmer@dkfz.de

Ramesh Ramakrishnan
Nanosphere, Inc.
4088 Commercial Avenue
Northbrook, IL 60062, USA
rramakrishnan@nanosphere.us

Bing Ren
University of California, San Diego
Department of Cellular and Molecular Medicine, School of Medicine
9500 Gilman Drive, La Jolla, CA
92093-0653, USA
biren@ucsd.edu

Jan Roegener
University of Bielefeld
Department of Applied Laser Physics
Universitaetsstrasse 25 D3
D-33615 Bielefeld, Germany
roegener@physik.uni-bielefeld.de

Simon Ronald
University of Basel
Institute of Pathology
Schoenbeinstrasse 40
4031 Basel, Switzerland
Ronald.Simon@unibas.ch

Matthew P. Rounseville
High Throughput Genomics, Inc.
6296 East Grant Road
Tucson, AZ 85712, USA
mrounseville@htgenomics.com

Guido Sauter
University of Basel Institute of Pathology Schoenbeinstrasse 40 4031 Basel, Switzerland
Guido.Sauter@unibas.ch

Peter Schellenberg
Institute for Molecular Biotechnology
Department of Single Cell and Single Molecule Techniques
Beutenbergstrasse 11
D-07745 Jena, Germany
schelli@imb-jena.de

Carsten Schwaenen
Medizinische Klinik der Universität Ulm
Innere Medizin III D-89081 Ulm, Germany
carsten.schwaenen@medizin.uni-ulm.de

Bruce E. Seligmann
High Throughput Genomics, Inc.
6296 East Grant Road
Tucson, AZ 85712, USA
bseligmann@htgenomics.com

Drew Smith
SomaLogic, Inc.
1745 38th Street
Boulder, CO 80301, USA
drew.smith@somalogic.com

James J. Storhoff
Nanosphere, Inc.
1818 Skokie Boulevard
Northbrook, IL 60062, USA
jstorhoff@nanosphere.us

Hans Martin. Striebel
Institute for Molecular Biotechnology,
Department of Single Cell and Single Molecule Techniques
Beutenbergstrasse 11 D-07745 Jena, Germany
hms@imb-jena.de

Ann-Christine Syvänen
Uppsala University Hospital
Dept Medical Sciences
S-751 85 Uppsala, Sweden
Ann-Christine.Syvanen@medsci.uu.se

Matthias Vaupel
Nanofilm Technologie
Anna-Vadenhoeck-Ring 5 D-37081 Göttingen, Germany
mv@nanofilm.de

Gerald Walter
Biorchard AS
Nedre Skogvei 14
N-0281 Oslo, Norway
gerald@biorchard.com

Brian L. Webb
Corning, Inc.
Biochemical Sciences, Science and Technology Division
Corning, NY 14870, USA
webbbl@corning.com

Hendrik Weiner
Max Planck Institute of Molecular Genetics
Ihnestrasse 73
D-14195 Berlin, Germany
weiner@molqen.mpg.de

Swen Wessendorf
Medizinische Klinik der Universität Ulm
Innere Medizin III
D-89081 Ulm, Germany
swen.wessendorf@medizin.uni-ulm.de

Part I

General Microarray Technologies

1

Array Formats

Ralph R. Martel, Matthew P. Rounseville, and Ihab W. Botros,
and Bruce E. Seligmann

1.1 Introduction

Arrays have become an increasingly diverse set of tools for biological studies; their use continues to expand rapidly. Likewise, the underlying array technologies, formats and protocols continue to evolve. Investigators can choose from a growing range of options when selecting an array technology that is appropriate for reaching their research objectives. Traditionally, arrays have consisted of collections of distinct capture molecules – typically cDNAs or oligonucleotides – attached to a substrate – usually a glass slide – at predefined locations within a grid pattern [1, 2]. However, today's formats are more diverse and can be grouped into several categories. Like any categorization effort, there will be exceptions, crossover technologies and tangential relations. The intent here is only to lay out some general trends.

The classes of capture molecules used in arrays include not only DNA, but also proteins [3], carbohydrates [4], drug-like molecules [5], cells [6], tissues [7] and the like. Array formats vary in their architecture. For closed architecture arrays, the analytes that can be measured are preselected and locked-in during the manufacturing process. In contrast open architecture array technologies allow the set of measured analytes to be modified or allow new analytes to be discovered. Regardless of the architecture, various manufacturing technologies and various substrate materials and coatings are available as are numerous means of attaching capture molecules to substrates. A broad variety of commercially prepared arrays can be purchased. In some instances, the pre-defined grid has been eliminated and replaced with 'virtual arrays' of optically encoded beads [8] or of analyte-specific detection labels (e.g. e-Tags; www.aclara.com). Coupled with the diversity of arrayed molecules and array formats is the diversity of detection schemes that include fluorescence, luminescence, electrochemical detection, mass spectrometry, surface plasmon resonance and others.

In spite of the diversity of formats, all arrays share a common feature: Arrays allow multiplexed analyses, that is, arrays allow multiple tests to be performed simultaneously. This is the case both when many analytes are measured simultaneously in an individual sample and also when many samples are tested at one time for an individual analyte. For instance, DNA arrays can be used to determine the expression levels of thousands of genes in an individual biological specimen, while tissue arrays can be used to determine the presence of a specific antigen in hundreds of specimens in a single experiment. Various 'array–of–arrays' technologies combine the measurement of numerous analytes across numerous samples.

The impact of array technologies on the life sciences has been important. In conjunction with bioinformatic tools to process and analyze the large amounts of data they generate, arrays have spawned new approaches to systems biology often described with the 'omics' suffix: genomics, transcriptomics and proteomics, to name a few.

This chapter will provide the rationales for using arrays to address various scientific questions and will outline some of the array technologies developed to fill specific needs. This is a series of examples to illustrate the range of available options and how one technology may be better suited than another to reach a specific research objective, not a comprehensive survey of available tools. The latter part of the chapter will discuss the ArrayPlate™ technology developed by High Throughput Genomics (HTG, Tucson, AZ) to bring the benefits of arrays to the high throughput screening phase of the drug discovery and development process. The procedure for a multiplexed ArrayPlate™ mRNA assay will be described and the results of an mRNA assay and a companion multiplexed ELISA will be presented.

1.2 Reasons to Use Arrays

There are three principle justifications for using array technologies. Arrays serve to discover unique patterns (of gene expression, protein synthesis or post-translational modification, etc.) associated with a particular physiological state. We use the term 'survey array' to describe the technologies that are employed for this purpose. 'Scan array' or 'focused array' refers to the array tools that measure a predefined pattern, previously established with survey arrays. Finally, 'efficiency array' refers to the techniques that do not require multiplexing per se, but that take advantage of the parallel processing common to arrays to provide savings of effort, time and materials or to improve data quality by incorporating internal controls that are measured in each sample. Most array technologies have been developed to achieve one of these three goals and may be inefficient for reaching the other two.

1.2.1 Arrays to Identify Patterns

The best-known array technology, the GeneChip® developed by Affymetrix (Santa Clara, California) is an excellent example of a 'survey array'. According to the company (www.affymetrix.com), the two arrays in the Human Genome U133 Set contain over one million distinct oligonucleotide features to monitor the expression of 39,000 transcript variants of 33,000 different human genes in a single sample. GeneChips® and their cDNA and oligonucleotide array counterparts are widely used to identify genes that are differentially expressed in diseased tissues or during development or upon treatment with a drug. In most instances, results obtained with DNA arrays show that the vast majority of genes are either not expressed or not affected by disease. Typically, a disease-specific pattern of gene expression or 'signature' is characterized that involves fewer than 50 genes [9–12]. Although well suited to initially define patterns based on the examination of a relatively small number of samples, survey arrays are generally too labor- and material-intensive and too costly to be used routinely thereafter in diagnostics or in drug discovery.

1.2.2 Arrays to Measure Patterns

'Scan arrays' that measure specific patterns are appropriate for clinical diagnostics and for drug discovery. While these techniques measure fewer analytes than do survey arrays, the analytes have been carefully selected and validated. Other attributes such as ease of use and throughput make various scan array technologies well-suited for particular niches.

Inexpensive readout equipment is a requirement for array-based diagnostic tests as such tests are performed at many different sites such as reference laboratories, hospital laboratories and physicians' offices but relatively infrequently at any given site. Cost per test however is less important since the results provide information that is of high value. Furthermore, most diagnostic testing is reimbursed by insurers. Hands-on manipulations must be simple as testing is frequently performed by inexperienced personnel. To gain approval from regulatory agencies, diagnostics tests must yield results that are robust and interpretable. For these reasons, various hand-held electronic array devices appear to be in the best position to make inroads in this arena.

In drug discovery, once targets are validated, throughput becomes an important criterion, that is, how rapidly collections of hundreds of thousands of chemical compounds can be tested to identify those compounds that elicit a desired effect. Efficiency in the high throughput screening laboratory is obtained with miniaturization (96–, 384– and 1536–well microplates) and with extensive automation and plate handling robotics. Besides performance criteria such as sensitivity and reproducibility, the success of a technology in this setting depends upon the development of automation-friendly protocols. While substantial expenditures on capital equipment are commonplace, cost per sample is an issue because of the large testing volumes. The ArrayPlate™ described later in this chapter was designed specifically for high throughput screening.

1.2.3 Arrays for Parallel Processing

Examples where the array format has been adopted for the efficiencies derived from parallel processing can be found in the combinatorial chemistry literature [13]. The synthesis of chemical compound libraries has been performed in an array format [14]. Indeed, the photolithographic process utilized by Affymetrix to manufacture its DNA chips had its origins in combinatorial chemistry [15]. Arrays of compounds have also been used in drug discovery screening [16]. Microtiter plate wells that contained individual compounds have been miniaturized to the point of vanishing with the compounds becoming elements of an array rather than contents of a well. Generally, using arrays leverages sample preparation efforts. In cell-based assays for instance, the effort of culturing cells and screening compounds is the same regardless of whether a single or multiple measurements are made.

1.3 Arrays for Nucleic Acid Analysis

Several review articles covering advances and applications of DNA microarray technology have recently been published [17, 18] hence, the same material will not be repeated here. Oligonucleotide and cDNA arrays have different strengths and weaknesses. There is more control over the design of oligonucleotide microarrays than there is for cDNA arrays. Consequently, oligonucleotide arrays tend to have more uniform physicochemical characteristics and fewer issues pertaining to cross–hybridization. For cDNA arrays, the capture probes are typically PCR amplicons of clones derived from the organism or the organ of interest. One advantage is that cDNA probes can be incorporated into arrays without further characterization of the underlying gene. For both types of microarrays however, the architecture is closed, albeit at times unknown for cDNA arrays. For illustrative purposes, several less conventional array technologies are described.

1.3.1 Arrays on Beads

The attachment of array moieties to small particles allows multiplexed assays to be performed in three–dimensions rather than on a flat surface. Luminex (Austin, TX) has developed fluorochrome-coded microspheres that can be coated with various classes of ligands. During an assay, a sample is incubated with the beads in solution, allowing the analytes of interest to be captured by their corresponding bead-bound ligands. A fluorescently tagged 'reporter molecule' then labels the analyte species. For readout, beads are passed, single file, through a flow cytometry device where the fluorescent tags are illuminated by laser excitation. The resulting fluorescence of both the bead and the reporter molecule are quantified and decoded to yield the identity and quantity of the captured molecule. The application of this method to RNA expression analysis has been described recently [8].

Illumina (San Diego, CA) has developed an alternative readout system for bead-based arrays. A manifold of 96 fiber optic bundles, each consisting of about 50,000 individual fibers, is manufactured to fit the standard microplate format. A dimple etched at the end of each fiber can accommodate one of the company's 3 μm beads. This enables fluorochrome excitation and emission of the beads and of fluorescently-labelled analytes through the fiber. The company claims that combinations of fluorescent dyes uniquely identify up to 1,500 beads that can be sampled with 30–fold redundancy to provide a statistical average readout. Presently, the method appears to be used mainly in single nucleotide polymorphism (SNP) genotyping of multiple samples, as reviewed by Oliphant [19].

1.3.2 Electronic Arrays

Array technologies have used electronics to program open architecture systems, to accelerate hybridization kinetics and control stringency, and to detect captured analytes. The NanoChip® (Nanogen, San Diego, CA) incorporates 100 electrode test sites that are coated with a hydrogel containing streptavidin. This system has an open architecture. Programming is with biotinylated target–binding probes that migrate to specific electrodes when a positive charge is applied and that remain bound to the streptavidin afterwards. An electric field is also used to concentrate target molecules at the electrodes to accelerate their hybridization and subsequently, to drive away non-specifically bound materials. Final detection of target is by fluorescence. The eSensor™ DNA detection system (Motorola, Pasadena, CA) uses a self-assembled monolayer (SAM) array of target-specific 22–mer oligonucleotides covalently bound to the gold electrodes of a circuit board [20]. Target nucleic acids hybridized to the array are detected with ferrocene-labelled signaling probes that hybridize with their target next to the capture probe. An applied potential causes the transfer of electrons from the ferrocene to the gold electrode with the measured current quantifying the ferrocene label. SNPs can be detected as perfect hybrids that generate signals at least twofold greater than do single–base mismatches. Both of these technologies have targeted diagnostic applications.

1.3.3 SAGE

Serial analysis of gene expression (SAGE) allows the simultaneous detection and quantification of multiple mRNA species [21, 22] although it is not an array technology per se. SAGE relies on the isolation of unique sequence tags from individual mRNA molecules via a process that includes mRNA isolation, reverse transcription, restriction enzyme digestion, ligation and PCR amplification. The tags are subsequently ligated to form concatamers that are sequenced to reveal both the identity and abundance of expressed genes. Unlike conventional arrays, SAGE can identify novel transcripts.

1.4 Protein Arrays

The development of protein arrays has lagged behind that of DNA arrays primarily because of the greater complexity of proteins. While DNA microarrays have become the tools of choice for characterizing patterns of gene expression, two–dimensional gel electrophoresis remains the standard method for generating 'protein fingerprints'.

Multiplexed immunoassays are the most developed application for protein arrays. Three strategies have emerged. One is the miniaturization and multiplexing of the standard enzyme linked immunosorbent assay (ELISA), in which capture antibodies are arrayed onto slides or microtiter plates. A variation on this method that requires only a single antibody for each antigen, is to label the proteins in a sample with one fluorochrome and the proteins in a reference sample with a second fluorochrome. The differentially labelled samples are mixed and incubated with an antibody microarray which is scanned. The ratio of the two fluorescent dyes at each spot in the array corresponds to the relative concentration of each protein in the two samples [23]. Improvements in sensitivity and signal–to–noise ratio will be required for this methodology to become useful for measuring protein changes in biologically relevant samples. A third strategy, which may be particularly useful for diagnostic assays, is to prepare arrays of antigens. Such arrays allow samples to be tested for the presence and the titer of antibodies to particular antigens. This approach lends itself to develop broad–spectrum tests for certain autoimmune diseases and for exposure to infectious agents. As for nucleic acids, bead arrays also lend themselves to proteomic applications.

The technological challenges that remain are the development of specific, high affinity ligands that can be produced on a large scale and in a relatively short time. Distinguishing between various post-translational modifications, such as phosphorylation and amidation, are also technical features that need to be addressed. It is likely that different types of protein arrays will be required for cataloging the proteome, detecting differences in expression, and for screening compounds. For a more extensive review on the development of protein-detecting microarrays and related devices see Kodadek [24] and Schweitzer [3].

The development of arrays of functionally active proteins such as enzymes and receptors is progressing rapidly and the significant advances in this area are the topic of Chaps. 14–16 in this book.

1.5 The ArrayPlate™

HTG developed the ArrayPlate™ as a platform technology with an open architecture to conduct a variety of multiplexed assays in microtiter plates. The goal was to extend the capabilities and information content of conventional drug discovery and development assays for two purposes. The first was to provide a technology to allow genomic and transcriptomic efforts to progress from target discovery to drug discovery, that is, from the description of disease-specific signature patterns of gene expression to the identification of signature-modulating compounds. How the multiplexed ArrayPlate™ mRNA assay achieves this is discussed. The second purpose was to provide screening laboratories with another means to increase their efficiency as multiplexing is synergistic with both automation and miniaturization to enhance productivity. The multiplexed ELISA serves as an example for this. ArrayPlate™ assays rely on a single hybridization to transition from an open to a closed architecture. The benefits of this hybridization step, termed "reagent programming", that modifies the binding specificity of each element in a universal array, will be outlined. For the mRNA assay, a multiplexed nuclease protection assay is combined with the capture of processed nuclease protection probes on the array. Enzyme-mediated chemiluminescent detection subsequently quantifies probes in the mRNA assay and antigens in the multiplexed ELISA.

1.5.1 Materials and Methods

ArrayPlate™ Manufacture

The 96–well ArrayPlates™ contained at the bottom of each well of flat-bottom poly-styrene microtiter plate (Falcon™) modified with N–oxysuccinimide ester, a four–by–four array of 16 distinct oligonucleotide elements 100 μm in diameter and spaced 800 μm on center. Each of the 16 anchor oligonucleotides incorporated a unique 25–mer sequence and was 3′-modified with heptylamine. Arrays were printed with a PixSys 3000 microarrayer equipped with 85 μm inner diameter ceramic dispensing tips (Cartesian Technologies, Irvine, CA) in an environmental chamber (26°C and 80% relative humidity).

Oligonucleotides and Antibodies

The 16 target human mRNA species each required three oligonucleotides: A nuclease protection probe, a programming linker and a detection linker. These oligonucleotides were designed using ArrayPlate™ Oligo v.3.0 software (HTG, Tucson, AZ) and synthesized (Epoch Biosciences, San Diego, CA and Sigma–Genosys, The Woodlands, TX) as detailed elsewhere [25]. The 16 genes examined were glyceraldehyde 3–phosphate dehydrogenase (GAPDH), interleukin–1β (IL–1β), tumor necrosis factor–α (TNF–α), tubulin, cathepsin G (catG), cyclooxygenase–2 (cox–2), granulocyte colony stimulating factor (G–CSF), granulocyte macrophage colony stimulating factor (GM–CSF), glutathione S–transferase Pi–1 (GST Pi–1), high mobility group 17 (HMG–17), cyclophilin (cyclo), β–thromboglobulin (bTG), lactate dehydrogenase (LDH), tissue inhibitor metalloprotease 1 (TIMP–1), matrix metaloproteinase 9 (MMP–9) and β–actin.

Briefly, each programming linker was a 50–mer comprising a 5′ 25–mer complementary to one of the 16 anchor oligonucleotides and a 3′ 25–mer complementary to one of the 16 target-specific nuclease protection probes. Each nuclease protection probe was a 65–mer composed of a 50–base sequence with 48% to 52% GC content, complementary to the target mRNA. Each protection probe also incorporated a target-independent 15–mer control sequence. Each detection linker oligonucleotide was a 50–mer designed with a common 3′ 25–mer sequence and a unique 5′ 25–mer complementary to the 5′–terminal 25–mer of the corresponding nuclease protection probe. Finally, a detection conjugate of horseradish peroxidase labelled with the 25–mer sequence complementary to the common 3′-end of all detection linkers was used to generate a luminescent signal.

All oligonucleotides were tested before use in an assay by means of a design of experiments protocol that ensured that each oligonucleotide hybridized as intended without showing unintended and interfering binding. The behavior of individual oligonucleotide species was deduced from the observed behavior of predefined oligonucleotide mixtures.

For the antibody assays, ELISA-ready antibody sets, recombinant antigen standards and streptavidin–peroxidase were obtained from R&D Systems (Minneapolis, MN).

Cell Culture and Treatments

The human THP–1 acute monocytic leukemia cell line (ATCC, Manassas, VA) was grown in either T–175 culture flasks or in 96–well V–bottom cell culture plates (Falcon) at 37°C with 4% (v/v) CO_2 and 80% relative humidity in RPMI 1640 medium supplemented with 10% (v/v) fetal bovine serum, 100 U/ml penicillin and 100 µg/ml streptomycin (Hyclone, Logan, UT). Phorbol merystil acetate (PMA) treatment (0.1 µg/ml in RPMI for 48 hours) caused the cells to differentiate to adherent monocytes.

Cells activation was induced with four hours of treatment with 1 µg/ml bacterial lipopolysaccharide (LPS) (Sigma, St. Louis, MO) in culture medium. Dexamethasone (Sigma, St. Louis, MO) treatments were with compound dissolved at various concentrations in culture medium. Cells growing in suspension in microtiter plates were harvested by centrifugation at 180× g for 5 minutes (GS15, Beckman Coulter, Fullerton, CA). Removal of culture medium from cell pellets and from adherent cells in wells was by aspiration.

Multiplexed Nuclease Protection Assay

All reagent additions were performed with a 96–channel Biomek FX automated pipettor (Beckman Coulter, Fullerton, CA). Media-free THP–1 cells in 96–well culture plates received in rapid succession 30 μl/well lysis solution (HTG, Tucson, AZ) that contained each of the 16 nuclease protection probes at 30 pM and 60 μl/well mineral oil (Sigma, St. Louis, MO). The plates were incubated for 10 minutes at 95°C, for 6 hours at 70°C and were allowed to cool to room temperature for 10 minutes. The plates received 20 μl/well S1 nuclease solution (50 S1 units in 1.4 M sodium chloride, 22.5 mM zinc sulfate, 250 mM sodium acetate, pH 4.5) (Promega, Madison, WI) and were incubated for 30 minutes at 50°C. The plates received 10 μl/well 1.6 M sodium hydroxide, 135 mM EDTA and were heated for 15 minutes at 95°C. After cooling at room temperature for 15 minutes, the plates received 10 μl/well Neutralizing Solution (1 M HEPES, pH 7.5, 1.6 M HCl, 6× SSC). For each well, 60 μl of the 70 μl aqueous subphase was transferred from the cell culture plate to a programmed (i.e. programming linker-modified) ArrayPlate™, followed immediately by the additional transfer of 60 μl of aqueous subphase and overlayering oil.

Reagent Modification of Universal Arrays

The washing of ArrayPlates™ was completed in 60 seconds with a 96–channel plate washer (ELx405 Auto Plate Washer, Bio–Tek Instruments, Minooski, VT) and consisted of six dispenses and aspirations of 300 μl/well 1× SSC (150 mM sodium chloride, 15 mM sodium citrate, pH 7) with 0.1% (v/v) Tween–20 (Sigma, St. Louis, MO).

Following a wash cycle, the ArrayPlates™ received 50 μl/well programming linker solution that consisted of each of the 16 programming linker oligonucleotides at 5 nM in SSCS (1× SSC, 0.1% (w/v) SDS). After a one-hour hybridization at 50°C, the ArrayPlates™ were washed again. These were programmed (i.e. programming linker-modified) ArrayPlates™.

Capture and Detection of Protection Probes on the ArrayPlate™

Programmed ArrayPlates™ containing nuclease protection-processed cell lysates were incubated overnight at 50°C and washed. The ArrayPlates™ received 50 μl/well detection linker solution that contained each of the 16 detection linker oligonucleotides 5 nM in SSCS. The plates were incubated for one hour at 50°C and washed. Next, the ArrayPlates™ received 50 μl/well detection enzyme conjugate solution and were incubated for 30 minutes at 37°C followed by a wash. Detection enzyme conjugate solution contained 10 nM detection enzyme conjugate in SSCS. The ArrayPlates™ received 50 μl/well chemiluminescent peroxidase substrate (Atto–PS™ Lumigen, Southfield, MI) and were imaged from the bottom with an Omix CCD imager (HTG, Tucson, AZ) for 30 seconds to 6 minutes, depending on signal intensity, within 30 minutes of substrate addition.

Image Analysis

Digital images of ArrayPlates^{TM} were analyzed with software (ArrayPlate^{TM} Fit v.3.31a, HTG, Tucson, AZ) that extracted luminescence intensity data for each array element in a plate. The resulting data were exported as comma-separated value (CSV) files that were processed further with software (ArrayPlate^{TM} Crunch, HTG, Tucson, AZ) that allowed manipulation of the intensity data, for instance, to normalize signals within arrays to any combination of array elements. Intensity data CSV files were also imported into Excel spreadsheets (Microsoft, Redmond, WA) for further analysis.

1.5.2 Results and Discussion

Reagent Programming of Universal Arrays

The 96–well ArrayPlates^{TM} contain the same universal array of 16 distinct elements printed at the bottom of each well. Each element consists of a position-specific, covalently bound 'anchor' species that incorporates an oligonucleotide 25–mer recognition feature. Since identical arrays are printed across all wells of all plates, the manufacture of ArrayPlates^{TM} is standardized and subject to rigorous quality control procedures.

In spite of this standardized production, ArrayPlates^{TM} provide an open architecture to allow customized assays: A 'reagent programming' hybridization immobilizes specific capture reagents at preselected positions in the universal array. This is achieved using a cocktail that contains 16 bifunctional 'programming linker' species. Each programming linker contains both an oligonucleotide complementary to a specific anchor and an analyte-specific region. Thus, the hybridization of linkers to anchors immobilizes analyte-specific reagents at predetermined positions within the array (Fig. 1.1, top left panel).

Reagent programming provides versatility. The analyte-specific region of a programming linker can be an oligonucleotide, a peptide, a protein or a chemical compound, depending upon the type of assay that is to be performed: Programming linkers that consist of antibody conjugated to anchor-binding oligonucleotide are suited for multiplexed ELISAs or for setting up arrays of antigens. Programming linkers that have two oligonucleotide regions serve to capture target RNA, DNA or oligonucleotides. Conjugates of anchor-binding oligonucleotide and substrate peptides can be used for instance, for multiplexed kinase and phosphatase assays. With reagent programming, different combinations of assay capacity versus content become possible. For example, the user can program all the wells in a plate identically to measure 16 targets per sample across 96 samples. Alternatively, by programming arrays in pairs and splitting samples across two wells, 32 targets (16×2) can be measured in 48 samples (96÷2).

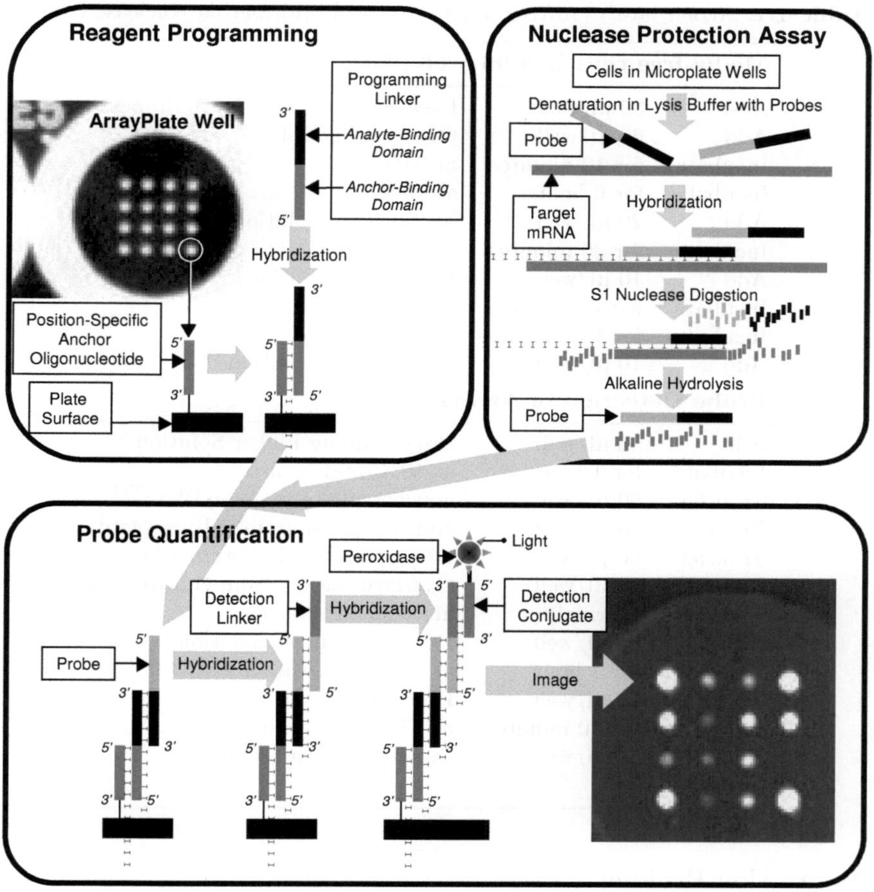

Fig. 1.1. ArrayPlate^{TM} mRNA Assay Principles. Upper left: Reagent programming modifies the binding specificity of each array element via the hybridization of a bifunctional programming linker to an anchor oligonucleotide. Upper right: A multiplexed nuclease protection assay preserves a stoichiometric quantity of oligonucleotide probe while destroying target mRNA. Bottom: Probe surviving the nuclease protection assay is immobilized by hybridization with its corresponding array-bound programming linker. Probe bound to the array in this manner is labelled in succession with detection linker oligonucleotide and peroxidase-containing detection conjugate. The light generated upon the addition of chemiluminescent peroxidase substrate is imaged with a CCD camera

Table 1.1. ArrayPlate^{TM} mRNA Assay Protocol Multiplexed Nuclease Protection

Media-free cells in a 96–well plate		
Add	30 µl/well	Lysis Solution with NPA Probes
Add	60 µl/well	Overlayering Oil
Incubate	for 10 minutes	at 95°C
Incubate	for 6 hours	at 70°C
Add	20 µl/well	S1 Nuclease Solution
Incubate	for 30 minutes	at 50°C
Add	10 µl/well	Hydrolysis Solution
Incubate	for 15 minutes	at 95°C
Incubate	for 15 minutes	at RT
Add	10 µl/well	Neutralizing Solution
Probe Detection in ArrayPlate^{TM}		
Add	50 µl/well	Programming Linker Solution
Incubate	for 1 hour	at 50°C and wash
Transfer	60 µl/well	aqueous phase to ArrayPlate^{TM}
Receive	60 µl/well	aqueous phase from culture plate
Transfer	60 µl/well	Overlayering Oil to ArrayPlate^{TM}
Receive	60 µl/well	Overlayering Oil from culture plate
Incubate	overnight	at 50°C and wash
Add	50 µl/well	Detection Linker Solution
Incubate	for 1 hour	at 50°C and wash
Add	50 µl/well	Detection Probe Solution
Incubate	for 30 minutes	at 37°C and wash
Add	50 µl/well	Luminescent Substrate
Image		

Expression Profiling

The multiplexed mRNA assay is a cell-based assay designed for the primary and follow-up screening of compound libraries. This required that the assay be capable of establishing structure–activity relationships (SAR) to correlate molecular features of screened compounds with their effects on the expression of target genes. Furthermore, assay protocols had to be automation-friendly. Both were achieved with a multiplexed solution–phase nuclease protection assay (NPA) that required only reagent additions and incubations and that avoided RNA isolation, reverse transcription, target amplification and fluorescent labelling.

The NPA served to convert labile target mRNA molecules to stoichiometric amounts of stable oligonucleotide probes (Fig. 1.1, top right panel); protocol details are provided in Table 1.1 Cells were grown in 96–well plates and treated with compounds. Following the treatment, culture media was removed and the cells were lysed with a solution that contained a large excess of nuclease protection probes complementary to each of the 16 target mRNA species. A heat denaturation step served to inactivate endogenous nucleases and to remove secondary structure in the target mRNA species. During a subsequent incubation, probe hybridized to mRNA. S1 nuclease, an enzyme that specifically cleaves single-stranded nucleic acids [26–28], was added to digest excess probes and unhybridized mRNA, leaving only duplexes of probe and mRNA intact. An alkaline hydrolysis simultaneously inactivated the S1 nuclease and destroyed the RNA component of the mRNA:probe duplexes. Upon neutralization of the samples, nuclease protection probes remained in amounts proportional to the concentration of the complementary target mRNA species that had been present in the original cell sample. These probes were subsequently quantified with an ArrayPlate™. Since all nuclease protection probes were designed to have similar lengths and GC content regardless of their target genes, various probes showed similar behaviors in the assay and consequently, a standardized NPA protocol could be used.

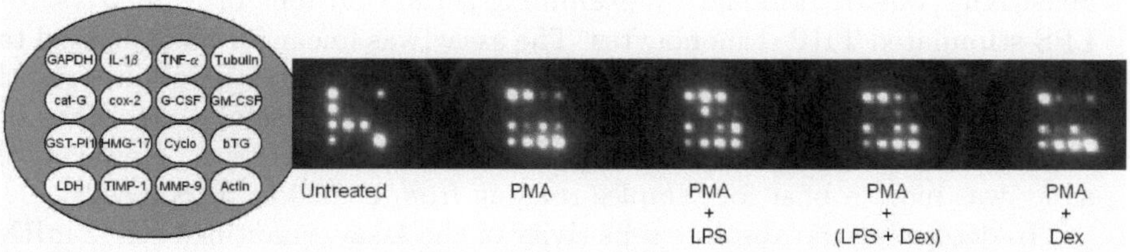

Fig. 1.2. Treatment-Dependent Gene Expression Patterns. The 16 genes that were measured are shown on the left. Five adjacent wells in an ArrayPlate™ are shown on the right. Each well contained sample from 30,000 THP–1 monocytes subjected to a particular regimen involving combinations of treatment with the phorbol ester PMA, with bacterial lipopolysaccharide (LPS) and with dexamethasone (Dex). Each treatment resulted in a distinct pattern of gene expression

The probe-containing hydrolysate resulting from the NPA was transferred from the cell culture plate to a reagent-programmed ArrayPlate™ (Fig. 1.1, lower panel). Array-bound programming linkers captured the various nuclease protection probes at specified elements within the array. Each 50–mer nuclease protection probe was bound by its 3′–terminal 25–mer to its complementary programming linker. The exposed 5′–terminal 25–mer of each probe was subsequently labelled by hybridization with a specific detection linker oligonucleotide. Each of the 16 different 50–mer detection linkers contained a common 3′ 25–mer in addition to a 5′ 25–mer specific to one of the probes. The common 3′ 25–mer of the detection linkers served to bind a final oligonucleotide that was conjugated to horseradish peroxidase. Thus, a five-layered sandwich hybridization took place at each element: Anchor to programming linker to nuclease protection probe to detection linker to peroxidase conjugate. The amount of peroxidase immobilized at a given array element was determined by the amount of nuclease protection probe bound there as this probe was the limiting reagent.

Upon the addition of chemiluminescent peroxidase substrate, light was generated at each array element in proportion to the amount of peroxidase immobilized there. Within 30 minutes of substrate addition, the entire ArrayPlate^{TM} was imaged for 30 seconds to 6 minutes with a high resolution CCD imager. The digital images of ArrayPlates^{TM} were analyzed with image analysis software that reported the signal intensity for each element in a plate after correcting the intensity for local background and, when applicable, for the contribution of adjacent elements.

Changes in the patterns of expression of 16 genes in THP–1 monocytes subjected to various treatment regimens are shown in Fig. 1.2. Various treatments were useful to establish performance characteristics for the assay.

Performance Characteristics

Sensitivity was determined by examining serial dilutions of a bulk lysate of LPS-stimulated THP–1 monocytes. The assay was linear for all expressed target genes over a broad range of sample sizes (Fig. 1.3a) and, more importantly, expression ratios between genes remained constant. Useful gene expression data could be obtained from samples of 1,000 cells or fewer. However, the assay was most robust for samples ranging from 25,000 to 50,000 cells.

To determine the absolute sensitivity of the assay, quantified cox–2 mRNA obtained by in vitro transcription was tested (Fig. 1.3b). Here too, assay response was linear over the entire range that was tested (up to nearly 6,000,000 molecules) with the best fit linear regression showing a coefficient of correlation greater than 0.99. As few as 150,000 cox–2 mRNA molecules were detectable. Similar sensitivities were observed with in vitro transcripts of other genes (data not shown). The reproducibility of the mRNA assay was determined for each target using 30,000 cells/well samples of untreated THP–1 cells (n=48) and cells treated with PMA and LPS (n=48). The data for each well were normalized to GAPDH (the housekeeping gene for these experiments) and the coefficient of variability (CV, i.e. standard deviation as a percentage of the average) was determined for each gene (Table 1.2). The average CV was 6.4% for untreated cells and 7.6% for treated cells, ranging from a low of 3% for cathepsin G in untreated cells to a high of 13% for GST Pi–1 and cyclophilin in treated cells.

Antibody Array

In a proof–of–principle study, a companion multiplexed ELISA was established to simultaneously quantify five antigens (IL–1β, TNFα, G-CSF, MCP–1 and IL–8). The antigens were selected based on the availability of an ArrayPlate^{TM} mRNA assay for the corresponding genes and of commercial ELISA reagents. The commercial kits contained capture antibody, biotinylated detection antibody, streptavidin–peroxidase conjugate and recombinant antigen standard.

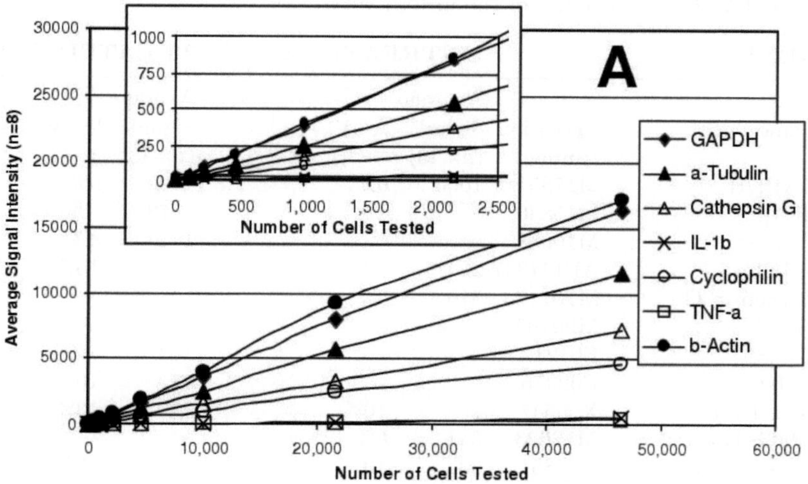

(a)

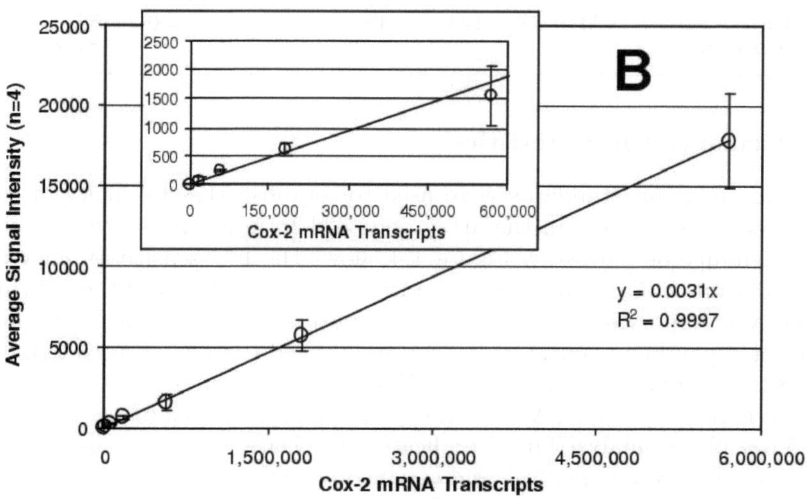

(b)

Fig. 1.3. Sensitivity of the mRNA Assay. **(a)** Serial dilutions of LPS-stimulated cells were analyzed. The linear response for seven of the target genes is shown with the low range enlarged in the insert. **(b)** Serial dilutions of cox–2 mRNA obtained by in vitro transcription were analyzed. The error bars show the standard deviation (n=4) of signal intensity at each concentration

Table 1.2. Reproducibility of the mRNA Assay

GENE		UNTREATED CELLS		TREATED CELLS	
Name	Accession Number	Average Signal (n=48)	%CV	Average Signal (n=48)	%CV
GAPDH	M17851	1000	6%	1000	9%
IL–1β	M15840	–	–	1778	5%
TNF–α	M10988	–	–	1416	4%
Tubulin	AF141347	224	7%	80	10%
Cathepsin G	M16117	510	3%	–	–
Cox 2	M90100	–	–	791	6%
G–CSF	E01219	–	–	103	8%
GM–CSF	E02975	–	–	77	10%
GST Pi–1	X06547	79	10%	35	13%
HMG–17	M12623	541	6%	–	–
Cyclophilin	X52851	333	10%	251	13%
β–Thromboglobulin	M17017	–	–	895	6%
LDH	X02152	228	5%	268	7%
TIMP–1	X03124	–	–	833	6%
MMP–9	J05070	–	–	1117	4%
Actin	M10277	1231	4%	1000	5%
AVERAGE:			6.4%		7.6%

Performance Characteristics

The recombinant standards were used to establish the specificity of each of the five antibody sets in the array and to determine the sensitivity and reproducibility of the assay. Figure 1.4 shows the five sensitivity curves that were obtained. For each of the five antigens, the sensitivity of the multiplexed assay was approximately the same as reported by the antibody supplier for the corresponding traditional ELISA and ranged from less than 0.5 pg/ml for IL–8 to approximately 2 pg/ml for G–CSF. To determine the reproducibility of the multiplexed ELISA, a solution that contained each of the five antigens at 5 pg/ml was analyzed in 36 replicate wells. Data were normalized to 10,000 luminescence counts per well and assigned to each of the five elements according to their relative intensities. CV values ranged from 7% for IL–8 to 15% for MCP–1 (Table 1.3).

Examples

To illustrate the high content that is achievable with multiplexed assays, samples of 30,000 THP–1 cells per well were treated with PMA and examined over time. Secreted and intracellular protein profiles were obtained with the multiplexed ELISA while the ArrayPlate^{TM} mRNA assay served to monitor gene expression. Secreted proteins were measured in the culture medium

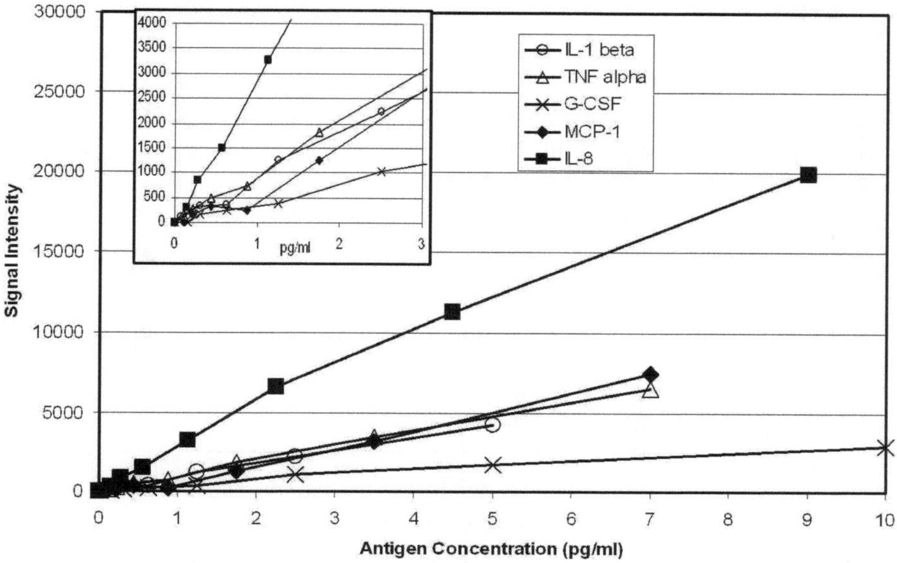

Fig. 1.4. Sensitivity of the Multiplexed ELISA. Serial dilutions of recombinant antigen standards were tested. The sensitivity curves are shown

Table 1.3. Reproducibility of the multiplexed ELISA

ANTIGEN	AVERAGE SIGNAL (Normalized)	S.D.	%C.V.
IL–1β	1,646	192	12%
TNF–α	1,685	129	8%
G–CSF	973	102	10%
MCP–1	1,415	214	15%
IL–8	4,281	280	7%

while mRNA and intracellular proteins were measured in cell lysate. Eight replicates (one column in a 96– well plate) were examined at each of six time points. Results for IL–1β are shown in Fig. 1.5. The induction of IL–1β mRNA, the intracellular accumulation IL–1β and the secretion of protein could all be measured for samples derived from individual wells. Additionally, similar data were obtained for four other proteins and 15 additional genes.

1.6 Conclusion

Arrays encompass a range of technologies to conduct multiplexed assays. The ArrayPlate^{TM} platform is aimed at bringing the benefits of arrays to the

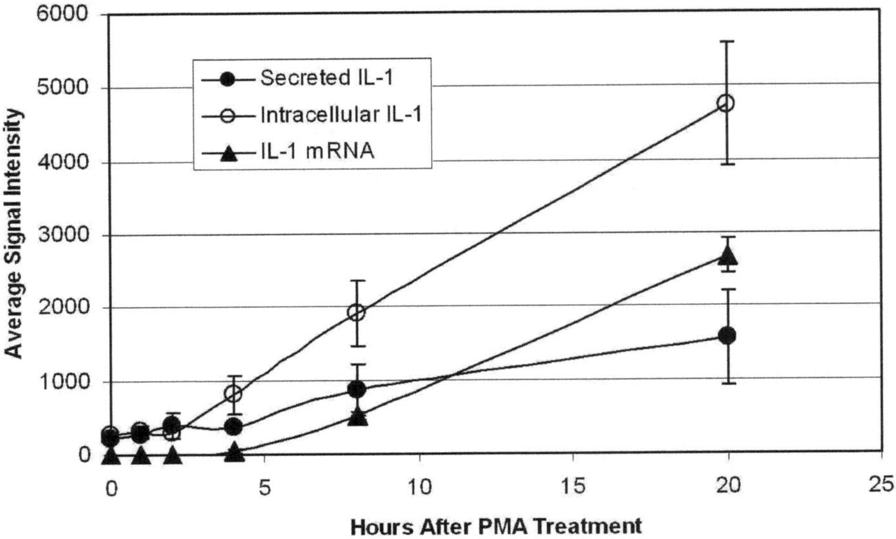

Fig. 1.5. mRNA and Protein Levels Following Treatment. THP–1 monocytes were examined at different intervals following treatment with PMA. mRNA and intracellular proteins were measured in cell lysate while secreted proteins were measured in the culture media. The results obtained for IL–1β are shown. The error bars show the standard deviations for eight replicates at each time point

drug discovery process. The ArrayPlate^{TM} mRNA assay is an automation-compatible method for quantifying 16 genes simultaneously with a sensitivity of 150,000 molecules and reproducibility of <10% average CV. The use of reagent-modifiable arrays and of whole–plate imaging of chemiluminescent read-out signals are features that will allow this multiplexed format to be applied to a variety of high throughput screening assays.

References

1. Lockhart DJ, Dong H, Byrne MC, Follettie MT, Gallo MV, Chee MS, Mittmann M, Wang C, Kobayashi M, Horton H, Brown EL (1996) *Expression monitoring by hybridization to high–density oligonucleotide arrays.* Nat Biotechnol 14:1675–80
2. Wodicka L, Dong H, Mittmann M, Ho MH, Lockhart DJ (1997) *Genome–wide expression monitoring in Saccharomyces cerevisiae.* Nat Biotechnol 15:1359–67
3. Schweitzer B, Kingsmore SF (2002) *Measuring proteins on microarrays.* Curr Opin Biotechnol 13:14–9
4. Love KR, Seeberger PH (2002) *Carbohydrate arrays as tools for glycomics.* Angew Chem Int Ed Engl 41:3583–6, 3513
5. Lam KS, Renil M (2002) *From combinatorial chemistry to chemical microarray.* Curr Opin Chem Biol 6:353–8

6. Wu RZ, Bailey SN, Sabatini DM (2002) *Cell–biological applications of transfected–cell microarrays.* Trends Cell Biol 12:485–8
7. Fejzo MS, Slamon DJ (2001) *Frozen tumor tissue microarray technology for analysis of tumor RNA, DNA, and proteins.* Am J Pathol 159:1645–50
8. Yang L, Tran DK, Wang X (2001) *BADGE, Beads Array for the Detection of Gene Expression, a high–throughput diagnostic bioassay.* Genome Res 11:1888–98
9. Tung WS, Lee JK, Thompson RW (2001) *Simultaneous analysis of 1176 gene products in normal human aorta and abdominal aortic aneurysms using a membrane–based complementary DNA expression array.* J Vasc Surg 34:143–50
10. Golub TR, Slonim DK, Tamayo P, Huard C, Gaasenbeek M, Mesirov JP, Coller H, Loh ML, Downing JR, Caligiuri MA, Bloomfield CD, Lander ES (1999) *Molecular classification of cancer: class discovery and class prediction by gene expression monitoring.* Science 286:531–7
11. Hedenfalk I, Duggan D, Chen Y, Radmacher M, Bittner M, Simon R, Meltzer P, Gusterson B, Esteller M, Kallioniemi OP, Wilfond B, Borg A, Trent J (2001) *Gene–expression profiles in hereditary breast cancer.* N Engl J Med 344:539–48
12. Heller RA, Schena M , Chai A, Shalon D, Bedilion T, Gilmore J, Woolley DE, Davis RW (1997) *Discovery and analysis of inflammatory disease–related genes using cDNA microarrays.* Proc Natl Acad Sci USA 94:2150–5
13. Blackwell HE, Perez L, Stavenger RA, Tallarico JA, Cope Eatough E, Foley MA, Schreiber SL (2001) *A one–bead, one–stock solution approach to chemical genetics: part 1.* Chem Biol 8:1167–82
14. LeProust E, Pellois JP, Yu P, Zhang H, Gao X, Srivannavit O, Gulari E, Zhou X (2000) *Digital light–directed synthesis. A microarray platform that permits rapid reaction optimization on a combinatorial basis.* J Comb Chem 2:349–54
15. Fodor SP, Read JL, Pirrung MC, Stryer L, Lu AT, Solas D (1991) *Light–directed, spatially addressable parallel chemical synthesis.* Science 251:767–773
16. David CA, Middleton T, Montgomery D, Lim HB, Kati W, Molla A, Xuei X, Warrior U, Kofron JL, Burns DJ (2002) *Microarray compound screening (microARCS) to identify inhibitors of HIV integrase.* J Biomol Screen 7:259–66
17. Heller JH (2002) *DNA microarray technology: devices, systems, and applications.* Annu Rev Biomed Eng 4:129–53
18. Shoemaker DD, Linsley PS (2002) *Recent developments in DNA microarrays.* Curr Opin Microbiol 5:334–337
19. Oliphant A, Barker DL, Stuelpnagel JR, Chee MS (2002) *BeadArray technology: enabling an accurate, cost–effective approach to high–throughput genotyping.* Biotechniques Suppl 32:56–61
20. Umek RM, Lin SW, Vielmetter J, Terbrueggen RH, Irvine B, Yu CJ, Kayyem JF, Yowanto H, Blackburn GF, Farkas DH, Chen YP (2001) *Electronic detection of nucleic acids: a versatile platform for molecular diagnostics.* J Mol Diagn 3:74–84
21. Velculescu VE, Zhang L, Vogelstein B, Kinzler KW (1995) *Serial Analysis of Gene Expression.* Science 270: 484–487
22. Bertelson AH, Velculescu VE (1998) *High–throughput Gene Expression Analysis Using SAGE.* Drug Discovery Today 3:152–159
23. Haab BB, Dunham MJ, Brown PO (2001) *Protein microarrays for highly parallel detection and quantitation of specific proteins and antibodies in complex solutions.* Genome Biol 2:research0004.1–0004.13

24. Kodadek T (2002) *Development of protein–detecting microarrays and related devices.* Trends Biochem Sci 27:295–300
25. Martel RR, Botros IW, Rounseville MP, Hinton JP, Staples RR, Morales DA, Farmer JB, Seligmann BE (2002) *Multiplexed screening assay for mRNA combining nuclease protection with luminescent array detection.* Assay Drug Dev Tech 1:61–71
26. Berk AJ, Sharp PA (1977) *Sizing and mapping of early adenovirus mRNAs by gel electrophoresis of S1 endonuclease–digested hybrids.* Cell 12:721–32
27. Maxwell IH, Van Ness J, Hahn WE (1978) *Assay of DNA–RNA hybrids by S1 nuclease digestion and adsorption to DEAE–cellulose filters.* Nucleic Acids Res 5:2033–8
28. Wittelsberger SC, Hansen JN (1977) *The specificity of S1 nuclease toward RNA–DNA hybrids as studied using isotopes of phosphorus–32 and phosphorus–33.* Nucleic Acids Res 4:1829–35

2

Biomolecules and Cells on Surfaces – Fundamental Concepts

Kristi L. Hanson, Luisa Filipponi, and Dan V. Nicolau

2.1 Introduction

In microarray technology, surfaces must be designed and prepared to optimize the immobilization of probe biomolecules and/or cells, but also to resist non-specific binding of target species. Further, the surface and type of immobilization technique selected will affect the concentration, bioactivity and target–binding ability of bound species. For any given probe molecule, there is likely to be an optimal surface and/or technique which will allow for attachment at the highest possible concentration and with preservation of required activity. However, for multi–probe array formats requiring a variety of probe molecules to be bound to the same type of surface, difficulties are encountered selecting a surface and immobilization method able to generate sufficient probe concentration, resolution and bioactivity for *all* probes. The resulting variability in probe concentration and activity within the array also leads to signal variability, causing difficulty in data interpretation. Thus, appropriate attachment methods are critical to the success of any array technology.

The aim of this chapter is to summarize the general knowledge and fundamental concepts underlying DNA, protein, small biomolecule and cell attachment to surfaces, and to highlight issues arising in the field of microarray fabrication. The section will provide background knowledge for the reader not familiar with general biomolecule immobilization techniques, while more specific protocols used in microarray technology will be discussed further in Chap. 3.

2.2 Types of Immobilization

Biomolecule attachment is dependent on the properties of the biomolecular surface, the solid surface, and the liquid medium. In most cases, the biomolecular surface will display a higher level of complexity than the attachment surface or the liquid medium, as biomolecules and cells exhibit not only an overall charge and hydrophobicity, but also a heterogeneous distribution of these, depending on the types and distributions of surface-exposed groups.

The biomolecules of interest can be broadly grouped into nucleic acids (DNA, RNA, PNA), proteins (antibodies, enzymes, receptors, affibodies), small molecules (e.g. peptides, metabolites) and other biomolecules (e.g. carbohydrates, lipids), of which the first two classes have been by far the most studied with respect to microarray applications. Figures 2.1 and 2.2 depict the distribution of charges and hydrophobicity on a single stranded oligonucleotide and a protein (lysozyme), respectively. The oligonucleotide shows more ordered and predictable patterns, with regularly-spaced negatively-charged phosphate groups in the backbone region, and hydrophobic base pairing regions. In contrast, proteins are characterized by both heterogeneous and irregular regions of positive charge, negative charge, and hydrophobicity. As we will see in the next section, the relative structural simplicity of DNA, as compared to protein, results in more predictable and controllable patterns of surface attachment.

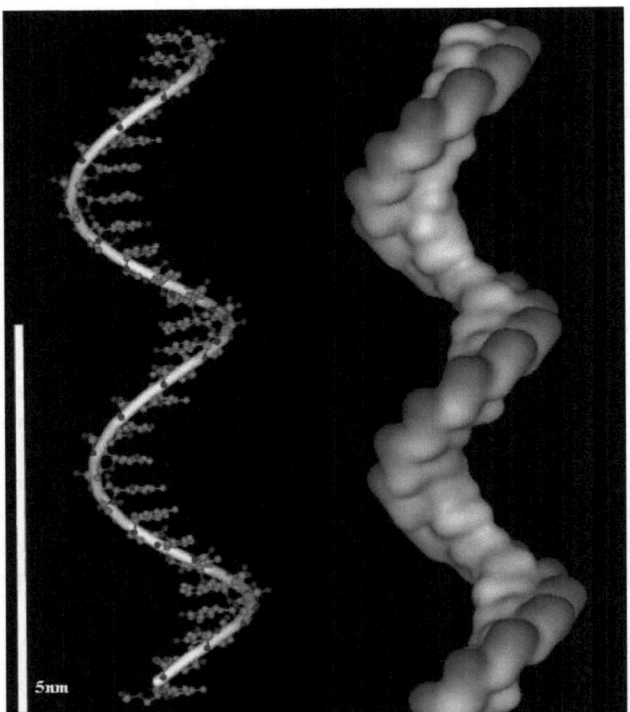

Fig. 2.1. Structure of a single stranded oligonucleotide (*left*) and the 3D map of the electrostatic potential (darker patches = negative charges)

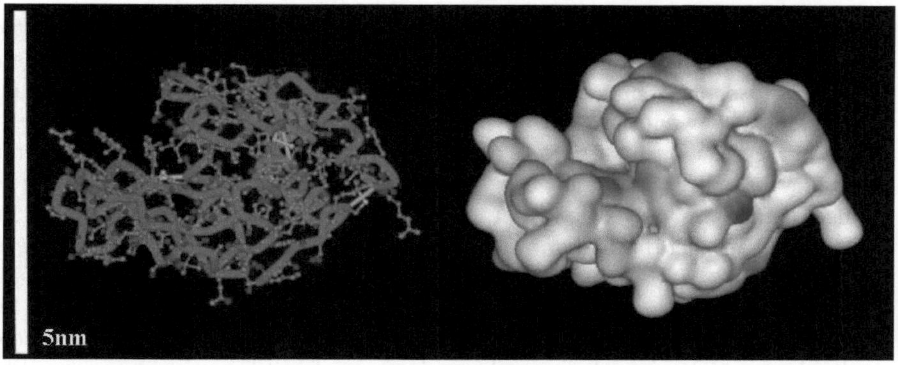

Fig. 2.2. Structure of a simple protein (lysozyme, *left*) and the 3D map of the electrostatic potential (red indicates negative charges; blue indicates positive charges)

Mechanisms of immobilization can be divided into two major categories: (i) adsorption, which relies on non-covalent interactions (mainly electrostatic, van der Waals, and dehydration of hydrophobic interfaces) and (ii) covalent binding of specific functional groups on the biomolecule to functionalized surfaces. The first mechanism is of a purely physical nature and therefore displays varying levels of reversibility, whereas covalent binding, by definition, involves the formation of essentially irreversible chemical bonds between biomolecule and surface.

2.2.1 Adsorption

In general, the extent of adsorption of any species at the solid–liquid interface will be the net result of several attractive and repulsive forces. For biomolecules, the most important of these include electrostatic interactions, van der Waals forces, energetically favorable dehydration of hydrophobic surfaces, structure rearrangement, and lateral interactions [1,2].

Electrostatic interactions result from the overlap of the electrical double layers around a charged protein molecule and a charged surface. These interactions generally depend on the net charge of the surface and the molecule, but heterogeneous surface charges distributed around a protein molecule (see Fig. 2.2) can also produce a dipole moment, thereby contributing to overall electrostatic interaction. The relatively weak character of these interactions renders them less appropriate for microarray technology, where strong and irreversible attachment is generally required. However, the possibility of charge control on the surface (e.g., using electrodes) and on the biomolecular surface (e.g. by variation of pH) make these interactions more versatile. For instance, control of surface charge of an electrode allows for the possibility of 'reusable' microarrays, where bound species can be desorbed, rinsed and re-arrayed.

Van der Waals forces can be broadly defined as other weak attractive forces contributing to intermolecular attraction, including dipole–dipole interactions, hydrogen bonding, and dispersion (London) forces. Where electrostatic interactions might be unfavorable due to like charges on molecule and surface, adsorption may still occur due to the strong effect of van der Waals forces at close range.

Classic DLVO theory [3, 4] models colloidal or protein interactions and stability based on the balance between the above forces (i.e. electrostatic repulsion and van der Waals attraction). In general, electrostatic forces are felt at longer distances than van der Waals forces, but both forces increase as molecules are brought closer together. At short distances, van der Waals attraction increases more rapidly than electrostatic repulsion, leading to adsorption (in the case of proteins and surfaces) or flocculation (in the case of colloidal particles in solution). Thus, in order for adsorption to occur, like-charged particles must have sufficient kinetic energy to overcome the energy barrier, which is dictated by the point of maximum repulsive energy on the net interaction curve (Fig. 2.3).

DLVO theory predicts strong adhesion between hydrophobic particles or molecules, due to the strong effect of van der Waals interactions at close range. These interactions are sometimes therefore referred to as hydrophobic interactions, but the driving force is considered to be the energetically favorable displacement of water molecules between two hydrophobic surfaces. Regardless of the details of theoretical explanation, there is no doubt that attractive interactions between proteins and hydrophobic surfaces are often very important, and in many cases dominate all other driving forces [1]. The application of these attractive forces to microarrays is complicated by the fact that hydrophobic interactions are often associated with conformational changes in molecular structure, as the hydrophobic interior of the biomolecule 'unfolds' to position itself against the hydrophobic interface.

Finally, of particular importance to microarray technology, but usually poorly characterized, *lateral interactions* will affect the density of surface-bound biomolecules. These interactions can result from either (1) electrostatic repulsion between molecules with like charges, or (2) dipole–dipole interactions, which can be repulsive or attractive, depending on the alignment and ordering of molecules on the surface.

In practice, it is difficult to predict or model the overall effects of the above interactions, and the nature of biomolecule adsorption on a particular surface is often investigated by determination of relevant adsorption isotherms (Fig. 2.4). Adsorption isotherms relate the quantity of adsorbed protein (relative to available surface area) to the concentration in solution at *equilibrium*. Typically, the amount of adsorbed protein increases sharply at low solution concentrations, and then eventually approaches a limiting value indicative of the saturated, or maximum possible loading. In the simplest case, the relationship can be modelled by the Langmuir equation, which assumes a single equilibrium constant for the reaction between adsorbed and dissolved protein. An alternative model, known as the Freundlich model, can be derived assuming a certain distribution function for multiple binding sites having different

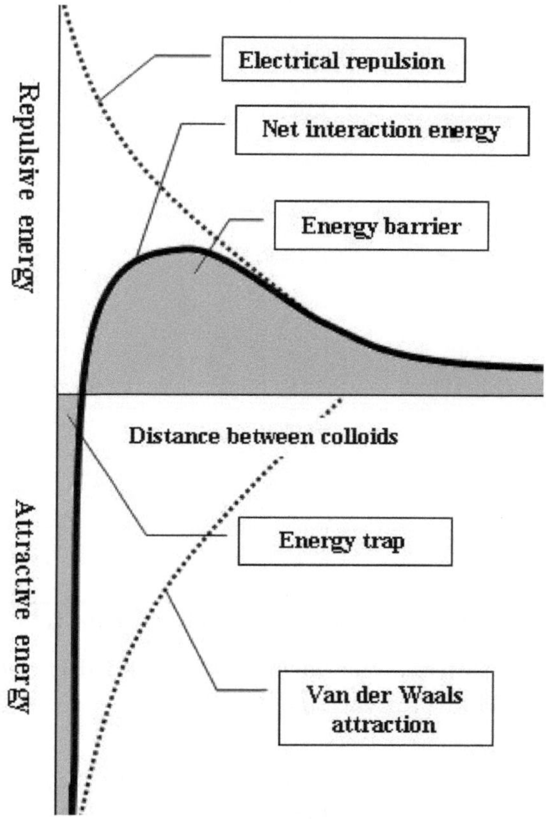

Fig. 2.3. Net interaction curve formed by subtracting the attraction curve (due to van der Waals forces) from the repulsion curve (due to electrostatic repulsion of like-charged particles)

equilibrium constants. In most cases, one or both of these models can be fit to protein adsorption data.

Evaluation of such isotherms is particularly useful when comparing different adsorption strategies, and can provide insight into maximum possible protein loading concentration, binding geometries and lateral interactions. For example, if the dimensions of the biomolecule are known, the maximum surface coverage achieved can be compared to theoretical monolayer coverage in all possible binding geometries, thereby allowing inference of attachment density and/or attachment orientation.

2.2.2 Covalent Attachment

The covalent binding of biomolecules allows for very strong attachment and in certain instances a positional linking at one end of a biomolecule. A variety

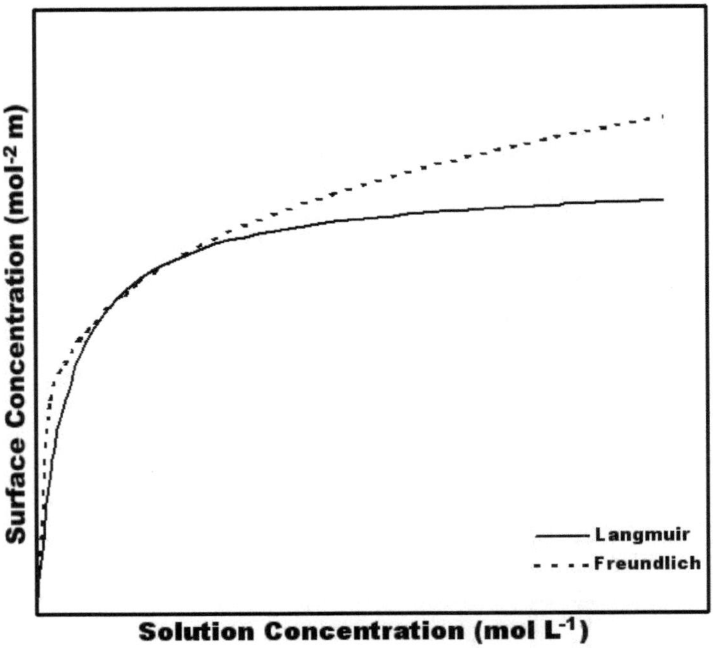

Fig. 2.4. Typical patterns of Langmuir (*solid line*) and Freundlich (*dashed line*) adsorption isotherms

of side groups are easily used for covalent binding, most commonly amino, carboxy, hydroxy, and thiol groups. Consequently similar groups on the surface are needed for a covalent interaction, and in many instances the covalent binding must be enabled by a functionalization of the surface and/or the biomolecule. Quite often the process is 'standardized' through the use of crosslinkers and associated protocols, many of these being reviewed in Chap. 3.

2.3 DNA Immobilization on Surfaces

Although DNA immobilization at the solid/liquid interface is not fully understood, especially with respect to molecular conformations at the surface, a wide variety of techniques have been successfully used for probe attachment.

At neutral pH, DNA molecules are charged negatively (Fig. 2.1), and the pattern of charges suggests that phosphates in the DNA backbone would be expected to bind strongly to a positively charged surface, leaving the bases facing towards the solution. As such, positively charged surfaces (e.g., aminopropyltriethoxylsilane [APTES] or poly–L–lysine coated glass) have commonly been used as DNA hybridization sensors [5, 6]. In contrast, hydrophobic and van der Waals interactions would be expected to bind base pairing regions to the hydrophobic surface, thus reducing the level of target hybridization in microarray format. These two possible conformations are illustrated in Fig. 2.5 (reprinted from [7]).

Despite the fact that in theory, DNA adsorption to a hydrophobic surface should not allow for efficient hybridization of DNA target, nitrocellulose and nylon supports have been widely used for many years as standard substrates for DNA hybridization [8]. It is interesting to note that both single stranded and double stranded DNA are able to bind by hydrophobic interactions [9], despite the fact that hydrophobic regions of double stranded DNA are presumably buried within the center of the helical structure. As a result, adsorbed double stranded DNA molecules overlap and superimpose through sticky end cohesions, forming complex lattices that are unstable and desorb easily from the surface. When a positive potential is applied to the surface, these lattices form coiled fibers with greatly increased stability due to the electrostatic attraction of phosphate backbone to the positively charged surface, but the DNA duplex is destabilized and stretched as a result of charge–charge repulsion on the unbound side of the DNA helix. Subsequent reorientation of the molecule forces DNA bases from inside the helix to be more exposed to solution. These processes demonstrate the relative simplicity of oligonucleotide

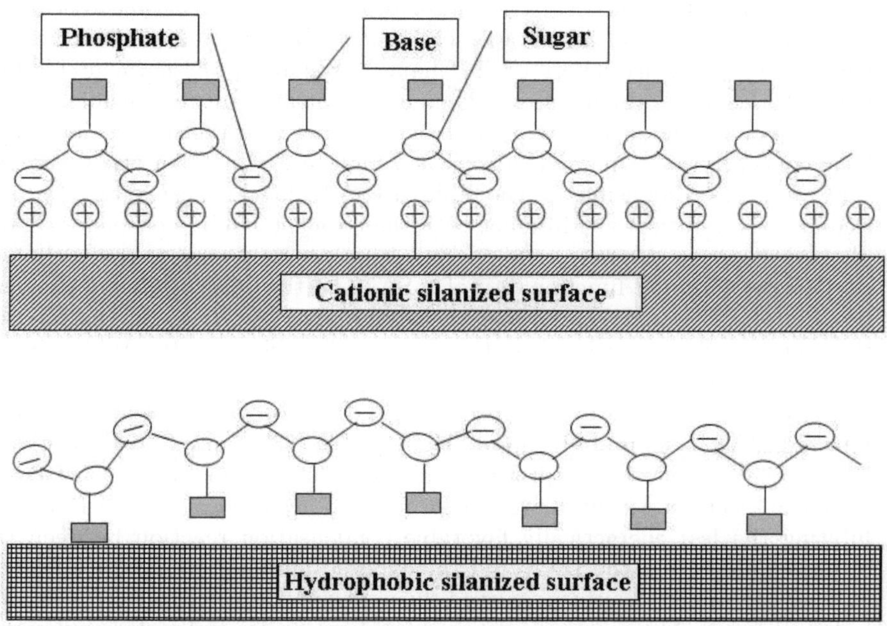

Fig. 2.5. Possible conformations of the DNA/oligonucleotide–surface complex on hydrophobic and cationic surfaces (Reprinted with permission from [7]. Copyright 1998 Academic Press Inc Elsevier Science)

behavior, as it appears to be dominated by electrostatic, van der Waals and hydrophobic interactions in a quasi–predictable manner.

For microarray applications, it appears that electrostatic interactions between negatively charged DNA and a positively charged surface will produce both higher concentrations of surface-bound probe DNA and more favorable orientation of the probe with respect to hybridization potential [7]. Moreover, the electrostatic adhesion has been found to result in significantly lower surface diffusion, which would be advantageous for maintaining high contrast areas of probe attachment.

Figure 2.6 shows patterns of oligonucleotide adsorption on hydrophobic and ionic substrates [7]. The chemical structures of the functionalized silanes coupled to the glass surface are shown along with adsorption isotherms for equilibrium oligonucleotide concentrations on the surfaces. Maximum adsorption densities reached $> 1 \times 10^{13}$ molecules cm^{-2} on cationic surfaces, approximately two times higher than on hydrophobic surfaces. The effect of such densities on hybridization signal were not evaluated in this study, but another study has specifically assessed the effects of array spot concentration on hybridization signal [10] by direct comparison of spot concentration to hybridization efficiency. With maximum hybridization signals (300–400 a.u.) were observed using a spot concentration of 0.25–1 ng nL^{-1}.

The simplicity of physical adsorption for DNA immobilization can be counterbalanced by several advantages of *covalent binding*, many strategies for which are specifically described in Chap. 3. Whatever the covalent binding method, the non-covalent interactions precede it and are responsible for the build–up of a high local concentration of molecules near the surface. This high local concentration is needed to achieve a high rate of reaction. However, the very processes responsible for the build–up of the local concentration (in particular electrostatics) can interact with the covalent binding efficiency. X-ray photoelectron spectrometry and cyclic voltametry were used to probe the impact of the terminal functionality of a SAM on the effectiveness of covalent binding of DNA to SAM-covered electrodes, shedding light on the interaction between electrostatic adsorption and covalent binding [10]. While the ratios of total immobilized DNA on hydroxyl-, carboxyl- and amino-terminated SAMs was (3–3.5):(1–1.5):1, respectively, the proportion of covalently immobilized DNA was found to be approximately 85%, 93%, and 25%, respectively. These results suggest that protonization of amino groups on the surface resulted in electrostatically driven adsorption of negatively charged DNA, inhibiting the less energetically favorable condensation reaction between the 5′ phosphate end of the DNA and the exposed amine group. Attachment to carboxyl-terminated surfaces showed the opposite effect, with electrostatic repulsion between like negatively charged DNA molecules and surface functional groups inhibiting adsorptive attachment, but higher covalent binding yields. However, the total amount of immobilized DNA on the carboxyl-terminated surfaces was low, due to inhibition of DNA movement towards the reactive surface by electrostatic repulsion. Optimal total attachment, with a high per-

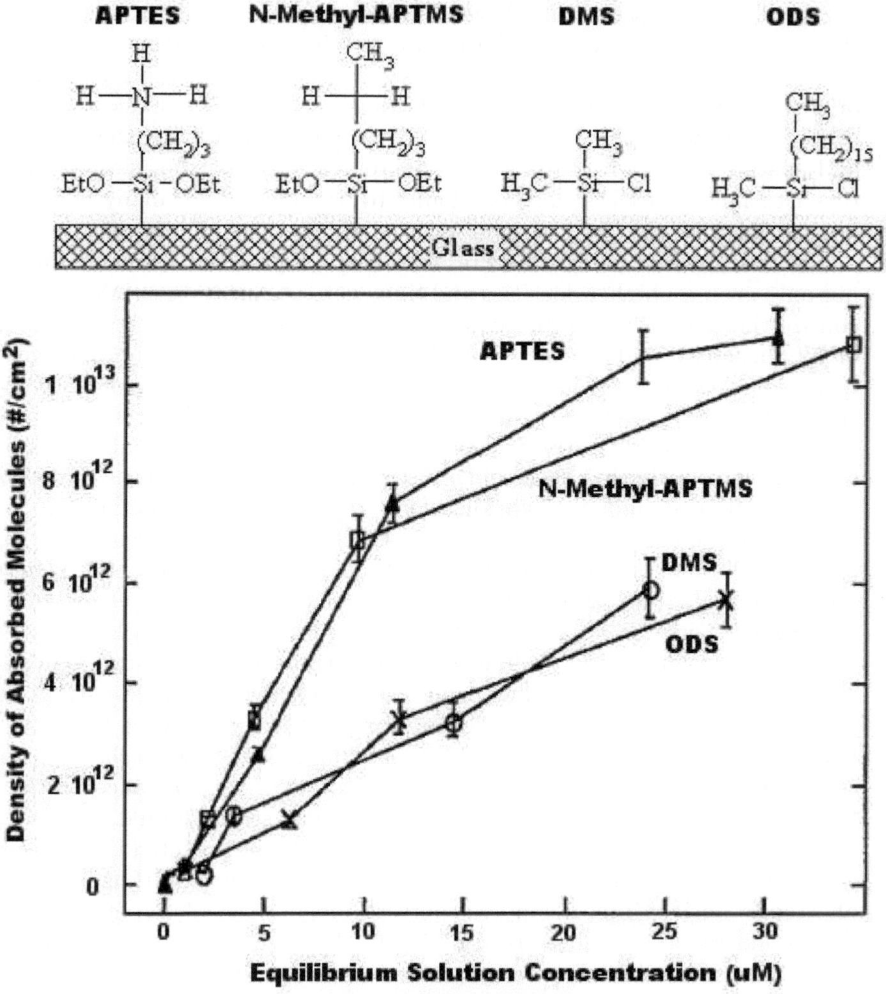

Fig. 2.6. Concentration of DNA molecules as a function of surface chemistry (Adapted from [7]. Copyright 1998 Academic Press Inc Elsevier Science)

centage of covalently bound DNA, was achieved on neutral or slightly negative hydroxyl-terminated surfaces.

Whatever the method of immobilization, the key performance criterion is the efficiency of hybridization. In theory, this should depend on (i) the surface characteristics, (ii) the surface density of probe molecules, (iii) probe orientation on the surface, and (iv) factors controlling transport of target molecules to the surface. While the DNA density can be controlled, the DNA conformation on the surface is more difficult to modulate. Studies discussed above [6, 7, 9] suggest that electrostatically driven DNA adsorption results in orientation of the molecule's backbone parallel to, rather than perpendicular to, the surface, but with base pairing sites exposed to the liquid medium. While this orientation is conducive to target hybridization, it does not allow for dense probe coverage on the surface, and will therefore limit both the sensitivity and spatial resolution of associated microarrays. That said, simple adsorptive attachment of DNA has been found to be sufficient for many microarray applications.

Where higher sensitivity and resolution are required, covalent binding can not only produce higher densities and tighter immobilization of DNA, but also control the orientation of molecular immobilization at either the 3′–hydroxyl or 5′–phosphate end of the DNA chain. However, increased probe density will also affect intra–strand interactions, surface interactions and charge density at the surface, which can in turn result in substantially different ionic strength, pH, and dielectric constant at the surface than in bulk electrolyte solution. It is likely that these differences will also impinge on the availability of immobilized strands for hybridization. For instance, the standard enthalpy change for the thermal denaturation of target bound DNA was found to be 2–3 times lower for immobilized DNA than for DNA in the bulk solution, and the melting temperature (T_m) was decreased by 6–10°C [11]. The lower melting temperature suggests that interstrand bonding is weaker on a surface than in solution, depending on bound strand density. Thus, depending on the ability to control immobilization density, there may be variations in sensitivity from case to case. Another study [12] suggests that this effect is likely to be more pronounced for shorter strands, which were shown to produce lower hybridization signals when spotted at the same concentration as longer molecules. Hybridization signals for shorter strands could, however, be improved by addition of a poly(A)tail. This would be expected if smaller target molecules are held more tightly and closer to the surface, thus being more affected by surface interactions. In contrast, longer molecules are likely to contain more free loops and ends available for hybridization further away from the surface. This effect has been observed elsewhere [13], where hybridization was found to be directly dependent on the length of immobilized strands.

2.4 Protein Immobilization on Surfaces

The fundamentals of protein attachment on surfaces have been widely reviewed (e.g., [14–16]), but the extreme diversity and complexity of proteins still make any prediction regarding attachment difficult. Technologies used for DNA microarray application have, to some degree, been adapted to protein microarrays, but the broader use of protein microarray technology is still limited, primarily due to the fact that surfaces and technology allowing uniform and global attachment of a wide variety of proteins are not currently available. This lag in technology stems from the fundamental structural difference between proteins and DNA. DNA is (i) uniformly structured with an ordered hydrophilic backbone, (ii) stable, (iii) does not lose binding activity easily, and (iv) has only one interaction site and geometry with target DNA. In contrast, proteins have (i) many different structures, (ii) contain heterogeneous hydrophobic and charged domains, (iii) are extremely fragile with activity dependent on retention of three–dimensional structure, and (iv) can have multiple interaction sites.

Additional complications arise with respect to microarray technology, where functional conservation and sufficient concentration of bound protein are critical to the success of the technology. Correct orientation of the bound protein is required to increase the exposure of functional domains to solvent/target, but protein adsorption mechanisms often result in random (or widely distributed) orientations on a surface.

In theory, random attachment is not likely to result in a high percentage of protein functional sites in the proper orientation for binding, but successful attachment and target detection have been achieved with random adsorptive attachment techniques (e.g., [17]). Further, a recent study indicates that there may be minimal effect on functionality between proteins immobilized by directed or random attachment [18]. In contrast, oriented attachment has also been found to increase array sensitivity up to 10–fold [19].

Overall, the need for directed orientation and choice of technique will depend on the specific proteins being used, and no single method is likely to work in all situations. It is clear, however, that as protein arrays become more comprehensive and as the number of proteins in a single array increases, the need for a technology that can accomplish immobilization across a wide range of proteins, or even an entire proteome, will become more desirable. Strategies used thus far can be broadly classified based on adsorptive or covalent binding mechanisms, and then further subdivided into methods resulting in random vs. directed orientation of the molecule.

2.4.1 Random Adsorptive Attachment

As proteins are charged biomolecules, it would be expected that *electrostatic interactions* could be used for efficient and controllable immobilization. However, electrostatic adsorption is often of a non-permanent nature and can be strongly affected by changes to solution pH and ionic activity, thereby allowing for the possibility that subsequent array processing might desorb the protein. Additionally, despite the simplicity of electrostatic interaction, which should make adsorption more predictable, such interactions are usually more difficult to predict than hydrophobic or covalent ones [20]. This is most likely due to the uneven spatial charge distribution on protein surfaces, which also varies with pH and ionic strength of the solution. Many chemical or physico–chemical schemes have been used to create charged surfaces that can adsorb proteins, for example polyelectrolyte multilayers [21] and sulfonated polymer surfaces [22,23]. However, because of the complexity of the electrostatic potential map as well as interference from other interactions (e.g. hydrophobicity), a generic 'magic' surface that can promote the electrostatically-driven adsorption of proteins has not been found.

Hydrophobic interactions are often stronger and less reversible than electrostatic attractions, but can result in loss of functional activity due to partial denaturation as the protein unfolds to expose hydrophobic interior portions to the hydrophobic surface [15].

Due to the complexity of proteins, a reasonable approach would be to explore combinatorially the level of adsorption versus descriptors of surface, solution, and protein characteristics. In an attempt to map the adsorption of virtually any protein on virtually any surface, Nicolau and co-workers have compiled a protein adsorption database [24]. The database contains about 500 cases of protein adsorption for approximately 30 proteins and approximately 100 surfaces in various solution conditions.

Molecular surface property algorithms developed to describe the proteins [25, 26], have been used to describe protein adsorption. A purely empirical approach using a linearly piecewise model with breakpoint was found to be capable of accounting for over 90% of the variance in the data [27]. Fundamentally, the model assumes that the protein concentration on the surface follows a piecewise linear regression conforming to a Langmuir relationship. The experimental data present in the database have been used to derive an empirical relationship that describes the correlation between protein adsorption (dependent variable) and process (independent) variables (i.e. protein concentration in solution; surface tension of the surface; ionic strength of the solution; and absolute value of the difference between pH and the isoelectric point of the protein), as follows:

$$\begin{aligned}
\Gamma &= f_1(\gamma, \text{ion_str}, \text{abs}(pH - pI), C) \cdot (1 - g(\Gamma)) \\
&\quad + f_2(\gamma, \text{ion_str}, \text{abs}(pH - pI), C) \cdot g(\Gamma), \\
f_1 &= a_{11}\gamma + a_{12} \cdot \text{ion_str} + a_{13} \cdot \text{abs}(pH - pI) + a_{14} \cdot C, \\
f_2 &= a_{21}\gamma + a_{22} \cdot \text{ion_str} + a_{23} \cdot \text{abs}(pH - pI) + a_{24} \cdot C, \\
&\Gamma \le \Gamma_{\text{breakpoint}} \Rightarrow g(\Gamma) = 0, \\
&\Gamma > \Gamma_{\text{breakpoint}} \Rightarrow g(\Gamma) = 1.
\end{aligned} \tag{2.1}$$

The parameters of the equations are: Γ – protein surface concentration (mg/m^2); C, protein concentration in solution (mg/ml); γ - surface tension of the polymer (dyne/cm); ion_str – ionic strength (M); pI – isoelectric point of the protein; $\Gamma_{\text{breakpoint}}$ – protein concentration at which the slope of the linear function $\Gamma = f(C)$ changes; and the rest of the parameters are constants (Table 2.1).

The level of fit using such a model is quite remarkable, especially considering that the adsorption data span over three orders of magnitude. This work

Table 2.1. Coefficients for the protein adsorption (Eq. 6)

	Surface tension (dyne/cm)	**Ionic strength (M)**	**Abs(pH-pI)**	**Protein concentration in solution (mg/ml)**	**Free term**
Coefficients in f_1	$a_{11} = 0.076$	a_{12}=-3.297	a_{13}=0.085	1.052	b_1=4.441
Coefficients in f_2	a_{21}=-0.014	a_{22}=3.701	a_{23}=-0.395	0.438	b_2=4.840
Mean	44.82892	0.07505	1.39033	0.49182	
Standard deviation	6.942584	0.071997	0.914402	0.760767	
Break point for the protein surface concentration (mg/m^2)					3

suggests that prediction of appropriate surface types for maximum protein adsorption may, in fact, be possible based on analysis of protein surface characteristics. Validation of the model with additional sets of data will assist in further applications to microarray technology.

Overall, it appears that random adsorptive attachment can be very effective for microarray purposes, and it is possible to make some general predictions about adsorption levels for a variety of proteins if the molecular surface characteristics are known or can be predicted. As mentioned above, the effects of non-oriented binding on array performance can be an issue with this type of attachment, but in many cases still result in acceptable levels of target binding.

2.4.2 Random Covalent Attachment

Covalent binding is quite commonly used in biochip fabrication. Proteins present a variety of functional groups, including amino–, carboxyl–, hydroxyl– and thiol–, which can readily be used for covalent binding to surfaces with complementary chemical groups. However, due to the relative lability of proteins as compared to DNA, more care is required to avoid chemically-induced protein denaturation during the attachment process. There are many strategies for crosslinking of available functional groups, most of which make use of specialized crosslinkers designed for both attachment and physical separation of protein from surface, thereby allowing for more of the protein functional domain to be exposed to the solvent [28].

Covalent binding generally produces a higher concentration of protein than does adsorption. For instance, a study [29] compared the effect of physical adsorption of a protein on Poly(tert–butyl–methacrylate), a highly hydrophobic surface, to the covalent binding of the same protein to a carboxylic-functionalized surface (derived from the former via e-beam photolysis). Covalent attachment resulted in significantly higher surface protein concentrations than adsorption, despite the fact that carboxylic functional groups result in a hydrophilic surface which tends to repel protein (Fig. 2.7).

2.4.3 Oriented Attachment

A variety of oriented immobilization techniques have been attempted, and have recently been summarized [30]. These techniques can be adsorptive, covalent or a combination of both. Some of the more common methods include:

1. use of antibody binding proteins to bind the Fc portion of antibodies leaving the binding sites exposed to solution [31, 32];
2. terminal biotinylation of genetically engineered proteins with subsequent end-specific attachment to a streptavidin coated surface [33];
3. terminal His–tag addition and subsequent attachment to a nitrilotriacetic acid-coated surface [34];
4. use of carbohydrate binding molecules to bind the carbohydrate moieties of antibodies [35]; and
5. cystine thiol production on the C–terminal (non-antigen binding) end of cleaved Fab regions, with subsequent attachment using the cystine thiol 'handle' [32, 33].

A recent study [19] explored the effect of some of the above methods of antibody attachment on analyte binding capacity, and found that orientation increases analyte binding capacity up to 10–fold. When Fab' fragments were specifically oriented in a dense monolayer, 90% of the adsorbed molecules were active, while randomly attached Fab fragments were packed at much lower density, and showed a much lower specific activity. Thus for applications requiring high sensitivity and low detection limits, such techniques are likely to greatly improve performance.

While the above discussion has outlined that there are, in fact, a variety of methods which are useful for the immobilisation of proteins and detection of target analytes in array format, this methodological variation also has a potential downside. Heterogeneous information from different laboratories may ultimately result in non-standardized datasets, difficult to compare and interpret, thus hindering the overall goal of a more complete understanding of proteomes.

2.5 Carbohydrate Immobilization

Carboydrate-based microarrays, which have appeared only recently, have recently been reviewed [36]. Applications of these arrays have enormous potential in microarray technology due to the structural diversity, specificity, and differential expression of carbohydrates [37]. Further, these molecules are

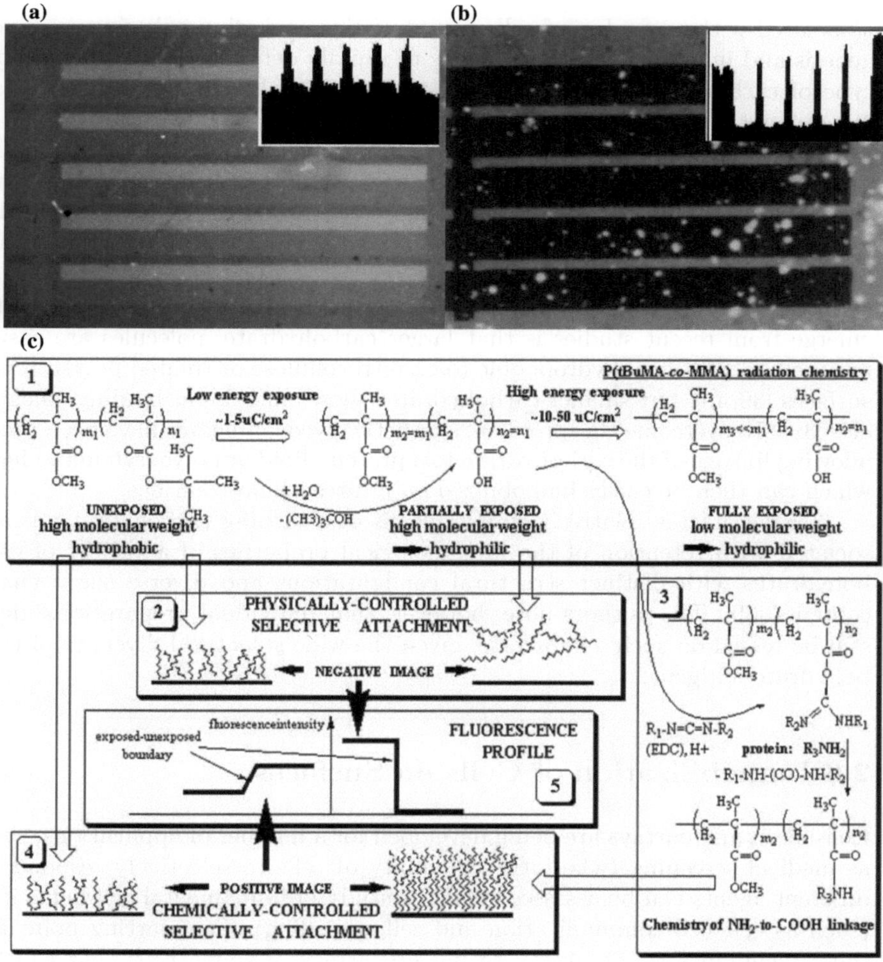

Fig. 2.7. Comparison of protein concentration on the surface, following adsorption on a hydrophobic and covalent attachment on a carboxylic-functionalized surface [29]. The image on the top left shows fluorescently-labelled (FITC) avidin adsorbed on the hydrophobic surface of Poly(tert–butyl–methacrylate), with dark regions showing lack of protein on the carboxylic-rich (hydrophilic) surface of Poly(methacrylic acid), obtained by deep–UV patterning. The image on the top right shows a similar patterned surface with protein covalently bound on the carboxylic-rich surface (lightest areas) and still adsorbed at lower concentrations on the hydrophobic surface (darker bands). (Reprinted with permission from 02Taguchi99. Copyright 1999 American Chemical Society Publications)

associated with a number of cell characteristics, including adhesion, carcinogenesis and immunity [38]. The ability to rapidly determine the presence and type of carboydrate molecules in a sample would therefore greatly increase our understanding of their in vivo functions.

Carbohydrates, like proteins, are structurally heterogeneous and require preservation of molecular conformation, 3–D structure, and topological configuration on a chip in order for molecular recognition to occur. As such, the existing array surfaces commonly used for DNA are generally not amenable to carbohydrate immobilization.

From an immobilization perspective, perhaps the most common theme to emerge from recent studies is that larger carbohydrate molecules are easily retained on relatively hydrophobic (e.g., nitrocellulose or treated polystyrene) surfaces [39,40], but smaller carbohydrates show much lower binding efficiencies [39]. To overcome this problem, synthetic glycoconjugates have been used, allowing linkage of the carbohydrate to a protein, lipid or polyacrylamide chain which can then be easily immobilized on a nitrocellulose surface.

In general, this relatively simple means of attaching carbohydrates is associated with retention of the immunological properties of a variety of carbohydrates with distinct structural configurations and diverse sugar chain contents [39]. The authors note, however, that individual preparations must still be tested on such a substrate, given the wide structural diversity of carbohydrate antigens.

2.6 Immobilization of Cells on Surfaces

Cell-based microarrays are being developed for a number of applications, such as medical screening (where the capability of cells to selectively respond to different agents can be assessed) and the study of fundamental cell behaviors (such as cell–cell communication and cell spreading). The starting point for these techniques is the ability to pattern arrays of single–cells that can be perturbed and monitored individually. As a consequence, the impact of cell confinement on microsized areas (i.e. areas that have dimensions comparable to that of a single cell – a technology generally referred as 'cell patterning' [41]) is of extreme importance in the context of microarray technology.

On a molecular level, the immobilization of cells is far more complex than the immobilization of single biomolecules, and may therefore require situation-specific studies to determine proper surfaces for a particular application. The difficulty of cell immobilization arises in the first instance from the complexity of the cell membrane, containing many types of molecules (membrane proteins, glycoproteins; lipid bilayer supramolecular structures; small molecules, etc.). These biomolecules could attach to a given surface based on the concepts described in the previous sections. Though each such interaction could be analyzed independently, it is likely that these interactions are cooperative or at least not fully independent. Furthermore, the cell is also very flexible, which makes the attachment of the respective molecular patches, independently and collectively, dynamic. Finally, and most importantly, the cell is a living entity that responds to the stimuli presented by the surface. One mechanism of response, and in fact the simplest from a panoply of responses, is to secrete chemical species that will extend the 'controlled' environment of the cell beyond its cell wall.

Simplistically speaking, cell attachment should follow the same rules that govern the non-covalent biomolecular immobilization. For instance, electrostatic interactions can be used for cell immobilization, if the surface of the cell is charged, as is the case for neuronal cells (negatively charged) or some bacteria (most negatively, but some positively charged). Because cells are normally surrounded by a sheath of proteins which presents the hydrophilic face towards the exterior, hydrophobicity-driven immobilization is generally less effective. The most powerful means of cell immobilization is by biomolecular recognition. Cells present proteins on the exterior of their walls that can be unique to a particular cell type or species and that can be recognized by complementary biomolecules (receptors). Alternatively, cells may present proteins with specific functions (including surface attachment) that can be supplied to immobilize cells. While the former mechanism has the propensity to be cell-specific, the latter is more general.

At the first instance, surfaces covered with cell specific proteins would be the natural technological path for cell immobilization. However, in the previous section we saw that the general behavior of proteins on surfaces is difficult to predict. The problems related to protein adsorption in the context of cell immobilization have been concisely described by Mrksich [42]. Briefly, it is difficult to know the density of ligands that are *effectively* available for binding to cellular receptors, due to the distribution in conformation and orientation of adsorbed protein. Many studies aimed at investigating the role of ligand density in cell adhesion and migration have improperly assumed a linear correlation between the density of adsorbed protein and the concentration of protein used to coat the substrates [43]. Also, as expected, the activity of protein-coated substrates can show a dramatic dependence on the choice of substrate. For instance, culturing of myoblasts on two different types of polystyrene resulted in completely different outcomes, namely proliferation or differentiation, even though both were coated with comparable densities of fibronectin (a cell adhesion protein) [44].

Because of the diversity of the response of different cells to surfaces, it is generally necessary to systematically and specifically test cell adhesion and preservation of bioactivity on substrates intended for microarray devices. For instance, one set of experiments [45,46] examined the attachment of neuronal cells on photosensitive polymers. The photoresists, when exposed to UV light, generate carboxylic-rich surfaces (with concentration modulated by exposure energy) that can be further functionalized with neuropeptides. Thermal processing was also used to manipulate the surface properties, either via polymer crosslinking and decarboxylation, or via diffusion of silicon-rich species. It was found that two pairs of partially independent antagonistic surface characteristics, namely (i) amino-rich vs. carboxylic-rich surfaces and (ii) hydrophilic vs. hydrophobic surfaces, controlled the cell attachment, with the former promoting adhesion (Fig. 2.8). This complex relationship means that one cannot predict the attachment of cells based only on hydrophobicity or hydrophilicity of the surface. However, surfaces designed for biomolecular recognition mechanisms (e.g. neuropeptide–functionalized) were the most effective for attachment of neuronal cells, and those designed with very high hydrophobicity were the most effective for repelling neuronal cells. This discussion is illustrative of the specific issues raised by cell attachment (i.e. neuronal cells) but these conclusions cannot, however, be extrapolated to other types of cells due to the diverse nature of cellular membranes types and receptors.

In addition to determining whether attachment will occur, it is also necessary to examine the effect that cell confinement will have on cell behavior. Several studies (e.g. [47]) have studied this relationship. Microcontact printing of SAMs has been used to fabricate substrates with micrometer–scale islands of bovine and human endothelial cell extracellular matrix separated by nonadhesive regions. The size and geometry of the islands were found to control cell shape, with immediate impact on the control of apoptosis as well as growth. Progressive restriction of cell extension by culturing cells on smaller and smaller micropatterned adhesive islands regulated a transition

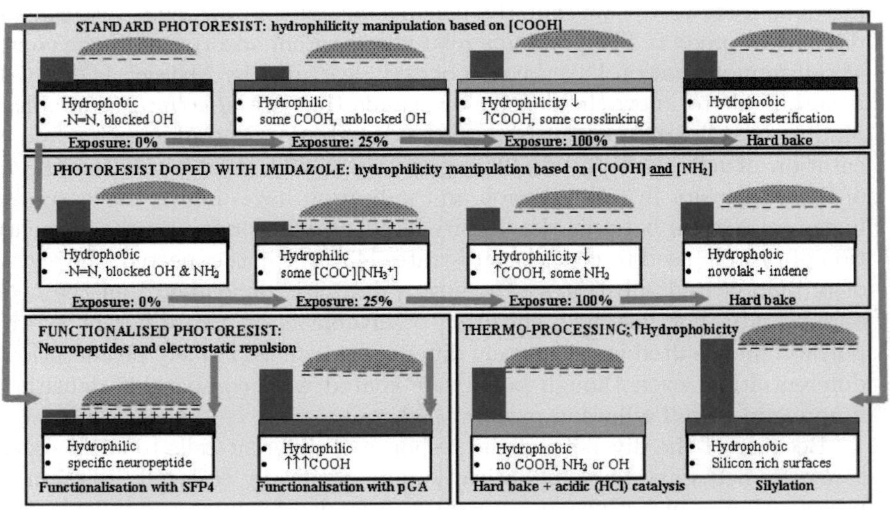

Fig. 2.8. Mechanisms of immobilization of neuronal cells on photoresist surfaces. The vertical bar on the left of each diagram represents the relative repelling effect of the respective surface. Neuropeptide-functionalized surfaces are found to be the most effective for immobilization of neuronal cells, while highly hydrophobic and negatively charged surfaces are the most repelling (Reprinted with permission from [46]. Copyright 1999 Academic Press Inc Elsevier Science)

from growth to apoptosis on a single continuum of cell spreading. This work showed that the size and geometry of the microarray pattern can have profound effects on cellular behavior, which in turn can influence the performance of a cell-based microarray.

In addition to surface chemistry, size and geometry, *topography* also influences cell physiology [48]. Substratum topography was found to influence a number of cell behaviors, such as spreading, secretion, attachment, shape, growth, polarity and differentiated functions [49]. The ability to effectively immobilize cells for the development of microarrays thus relies on the ability to accurately design and control the microarray surface properties at the micro– and nano– scale level.

Another fundamental problem is the intrinsic limitation of culturing cells in an environment that lacks cells' natural three–dimensional organization. The question of whether a cell patterned on a flat surface will behave the same way as when in a three–dimensional matrix (e.g. a gel) is still a matter of investigation, but evidence points to important differences in cell behavior when grown in 2D versus 3D cultures [50–52]. This could mean that unpredictable and different behaviors might be obtained when cells are patterned over a 2D environment, such as a microarray. Moreover, cell behavior in vivo is modulated by interaction with the surrounding cells and by the environment, a *heterogeneous medium* which comprises gradients of nutrients and secreted factors. As a result, use of isolated cell populations in vitro may trigger different behavior from the 'natural' state. These issues are currently under extensive research and should be taken in great consideration when designing a cell-based microarray and evaluating its performance.

2.7 Conclusions

The immobilization of biomolecules and cells for microarray technology has three 'dimensions': (i) a biomolecule or cell to be immobilized on a surface, of which we have limited information and which generally cannot, and/or should not, be altered; (ii) an immobilization surface which can be partially tuned, and of which we have quasi-complete information, and (iii) a liquid environment which is fully controllable and of which we have complete information. This chapter addressed the fundamental concepts dictating the likely response of biomolecules and cells immobilized on surfaces, and their resulting bioactivity. Our approach was to present the general interrelationships between the input and output technological parameters, and then to qualify these general rules on several specific situations. The only certainty that we hope we transmitted to the reader, in a field littered with more exceptions than rules, is that, although general rules are relevant in all situations, nothing can replace experience and innovation.

References

1. Norde, W. 2003. *Colloids and interfaces in life sciences.* Marcel Dekker, Monticello, NY
2. Haynes, C. A. and W. Norde. 1994. *Globular proteins at solid/liquid interfaces.* Colloids and surfaces B: Biointerfaces 2:517–566
3. Derjaguin, B. V. and D. L. Landau. 1941. *A theory of the stability of strongly charged lyophobic sols and the coalescence of strongly charged particles in electrolytic solution.* Acta Physicochimica USSR 14:633–662
4. Verwey, E. J. W. and J. Th. G. Overbeek. 1948. *Theory of stability of lyophobic colloids.* Elsevier, Amsterdam
5. Schena, M., D. Shalon, R. W. Davis, and P. O. Brown. 1995. *Quantitative monitoring of gene expression patterns with a complementary DNA microarray.* Science 270:467–470
6. Chan, V., D. J. Graves, P. Fortina, and S. E. McKenzie. 1997. *Adsorption and surface diffusion of DNA oligonucleotides at liquid/solid interfaces.* Langmuir 13:320–329
7. Chan, V., S. E. McKenzie, S. Surrey, P. Fortina, and D. J. Graves. 1998. *Effect of Hydrophobicity and Electrostatics on Adsorption and Surface Diffusion of DNA Oligonucleotides at Liquid/Solid Interfaces.* Journal of Colloid & Interface Science 203:197–207
8. Ausubel, F. M., R. Brent, R. E. Kingston, D. D. Moore, J. G. Seidman, J. A. Smith, and K. Struhl. 1997. *Current protocols in molecular biology.* John Wiley and Sons, New York
9. Brett, A. M. O. and A.-M. Chiorcea. 2003. *Atomic Force Microscopy of DNA Immobilized onto a Highly Oriented Pyrolytic Graphite Electrode Surface.* Langmuir 19:3830–3839
10. Zhao, Y.-D., D.-W. Pang, S. Hu, Z.-L. Wang, J.-K. Cheng, and H.-P. Dai. 1999. *DNA–modified electrodes; part 4: optimization of covalent immobilization of DNA on self–assembled monolayers.* Talanta 49:751–756
11. Watterson, J. H., P. A. E. Piunno, C. C. Wust, and U. J. Krull. 2000. *Effects of Oligonucleotide Immobilization Density on Selectivity of Quantitative Transduction of Hybridization of Immobilized DNA.* Langmuir 16:4984–4992
12. Franssen-van Hal, N. L. W., O. Vorst, E. Kramer, R. D. Hall, and J. Keijera. 2002. *Factors in sequencing cDNA microarray hybridization on silylated glass slides.* Analytical Biochemistry 308:5–17
13. Stillman, B. A. and J. L. Tonkinson. 2001. *Expression microarray hybridization kinetics depend on length of the immobilized DNA but are independent of immobilization substrate.* Analytical Biochemistry 295:149–157
14. Norde, W. 1986. *Adsorption of Proteins from Solution at the Solid–Liquid Interface.* Advances in Colloid and Interface Science 25:267–340
15. Brash, J. L. and T. A. Horbett. 1987. *Protein at interfaces: physicochemical and biochemical studies.* American Chemical Society, Washington, D.C.
16. Horbett, T. A. and J. L. Brash. 1995. *Protein at interfaces II: fundamentals and applications.* American Chemical Society, Washington, D.C.
17. Angenendt, P., J. Glokler, D. Murphy, H. Lehrach, and D. J. Cahill. 2002. *Toward optimized antibody microarrays: a comparison of current microarray support materials.* Analytical Biochemistry 309:253–260
18. Wilchek, M. and T. Miron. 2003. *Oriented versus random protein immobilization.* Journal of Biochemical and Biophysical Methods 55:67–70

19. Peluso, P., D. S. Wilson, D. Do, H. Tran, M. Venkatasubbaiah, D. Quincy, B. Heidecker, K. Poindexter, N. Tolani, M. Phelan, K. Witte, L. S. Jung, P. Wagner, and S. Nock. 2003. *Optimizing antibody immobilization strategies for the construction of protein microarrays.* Analytical Biochemistry 312:113–124
20. Asthagiri, D. and A. M. Lenhoff. 1997. *Influence of Structural Details in Modeling Electrostatically Driven Protein Adsorption.* Langmuir 13:6761–6768
21. Ladam, G., P. Schaaf, F. J. G. Cuisinier, G. Decher, and J.-C. Voegel. 2001. *Protein adsorption onto auto-assembled polyelectrolyte films.* Langmuir 17:878–882
22. Lestelius, M., B. Liedberg, and P. Tengvall. 1997. *In Vitro Plasma Protein Adsorption on -Functionalized Alkanethiolate Self-Assembled Monolayers.* Langmuir 13:5900–5908
23. Sundaram, S., F. Lim, S. L. Cooper, and R. W. Colman. 1996. *Role of leucocytes in coagulation induced by artificial surfaces: investigation of expression of Mac-1, granulocyte elastase release and leucocyte adhesion on modified polyurethanes.* Biomaterials 17
24. Nicolau, D. V. Biomolecular adsorption database. www.bionanoeng.com/BAD/
25. Nicolau, D. V. Jr. and D. V. Nicolau. 2002. *A database comprising biomolecular descriptors relevant to protein adsorption on microarray surfaces.* SPIE Proceedings 3:109–116
26. Nicolau, D. V. Jr. and D. V. Nicolau. 2002. *A model of protein adsorption to solid surfaces from solution.* In Biomedical Nanotechnology Architectures and Applications, Darryl J. Bornhop; David A. Dunn; Raymond P. Mariella; Catherine J. Murphy; Dan V. Nicolau; Shuming Nie; Michelle Palmer; Ramesh Raghavachari; Eds, Proc. SPIE Vol. 4626, 1–8, 2002
27. Nicolau, D. V. Jr., Fulga, F. and Nicolau, D. V. 2003 *Impact of Protein Adsorption on the Geometry of Microfluidics Devices.* Biomedical Microdevices 5:3, 227–233
28. Wong, S. H. 1991. *Chemistry of protein conjugation and cross-linking.* CRC Press, Boca Raton, FL
29. Nicolau, D. V., T. Taguchi, H. Taniguchi, and S. Yoshikawa. 1999. *Positive and negative tone protein patterning using conventional deep-UV/e-beam resists.* Langmuir 15:3845–3851
30. Seong, S. and C. Choi. 2003. *Current status of protein chip development in terms of fabrication and application.* Proteomics 3:2176–2189
31. Vijayendran, R. A. and D. E. Leckband. 2001. A *Quantitative Assessment of Heterogeneity for Surface-Immobilized Proteins.* Analytical Chemistry 73:471–480
32. Nakanishi, K., H. Muguruma, and I. Karube. 1996. *A novel method of immobilizing antibodies on a quartz crystal microbalance using plasma-polymerized films for immunosensors.* Analytical Chemistry 68:1695–1700
33. Rowe, C. A., S. B. Scruggs, M. J. Feldstein, J. P. Golden, and F. S. Ligler. 1999. *An Array Immunosensor for Simultaneous Detection of Clinical Analytes.* Analytical Chemistry 71:433–439
34. Lu, Y. J., F. Zhang, and S. F. Sui. 2002. *Specific binding of integrin to RGD peptide immobilized on a nitrilotriacetic acid chip: a surface plasmon resonance study.* Biochemistry (Moscow) 67:1122–1129
35. Hoffman, W. L. and D. J. O'Shannessy. 1988. *Site-specific immobilization of antibodies by their oligosaccharide moieties to new hydrazide derivatized solid supports.* Journal of Immunological Methods 112:113–120

36. Wang, D. 2003. *Carbohydrate microarrays.* Proteomics 3:2167–2175
37. Wang, D. and E. A. Kabat . 1996. *Carbohydrate Antigens (Polysaccharides)*, p. 247–276. In M. H. V. Van Regenmortal (ed.), Structure of Antigens, Volume Three. CRC Press, Boca Raton, New York, London, Tokyo
38. Hirabayashi, J., Y. Arata, and K. Kasai. 2001. *Glycome project: Concept, strategy and preliminary application to Caenorhabditis elegans.* Proteomics 1:295–303
39. Wang, D., S. Liu, B. J. Trummer, C. Deng, and A. Wang. 2002. *Carbohydrate microarrays for the recognition of cross–reactive molecular markers of microbes and host cells.* Nature biotechnology 20:275–281
40. Willats, W. G. T., S. E. Rasmussen, T. Kristensen, J. D. Mikkelsen, and J. P. Knox. 2002. *Sugar–coated microarrays: A novel slide surface for the high–throughput analysis of glycans.* Proteomics 2:1666–1671
41. Folch, A. and M. Toner. 2000. *Microengineering of cellular interactions.* Annual review of biomedical engineering 2:227-256
42. Mrksich, M. 2000. *A surface chemistry approach to studying cell adhesion.* Chemical Society Reviews 29
43. Palecek, S. P., J. C. Loftus, M. H. Ginsberg, D. A. Lauffenburger, and A. F. Horwitz. 1997. *Integrin–ligand binding properties govern cell migration speed through cell–substratum adhesiveness.* Nature 385:537–540
44. Garcia, A. J., M. D. Vega, and D. Boettiger. 1999. *Modulation of cell proliferation and differentiation through substrate–dependent changes in fibronectin.* Molecular biology of the cell 10:785–798
45. Nicolau, D. V., T. Taguchi, H. Tanigawa, and S. Yoshikawa. 1996. *Control of the neuronal cell attachment by functionality manipulation of diazo–naphthoquinone/novolak photoresist surface.* Biosensors and Bioelectronics 11:1237–1252
46. Nicolau, D. V., T. Taguchi, H. Taniguchi, H. Tanigawa, and S. Yoshikawa. 1999. *Patterning neuronal and glia cells on light–assisted functionalized photoresists.* Biosensors & Bioelectronics 14:317–325
47. Chen, C. S., M. Mrksich, S. Huang, G. M. Whitesides, and D. E. Ingber. 1998. *Micropatterned surfaces for control of cell shape, position, and function.* Biotechnology Progress 14
48. Flemming, R. G., C. J. Murphy, G. A. Abrams, S. L. Goodman, and P. F. Nealey. 1999. *Effects of synthetic micro– and nano–structured surfaces on cell behavior.* Biomaterials 20:573–588
49. Singhvi, R., G. Stephanopoulos, and D. I. C. Wang. 1994. *Review: Effects of substratum morphology on cell physiology.* Biotechnology and Bioengineering 43:764–771
50. Abbott, A. 2003. *Biology's new dimension.* Nature 424:870–872
51. Weaver, V. M., O. W. Petersen, F. Wang, C. A. Larabell, P. Briand, C. Damsky, and M. J. Bissell. 1997. *Reversion of the malignant phenotype of human breast cells in three–dimensional culture and in vivo by integrin blocking antibodies.* The Journal of Cell Biology 137:231–245
52. Wang, F., V. M. Weaver, O. W. Petersen, C. A. Larabell, S. Dedhar, P. Briand, R. Lupu, and M. J. Bissell. 1998. *Reciprocal interactions between 1–integrin and epidermal growth factor receptor in three–dimensional basement membrane breast cultures: A different perspective in epithelial biology.* Proceedings of the National Academy of Science 95:14821–14826

3

Surfaces and Substrates

Alvaro Carrillo, Kunal V. Gujraty, and Ravi S. Kane

3.1 Introduction

This chapter describes several approaches that have been used to fabricate DNA and protein microarrays . These microarrays may be used to perform highly miniaturized assays, in parallel, for numerous research, clinical, and diagnostic applications.

The composition and morphology of the substrate and the choice of surface chemistry influence several critical requirements for the successful implementation of microarray technology. These requirements include the controlled and reproducible spatial deposition of microliter or nanoliter amounts of sample on a surface, the stable attachment of biomolecules to the surface without denaturation, the immobilization of biomolecules at high density and at a high and consistent surface concentration, and detection methods that will provide a quantitative measure of the interaction. The ability to reuse the microarray surface is also desirable.

The surface modification technique should be easy, fast, reliable and form stable surfaces; surface chemistry is a major determinant of the stability of attachment of biomolecules. The surface must allow biomolecule attachment without denaturation or deactivation. In order to guarantee that only relevant interactions are measured, it is also necessary for the surface to be resistant to the non-specific adsorption of biomolecules and other analytes present in solution. The substrate must be compatible with the measurement method, that is, depending on the case, it must offer low fluorescence [1,2] or chemiluminescence background, or should be compatible with surface plasmon resonance or mass spectrometry. Minimum interference from the substrate in the detection stage is critical for generating microarrays with high sensitivity.

3.2 DNA Microarrays

Substrates to be used in DNA microarrays are required to have thermal and chemical stability, flatness and homogeneity, and need to be amenable to biochemical manipulation. A variety of techniques have been developed to attach probes – cDNA or oligonucleotides – to different substrates. There are two major strategies that are used: (1) in situ synthesis, which involves the synthesis of oligonucleotides on the substrates, base by base, and (2) the attachment of cDNA or presynthesized oligonucleotides to the substrate, either covalently or non-covalently. The surface modification techniques that have been used to fabricate DNA microarrays on a variety of substrates are summarized below. Table 3.1 lists several commercial suppliers of DNA arrays.

Table 3.1. Summary of commercially available surfaces for DNA microarrays

Provider	Technology	Web site
Affymetrix	In–situ synthesis using photolithographic method	`www.affymetrix.com`
Corning Inc.	GAPS™ derivatized surface	`www.corning.com/lifesciences`
BD Biosciences	Nylon, glass and plastic based arrays	`www.clontech.com`
Erie Scientific Company	Aminopropylsilane coated slides, 3D APS, poly(L–lysine) coated slides, epoxy coated substrates	`www.eriesci.com`
Metrigenix	Flow–Thru Chip™ (4D array) substrate comprising of a network of microchannels	`http://www.metrigenix.com`
Apogent Discoveries	Aminosilane derivatized slides, proprietary modified oligonucleotides technology for attaching Acrydite™	`http://www.apogent discoveries.com`
Surmodics	Code–Link™ slides designed to covalently attach amino-modified oligonucleotides	`http://www.surmodics.com`
Xenopore	Amino, aldehyde, epoxy, maleimide, thiol, biotin and streptavidin coated slides	`http://www.xenopore.com`

3.2.1 Glass Substrates

Glass is the most widely used substrate for DNA arrays as it is flat, transparent, resistant to high temperatures, easy to handle, and has low fluorescence. Techniques for modifying glass substrates are also well developed.

In Situ Synthesis

The Affymetrix method [3–5] uses solid–phase chemistry, photolabile protecting groups, and photolithography to synthesize oligonucleotides base–by–base. The surface is reacted with a linker having a photolabile group at its free end. Light is then directed to specific regions of the substrate by using a photolithographic mask, resulting in the removal of the photolabile groups and the activation of the linkers in these regions. The 'activated' ends react with nucleotides forming a covalent bond, and the process is repeated to build up different sequences at different sites on the substrate. The photolithographic fabrication method allows the construction of dense arrays containing many different probes in a small area. More than 400,000 different square probe regions can be packed into an area of about 1 cm^2 [6]. The major disadvantage of this method stems from the fact that the yield per cycle (i.e. per nucleotide attachment step) is $\sim$ 95% [5], which limits the probe length that can be synthesized with high fidelity. An alternate method for the in situ synthesis of oligonucleotide arrays using photogenerated acids (PGAs) has been reported by Gao et al. [7].

Covalent Attachment of Probes to Substrates Functionalized with Amino Groups

This method is among the most widely used techniques for immobilizing probes onto glass substrates. Glass slides can be silanized by immersing them in a 2% solution of 3–aminopropyl–triethoxysilane (APTES) in acetone for 40 minutes at room temperature followed by three acetone washes [8]. The silanization may also be carried out using p–aminophenyl–trimethoxysilane [8]. Aminosilane-coated slides may also be purchased commercially [9]. Presynthesized oligonucleotides having amino–modifiers can be attached to the aminosilane coated slides using bi-functional linkers [6]; alternatively, the amino-modified oligonucleotides can be succinylated and then covalently attached to the slides by amide bond formation using 1–(3–dimethylamino–propyl)–3–ethylcarbodiimide hydrochloride (EDC). Free amine groups on the substrate may be blocked chemically, in order to minimize the non-specific adsorption of the negatively charged oligonucleotides during the hybridization step [9]. Non-specific binding may also be minimized by prehybridization in a solution containing 1% bovine serum albumin [9].

Attachment of Probes to Poly(L–lysine)-coated Glass Substrates

This technique makes use of the adsorption of the polyanionic probes onto the polycation-coated glass substrate via electrostatic interactions [6]. Poly(L–lysine)-coated glass slides are obtained by immersing cleaned glass slides in an aqueous buffered solution of poly(L–lysine) [10]. The slides are dried, and then stored at room temperature for a month, to allow the surface to become sufficiently hydrophobic [10]. The hydrophobicity of the surface is critical for obtaining printed DNA spots of small size, and hence for generating high density arrays. An arraying robot is used to deposit the probes onto the slides from solutions in aqueous buffer. This step is followed by four post-processing steps: rehydration and drying, crosslinking of the DNA to the slide by UV irradiation, blocking of the free amine groups by acylation with succinic anhydride, and denaturation [10].

Other Techniques

There are several other methods for immobilizing DNA onto glass substrates. Silanized DNA can be attached to unmodified glass surfaces covalently [11]. Chrisey et al. have described methods for the formation of patterned single or multiple DNA species on glass microscope slides using photolithographic technique [12]. The covalent attachment of disulfide-modified oligonucleotides to mercaptosilane-modified glass [13], amine-modified oligonucleotides to aldehyde-modified surfaces [14] or epoxy-modified surfaces [15], aldehyde-modified oligonucleotides to semicarbazide-coated surfaces, and oligonucleotides to diazotized surfaces [16] are other approaches that may be used for fabricating microarrays.

3.2.2 Silicon Substrates

Oxidized silicon substrates can be modified by silanization, and by the adsorption of polycations such as poly(L–lysine). Consequently, the techniques used to attach probes to glass substrates may also be used to attach probes to oxidized silicon substrates [12,15,17,18].

Unoxidized crystalline silicon offers several advantages as a substrate for DNA microarrays including high purity, a highly organized and defined crystalline structure, robustness, and thermal and chemical stability [19]. Strother et al. [19] developed a technique for attaching oligonucleotides to unoxidized silicon substrates. Hydrogen terminated silicon wafers are generated by exposing wafers to a 2% solution of HF in water. The wafers are then covered with tertbutyloxycarbonyl (t–BOC)-protected 10–aminodec–1–ene and exposed to UV light for 2 hours. The surfaces are then treated with 25% trifluoroacetic acid in dichloromethane and rinsed with 10% ammonium hydroxide to remove the t–BOC protecting group and form surfaces terminated with primary amines. Thiol-modified probes can then be covalently attached to the amine-functionalized surfaces using the heterobifunctional crosslinker sulfo–succinimidyl 4–(N–maleimidomethyl) cyclohexane–1–carboxylate (SSMCC). The DNA-modified surfaces are rinsed with distilled water and stored at 37°C for 1 hour in a buffer containing sodium dodecyl sulfate to remove nonspecifically bound strands.

3.2.3 Gold Substrates

Gold-coated substrates have been used for immobilizing oligonucleotides to form an array. The primary advantage of gold-coated substrates is that they can be functionalized by forming self-assembled monolayers (SAMs) of alkanethiolates. The use of ω-functionalized alkanethiolates allows the chemistry of the interface to be controlled at the molecular level. Patterned SAMs may be generated by using photolithographic, soft lithographic, and other techniques [20–22]. Gold-coated substrates are also compatible with surface plasmon resonance (SPR) imaging techniques. SPR can be used to investigate the thermodynamics and the kinetics of binding interactions between unlabelled biomolecules in real time.

Gilmor et al. [21] have developed a technique for attaching probes onto patterned gold substrates. Substrates are prepared by evaporating chromium (an adhesion layer) followed by gold onto glass slides or silicon wafers. The gold coated slides are dipped into a solution of 11–mercaptoundecanoic acid (1 mM in ethanol) for $\sim$ 18 hours to form a SAM. Poly(L–lysine) is adsorbed onto the SAM from an aqueous solution (1 mg/ml, pH 8). The surface is then exposed to UV light through a quartz mask, resulting in the oxidation of the gold–sulfur bond in the exposed regions; rinsing the surface with ethanol completely removes the alkanethiol in these regions. Immersion of the substrate into a solution of octadecanethiol generates a substrate having a pattern of hydrophobic (methyl-terminated) and hydrophilic (poly(L–lysine)-terminated) domains. Thiol-modified oligonucleotides can be covalently attached to poly(L–lysine)-terminated regions of the array by using the heterobifunctional linker SSMCC. Corn et al. [23] have also developed a multistep chemical modification procedure to create DNA arrays on gold surfaces. They used SPR imaging to measure the adsorption of single stranded DNA–binding protein onto the oligonucleotide array.

3.2.4 Gels

Probes have been immobilized in gels on substrates like glass. 'Three–dimensional' gels can provide more than 100 times greater capacities for immobilization than two–dimensional substrates, and can provide higher sensitivities [24–26]. Gels provide a stable support with low fluorescence background and a high shelf life.

Mirzabekov and co-workers have developed procedures for immobilizing oligonucleotide probes in polyacrylamide gels [24–26]. The gel micromatrices, prepared by the photopolymerization of acrylamide, are activated by treatment with 100% hydrazine hydrate at 18 $\pm$ 2°C for 40 minutes, resulting in the incorporation of hydrazide groups into the gel. The space between gel elements on the glass slide is made hydrophobic by treatment with Repel–Silane. Activated oligonucleotides are immobilized by coupling with the hydrazide groups of the gel. Alternatively, amine-modified oligonucleotides can be immobilized onto micromatrix gel pads containing aldehyde groups [25,27]. Oligonucleotides have also been immobilized on glass slides coated with an activated agarose film [28]; the outcome of hybridizations with longer labelled fragments was less reliable on these slides than on conventional aldehyde-functionalized glass slides.

3.2.5 Fiber Optic Arrays

This technique utilizes probes that are immobilized onto microspheres using well established procedures [29–32]. Probe-functionalized microspheres are coupled to high density fiber optic arrays; the optical fiber substrate allows simultaneous and repetitive monitoring of the microsphere array [30,33]. Amine-modified oligonucleotides are activated by treatment with cyanuric chloride, and then reacted with polyethyleneimine (PEI)-coated microspheres. The beads are then rinsed with a sodium borate buffer, and the unreacted amine groups on the beads are capped using succinic anhydride to prevent non-specific binding of DNA.

3.2.6 Polymers

Polymers like nylon and polypropylene have been used for arraying oligonucleotides. Oligonucleotides have been immobilized onto nylon supports using UV crosslinkers [34–36]. High density arrays have been constructed on aminated polypropylene supports using phosphoramidite chemistry [37–39]. Oligonucleotides have been immobilized onto polypropylene supports covalently, using bifunctional crosslinkers or EDC-mediated amide bond formation between amine-terminated oligonucleotide and carboxylate-modified polypropylene plates [40, 41], and non covalently [42]. Non-covalently immobilized oligonucleotides are, however, susceptible to removal under high salt/high temperature conditions.

3.3 Protein Microarrays

The commercial development of protein microarrays has been difficult in great part due to the increased complexity that comes with dealing with proteins (compared to oligonucleotides or cDNA). Proteins tend to denature on surfaces [43,44]; this denaturation can result in a loss of their activity. Proteins also tend to adsorb non-specifically on a wide variety of surfaces [45]; this non-specific adsorption can lead to the misinterpretation of the results of microarray experiments. On account of these challenges, surface functionalization and protein immobilization procedures are very important for the successful implementation of protein microarrays. A wide variety of substrates and surface chemistries have been used in academic research and some have been developed commercially. Some of these methods are described below; Table 3.2 lists several surfaces available commercially.

Table 3.2. Summary of commercially available surfaces for protein microarrays

Provider	Technology	Web site
TeleChem International Inc.	Aldehyde-modified glass substrates.Epoxy–derivatized glass substrates	www.arrayit.com
Zyomyx Inc.	Titanium dioxide substrates modified with copolymers of poly(L–lysine)–g–poly(ethylene glycol)	www.zyomyx.com
PerkinElmer Inc.	Polyacrylamide gel-coated glass slides. HydroGel™.	http://lifesciences.perkinelmer.com
Biocept Inc.	Polyisocyanate-modified PEG gel on glass substrate.	www.biocept.com
Accelr8 Technol–ogy Corproation	Substrate covered by a three–dimensional polymer matrix	www.accelr8.com
Corning Inc.	GAPS™ coated glass slides	www.corning.com/lifesciences
BD Biosciences Clontech	Antibodies covalently bound to glass surface in ordered array, ready for protein detection. Ab Microarray™.	www.clontech.com
Ciphergen Biosystems Inc.	Surfaces are modified so that they bind proteins by hydrophobic attraction, anion exchange, cation exchange, or metal affinity. Afterwards, proteins are analyzed by MS technology. ProteinChip™ Arrays.	www.ciphergen.com
Panomics Inc.	SH3 domain arrays interact with proline-rich peptides. Ready to be used for investigation of protein function.	www.panomics.com
HTS Biosystems	Complete system for protein detection on gold substrates via Surface Plasmon Resonance.	www.htsbiosystems.com

3.3.1 Glass Substrates

Due to easy availability, flatness and the possibility of chemical modification, the use of glass substrates, especially in the form of microscope slides, has been common.

Peptide arrays have been prepared in situ on amino-modified glass substrates. The amino groups at the ends of linkers attached to glass substrates were protected with the photolabile nitroveratryloxycarbonyl (NVOC) protecting group. Illumination of the substrate through a patterned mask resulted in the removal of the protecting groups in selected regions of the substrate; the free amino groups were reacted with an NVOC-protected amino acid. This process was repeated several times to generate different peptide sequences at different locations on the substrate [3].

MacBeath and Schreiber described a procedure for fabricating protein microarrays on glass slides modified with an aldehyde–containing silane reagent [46,47]. This approach was used to screen protein–protein interactions, identify substrates of protein kinases, and identify protein targets of small molecules. A high precision robot was used to print proteins in phosphate-buffered saline containing 40% glycerol. The aldehydes react with primary amines on the protein to form a Schiff's base linkage. The slides were then immersed in a buffer containing bovine serum albumin (BSA) to quench unreacted aldehydes and prevent the non-specific binding of proteins in subsequent steps. The aldehyde-modified substrates used in this study were obtained commercially from TeleChem International under the trade name SuperAldehyde Substrates [48]. This vendor also offers epoxy-derivatized glass surfaces under the trade name SuperEpoxy Substrates; the reaction of the epoxy groups with primary amines of the protein can also be used to attach proteins to surfaces covalently. Protein microarrays fabricated using the aldehyde-based protein immobilization strategy have been used to study protein–protein interactions in the yeast proteome [49], and to study protein expression in cancer cells [50,51].

Amino-derivatized surfaces have been used to covalently immobilize proteins in a microarray [52–54]. Optically flat, 96–well glass plates were functionalized with amine groups by immersing them in a solution of aminopropyltrimethoxysilane (APTMS). Reaction of the amino groups with bis–sulfo–succinimidyl suberate generated an N–hydroxysuccinimide (NHS)-activated surface. Proteins were printed onto the activated substrates robotically, resulting in their covalent attachment to the surface. After washing excess unbound protein, the substrates were incubated with a solution of casein in phosphate buffered saline (PBS) to minimize the non-specific adsorption of proteins in subsequent steps [53,54].

Peptide arrays have also been fabricated by the site-specific ligation of glyoxylyl peptides onto glass surfaces functionalized with semicarbazide groups [55]. Cleaned glass slides were silanized with APTMS. The amino-functionalized surfaces were treated with triphosgen/diisopropylethylamine and 9–Fluorenylmethyl-protected hydrazine (Fmoc–$NHNH_2$); the semicarbazide groups were obtained on removal of the Fmoc groups. These arrays allowed the highly sensitive and specific detection of antibodies in very small blood samples from infected individuals [55].

Poly(L–lysine)-derivatized glass slides have been used to create protein microarrays [51,56]. Proteins were immobilized onto the slides non-covalently, by spotting solutions of the proteins in PBS [56]. The arrays were rinsed to remove unbound protein, and were then incubated overnight at 4°C in a blocking solution containing non-fat milk to minimize the non-specific adsorption of proteins. A further reduction in the extent of non-specific adsorption was deemed to be necessary in order to detect specific target proteins at concentrations below 1 ng/ml [56]. Other protein immobilization techniques have also been reported; for instance, the binding of histidine-tagged proteins to nickel coated slides was used to form a yeast proteome microarray [49].

The analysis of membrane proteins is important, since these proteins represent the most important class of drug targets; approximately 50% of current molecular targets are membrane-bound [57]. The application of microarray technology to membrane proteins has been complicated by the need to immobilize the accompanying lipid membranes in addition to the proteins themselves in order to maintain bioactivity [58, 59]. Fang et al. [57, 60] fabricated microarrays of G protein-coupled receptors (GPCR). Membrane preparations were printed onto ultraflat glass slides modified with γ–aminopropylsilane (GAPS^{TM}). Assays for the screening of ligands on membrane protein microarrays were also described [57].

The techniques described above allow the formation of microarrays of proteins; the immobilization of small molecules in microarrays is also useful for the identification of small molecule ligands for proteins [61, 62]. Macbeath et al. reacted aminosilane-functionalized glass slides with N–succinimidyl 3–maleimido propionate to obtain a surface presenting maleimide groups at high density [62]. Thiol–containing small molecules are covalently attached to the surface on printing, presumably due to the formation of a thioether linkage [62, 63]. No non-specific protein binding was observed in aqueous buffer [62]. Kuruvilla et al. fabricated small molecule microarrays by covalently attaching alcohol–containing small molecules to chlorinated glass surfaces [64]. They used these microarrays to identify compounds that bind the yeast protein Ure2p; one of these compounds was found to specifically activate a glucose-sensitive transcriptional pathway downstream of Ure2p [64].

3.3.2 Silicon Substrates

The techniques described above for immobilizing proteins on silanized or polycation-derivatized glass substrates may also be used for immobilization on oxidized silicon substrates. Mooney et al. have described another technique for immobilizing proteins non-covalently on glass or oxidized silicon substrates [65]. Substrates were functionalized with a monolayer of n–octadecyltrimethoxysilane (OTMS). UV photolithography was used to remove the monolayer in selected regions of the substrate creating a pattern, and biotinylated BSA was then allowed to adsorb onto the substrate. Significantly greater amounts of biotinylated BSA adsorbed in the OTMS-coated regions on the substrate. Streptavidin could be captured on the biotinylated regions; an additional layer of biotinylated protein could then be deposited in these regions [65].

3.3.3 Gold Substrates

Several studies have described the immobilization of proteins on SAMs of alkanethiolates on gold [66–68]. SAMs allow the investigation of biospecific interactions while minimizing background due to the non-specific adsorption of proteins. In a recent study, dip–pen nanolithography^{TM} (DPN^{TM}) was used to generate protein nanoarrays [68]. DPN^{TM} was used to pattern 16–mercaptohexadecanoic acid (MHA) on gold-coated substrates in the form of dots or grids having features ranging from 100 to 350 nanometers. The surrounding areas were passivated with a SAM of a protein-resistant triethylene glycol-terminated alkanethiol. Proteins such as lysozyme and IgG adsorbed selectively on the MHA-coated regions of the substrate. Proteins also retained their biological activity after adsorption [68]. Yang et al. have also described a technique, which they call light-activated micropatterning of proteins (LAMP), for the spatially resolved micropatterning of proteins on SAM-coated substrates [69].

Hodneland et al. described a method for the selective and covalent immobilization of proteins on gold substrates with control over the density and orientation of the protein [66]. The method is based on the active–site directed covalent immobilization of fusion proteins to mixed SAMspresenting phosphonate ligands in a background of protein-resistant triethylene glycol groups. The fusion proteins are comprised of the capture protein (cutinase) and the protein of interest; cutinase forms an active site-specific covalent adduct with phosphonate ligands. SPR spectroscopy showed that cutinase binds irreversibly to the mixed SAM and that the triethylene glycol groups prevent the non-specific adsorption of proteins [66].

Bieri et al. reported a study dealing with G protein-coupled receptors (GPCR) in which biotinylated membranes containing the protein rhodopsin in a specific orientation were immobilized in micrometer-sized patterns onto gold-coated substrates, and SPR was used to follow the process of ligand binding, G protein activation and receptor deactivation [70].

3.3.4 Titanium Dioxide Substrates

Titanium dioxide substrates can be functionalized by the adsorption of polycations such as poly(L–lysine) (PLL). Copolymers of poly(L–lysine)–g–poly(ethylene gly–col) (PLL–g–PEG) also spontaneously adsorb on these substrates and generate a comb-like structure in which the PEG side chains extend into the solution [71,72]. The PEG chains resist the non-specific adsorption of proteins on the underlying substrate [73, 74]. By modifying the PEG side chains with biotin, it is possible to adsorb streptavidin specifically; the streptavidin layer can be used to capture biotinylated proteins in a microarray format [75]. This technology is being commercialized by Zyomyx Inc. [76].

3.3.5 Gels and Membranes: 3D Immobilization

An alternative to printing proteins on flat surfaces is to immobilize proteins in three–dimensional gels. Gels greatly increase the capacity for the immobilization of proteins [77]. Polyacrylamide gels have been produced by persulfate– [78] and photo-induced [24, 78] polymerization; proteins are bound to the gel either by the reaction of amine groups on proteins with the glutaraldehyde-activated gel [24], by copolymerization of acrylamide and bisacrylamide with acryloyl-modified proteins [78], or by the reaction of antibodies that contain aldehyde groups after periodate oxidation with polyacrylamide gels previously activated by partial substitution of amide groups by hydrazide groups [79]. Gel formulations can be tuned to accommodate proteins having a molecular weight as high as 400 kDa [79]. Commercial offerings of gel technology include polyacrylamide [80] and polyisocyanate-modified PEG gels [81].

Proteins also adsorb to hydrophobic nitrocellulose membranes. The binding capacity per unit area is higher than that for flat surfaces, resulting in a greater sensitivity than that achieved on amine- and aldehyde-modified glass surfaces [50]. Nitrocellulose membrane microarrays have been used to study protein–protein, protein–DNA [82], and antibody–antigen interactions [83], and also to monitor the phosphorylation of proteins during cancer progression [84]. Polyvinylidene difluoride filter membranes reportedly offer superior protein binding capacity and mechanical resistance than nitrocellulose membranes, and have also been used to generate protein microarrays using a robotic arrayer. BSA was used as a blocking agent to minimize the non-specific adsorption of proteins [85].

3.3.6 Polymers

Poly(dimethylsiloxane) (PDMS) has been used as a substrate to immobilize proteins both covalently and non-covalently. Yeast kinases were immobilized covalently in arrays of PDMS microwells, by using the crosslinker 3–glycidoxypropyltrimethoxysilane (GPTS) [86]. Microfluidic networks have also been used to form patterns of proteins, adsorbed non-covalently, onto hydrophobic PDMS substrates [87].

Electrospray deposition has been used to fabricate protein microarrays on aluminized plastic substrates. The proteins were administered in a mixture with sucrose, and were attached to the surfaces either non-covalently or covalently by the reaction of amine groups of the proteins with aldehyde-modified substrates [88, 89].

3.4 Conclusion

The choice of substrate and surface chemistry has a major impact on the performance of DNA and protein microarrays. A wide variety of approaches have been used to fabricate these arrays, involving the covalent and non-covalent attachment of probes (oligonucleotides, cDNA, oligopeptides, proteins, and small molecules) to glass, silicon and gold substrates, gels and membranes. Future challenges include the fabrication of microarrays with increased density, lower background, higher immobilization yield, and higher sensitivity.

References

1. Che DP, Bao YJ, Muller UR (2001) *Novel surface and multicolor charge coupled device-based fluorescent imaging system for DNA microarrays.* Journal of Biomedical Optics 6: 450–456
2. Raghavachari N, Bao YP, Li G, Xie X, Muller UR (2003) *Reduction of autofluorescence on DNA microarrays and slide surfaces by treatment with sodium borohydride.* Anal Biochem 312: 101–105
3. Fodor SPA, Read JL, Pirrung MC, Stryer L, Lu AT, Solas D (1991) *Light-directed, spatially addressable parallel chemical synthesis.* Science 251: 767–773
4. Lipshutz R, Fodor SPA, Gingeras TR, Lockhart DJ (1999) *High Density synthetic oligonucleotide arrays.* Nature 21: 20–24
5. McGall GH, Barone AD, Duggelmann M, Fodor SPA, Gentalen E, Ngo N (1997) *The Efficiency of light directed synthesis of DNA arrays on glass substrates.* J Am Chem Soc 119: 5081–5090
6. Graves DJ (1999) *Powerful tools for genetic analysis come of age.* Trends in Biotechnology 17: 127–134
7. Gao X, LeProust E, Zhang H, Srivannavit O, Gulari E, Yu P, Nishiguchi C, Xiang Q, Zhou X (2001) *A flexible light-directed DNA chip synthesis gated by deprotection using solution photogenerated acids.* Nucleic Acids Research 29: 4744–4750
8. Joos B, Kuster H, Cone R (1997) *Covalent attachment of hybridizable oligonucleotides to glass supports.* Anal Biochem 247: 96–101
9. Hegde P, Qi R, Abernathy K, Gay C, Dharap S, Gaspard R, Hughes JE, Snesrud E, Lee N, Quakenbush J (2000) *A concise guide to cDNA microarray analysis.* BioTechniques 29: 548–562
10. Eisen MB, Browm PO (1999) *DNA arrays for analysis of gene expression.* Methods in Enzymology 303: 179–205
11. Kumar A, Larsson O, Parodi D, Liang Z (2000) *Silanized nucleic acids: a general platform for DNA immobilization.* Nucleic Acids Research 28: e71
12. Chrisey LA, O'Ferrall CE, Spargo BJ, Dulcey CS, Calvert JM (1996) *Fabrication of patterned DNA surfaces.* Nucleic Acids Research 24: 3040–3047
13. Rogers Y-H, Jiang-Baucom P, Huang Z-J, Bogdanov V, Anderson S, Boyce-Janico MT (1999) *Immobilization of oligonucleotides onto a glass support via disulfide bonds: a method for preparation of DNA arrays.* Anal Biochem 266: 23–30
14. Zammatteo N, Jeanmart L, Hamels S, Courtois S, Louette P, Hevesi L, Remacle J (2000) *Comparison between different strategies of covalent attachment of DNA to glass surfaces to build DNA microarrays.* Anal Biochem 280: 143–150
15. Lamture JB, Beattie KL, Burke BE, Eggero HD, Ehrlich DJ, Fowler R, Hollis MA, Kosicki BB, Reich RK, Smith SR, Varma RS, Hogan ME (1994) *Direct*

detection of nucleic acid hybridization on the surface of a charge coupled device. Nucleic Acids Research 22: 2121–2125
16. Dolan P, Wu Y, Ista LK, Metzenberg RL, Nelson MA, Lopez GP (2001) *Robust and efficient synthetic method for forming DNA microarrays.* Nucleic Acids Research 29: e107
17. Blanchard AP, Kaiser RJ, Hood LE (1996) *High density oligonucleotide arrays.* Bio-sensors and Bioelectronics 11: 687–690
18. Gray DE, Case-Green SC, Fell TS, Dobson PJ, Southern EM (1997) *Ellipsometric and interferometric characterization of DNA probes immobilized on a combinatorial array.* Langmuir 12: 2833–2842
19. Strother T, Cai W, Zhao X, Hamers RJ, Smith LM (2000) *Synthesis and characerization of DNA–modified silicon (111) surfaces.* J Am Chem Soc 122: 1205–1209
20. Demers LM, Ginger DS, Park SJ, Li Z, Chung SW, Mirkin CA (2002) *Direct patterning of modified oligonucleotides on metals and insulators by dip–pen nanolithography.* Science 296: 1836–1841
21. Gilmor SD, Thiel AJ, Strother TC, Smith LM, Lagally MG (2000) *Hydrophilic/hydrophobic patterned surfaces as templates for DNA arrays.* Langmuir 16: 7223–7228
22. Zhang H, Li Z, Mirkin CA (2002) *Dip–pen nanolithography–based methodology for preparing arrays of nanostructures functionalized with oligonucleotides.* Advanced Materials 14: 1472–1474
23. Brockman JM, Frutos AG, Corn RM (1999) *A multistep chemical modification procedure to create DNA arrays on gold surfaces for the study of protein–DNA interactions with surface plasmon resonance imaging.* J Am Chem Soc 121: 8044–8051
24. Guschin D, Yershov G, Zaslavsky A, Gemmell A, Shick V, Proudnikov D, Arenkov P, Mirzabekov A (1997) *Manual manufacturing of oligonucleotide, DNA, and protein microchips.* Anal Biochem 250: 203–211
25. Proudnikov D, Timofeev E, Mirzabekov A (1998) I*mmobilization of DNA in polyacrylamide gel for the manufacture of DNA and DNA–oligonucleotide microchips.* Anal Biochem 259: 34–41
26. Timofeev EN, Kochtkova SV, Mirzabekov AD, Florentiev VL (1996) *Regioselective immobilization of short oligonucleotides to acrylic copolymer gels.* Nucleic Acids Research 24: 3142–3148
27. Dubiley S, Kirillov E, Lysov Y, Mirzabekov A (1997) *Fractionation, phosphorylation and ligating on oligonucleotide microchips to enhance sequencing by hybridization.* Nucleic Acids Research 25: 2259
28. Afanassiev V, Hanemann V, Wölfl S (2000) *Preparation of DNA and protein microarrays on glass slides coated with an agarose film.* Nucleic Acids Research 28: e66
29. Egner BJ, Rana S, Smith H, Bouloc N, Frey JG, Brocklesby WS, Bradley M (1997) *Tagging in combinatorial chemistry: the use of coloured and fluorescent beads.* Chemical Communications 8: 735–736
30. Ferguson JA, Steemers FJ, Walt DR (2000) *High density fiber optic DNA random microsphere array.* Anal Chem 72: 5618–5624
31. Needels MC, Jones DG, Tate EH, Heinkel GL, Kochersperger LM, Dower WJ, Barrett RW, Gallop MA (1993) *Generation and screening of an oligonucleotide–encoded synthetic peptide library.* Proc Natl Acad Sci USA 90: 10700–10704

32. Ness JV, Kalbfleish S, Petrie CR, Reed MW, Tabone JC, Vermeulen NMJ (1991) *A versatile solid support system for oligodeoxynucleotide probe–based hybridization assays.* Nucleic Acids Research 19: 3345–3350
33. Walt DR (2000) *Molecular biology – bead-based fiber–optic arrays.* Science 287: 451–452
34. Church GM, Gilbert W (1984) *Genomic sequencing.* Proc Natl Acad Sci USA 81: 1991–1995
35. Saiki RK, Walsh PS, Levenson CH, Erlich HA (1989) *Genetic analysis of amplified DNA with immobilized sequence–specific oligonucleotide probes.* Proc Natl Acad Sci USA 86: 6230–6234
36. Zhao N, Hashida H, Takahashi N, Misumi Y, Sakaki Y (1995) *High–density cDNA filter analysis: a novel approach for large scale quantitative analysis of gene expression.* Gene 156: 207–213
37. Matson RS, Rampal J, Pentoney SL, Anderson PD, Coassin P (1995) *Biopolymer synthesis on polypropylene supports: oligonucleotide arrays.* Anal Biochem 224: 110–116
38. Shchepinov MS, Case-Green SC, Southern EM (1997) *Steric factors influencing hybridization of nucleic acids to oligonucleotide arrays.* Nucleic Acids Research 25: 1155–1161
39. Weiler J, Hoheisel JD (1996) *Combining the preparation of oligonucleotide arrays and synthesis of high–quality primers.* Anal Biochem 243: 218–227
40. Kohsaka H, Taniguchi A, Richman DD, Carson DA (1993) *Microtiter format gene quantification by covalent capture of competitive PCR products–application to HIV1 detection.* Nucleic Acids Research 21: 3469–3472
41. Running JA, Urdea MS (1990) *A procedure for productive coupling of synthetic oligonucleotides to polystyrene microtiter wells for hybridization capture.* BioTechniques 8: 276–277
42. Nikiforov TT, Rogers Y-H (1995) *The use of 96–well polystyrene plates for DNA hybridization–based assays : An evaluation of different approaches to oligonucleotide immobilization.* Anal Biochem 227: 201–209
43. Butler JE, Ni L, Nessler R, Joshi DS, Suter M, Rosenberg B, Chang J, Brown WR, Cantarero LA (1992) *The physical and functional behavior of capture antibodies adsorbed on polystyrene.* J Immunol Meth 150: 77–90
44. Butler JE, Ni L, Brown WR, Joshi KS, Chang J, Rosenberg B, Voss EW (1993) *The immunochemistry of sandwich ELISAs – VI. Greater than 90% of monoclonal and 75% of polyclonal anti-fluorescyl capture antibodies (CAbs) are denatured by passive adsorption.* Molecular Immunology 30: 1165–1175
45. Chapman RG, Ostuni E, Liang MN, Meluleni G, Kim E, Yan L, Pier G, Warren HS, Whitesides GM (2001) *Polymeric thin films that resist the adsorption of proteins and the adhesion of bacteria.* Langmuir 17: 1225–1233
46. www.cgr.harvard.edu/macbeath/research/protein_microarrays/protein_microarrays.html
47. MacBeath G, Schreiber SL (2000) *Printing proteins as microarrays for high–throughput function determination.* Science 289: 1760–1763
48. http://arrayit.com/products/substrates/substrates.html
49. Zhu H, Bilgin M, Bangham R, Hall D, Casamayor A, Bertone P, Lan N, Jansen R, Bidlingmaier S, Houfek T, Mitchell T, Miller P, Dean R, Gerstein M, Snyder M (2001) *Global analysis of protein activities using proteome chips.* Science 293: 2101–2105

50. Madoz-Gúrpide J, Wang H, Misek DE, Brichory F, Hanash SM (2001) *Protein based microarrays: a tool for probing the proteome of cancer cells and tissues.* Proteomics 1: 1279–1287
51. Sreekumar A, Nyati MK, Varambally S, Barrette TR, Ghosh D, Lawrence TS, Chinnaiyan AM (2001) *Profiling of cancer cells using protein microarrays: discovery of novel radiation–regulated proteins.* Can Res 61: 7585–7593
52. Benters R, Niemeyer CM, Wohrle D (2001) *Dendrimer–activated solid supports for nucleic acid and protein microarrays.* Chembiochem 2: 686–694
53. Mendoza LG, McQuary P, Mongan A, Gangadharan R, Brignac S, Eggers M (1999) *High–throughput microarray–based enzyme–linked immunosorbent assay (ELISA).* BioTechniques 27: 778–788
54. Wiese R, Belosludtsev Y, Powdrill T, Thompson P, Hogan M (2001) *Simultaneous multianalyte ELISA performed on a microarray platform.* Clin Chem 47: 1451–1457
55. Melnyk O, Duburcq X, Olivier C, Urbes F, Auriault C, Gras-Masse H (2002) *Peptide arrays for highly sensitive and specific antibody–binding fluorescence assays.* Bioconjugate Chem 13: 713–720
56. Haab BB, Dunham MJ, Brown PO (2001) *Protein microarrays for highly parallel detection and quantitation of specific proteins and antibodies in complex solutions.* Genome Biology 2: research0004.1–0004.13
57. Fang Y, Frutos AG, Lahiri J (2002) *Membrane protein microarrays.* J Am Chem Soc 124: 2394–2395
58. Cremer PS, Boxer SG (1999) *Formation and spreading of lipid bilayers on planar glass supports.* J Phys Chem B 103: 2554–2559
59. Lahiri J, Kalai P, Frutos A, Jonas SJ, Schaeffler R (2000) *Method for fabricating supported bilayer lipid membranes on gold.* Langmuir 16: 7805–7810
60. Fang Y, Frutos AG, Lahiri J (2002) *G–protein–coupled receptor microarrays.* Chem–biochem 3: 987–991
61. Hergenrother PJ, Depew KM, Schreiber SL (2000) *Small–molecule microarrays: covalent attachment and screening of alcohol–containing small molecules on glass slides.* J Am Chem Soc 122: 7849–7850
62. MacBeath G, Koehler AN, Schreiber SL (1999) *Printing small molecules as microarrays and detecting protein–ligand interactions en masse.* J Am Chem Soc 121: 7967–7968
63. http://www.schreiber.chem.harvard.edu/home/protocols/SMP.html
64. Kuruvilla FG, Shamji AF, Sternson SM, Hergenrother PJ, Schreiber SL (2002) *Dissecting glucose signalling with diversity–oriented sysnthesis and small–molecule microarrays.* Nature 416: 653–657
65. Mooney JF, Hunt AJ, McIntosh JR, Liberko CA, Walba DM, Rogers CT (1996) *Patterning of functional antibodies and other proteins by photolithography of silane monolayers.* Proc Natl Acad Sci USA 93: 12287–12291
66. Hodneland CD, Lee Y, Min D, Mrksich M (2002) *Selective immobilization of proteins to self–assembled monolayers presenting active site–directed capture ligands.* Proc Natl Acad Sci USA 99: 5048–5052
67. Lahiri J, Isaacs L, Tien J, Whitesides GM (1999) *A strategy for the generation of surfaces presenting ligands for studies of binding based on an active ester as a common reactive intermediate: a surface plasmon resonance study.* Anal Chem 71: 777–790
68. Lee K, Park S, Mirkin C, Smith J, Mrksich M (2002) *Protein nanoarrays generated by dip–pen nanolithography.* Science 295: 1702–1705

69. Yang Z, Frey W, Oliver T, Chilkoti A (2000) *Light–activated affinity micropatterning of proteins on self–assembled monolayers on gold.* Langmuir 16: 1751–1758
70. Bieri C, Heyse OP, Hofmann KP, Vogel H (1999) *Micropatterned immobilization of a G protein–coupled receptor and direct detection of G protein activation.* Nat Biotechnol 17: 1105–1108
71. Kenausis GL, Vörös J, Elbert DL, Huang N, Hofer R, Ruiz-Taylor L, Textor M, Hubbell JA, Spencer ND (2000) *Poly(L–lysine)–g–Poly(ethylene glycol) layers on metal oxide surfaces: attachment mechanism and effects of polymer architecture on resistance to protein adsorption.* J Phys Chem B 104: 3298–3309
72. Ruiz-Taylor LA, Martin TL, Wagner P (2001) *X-ray photoelectron spectroscopy and radiometry studies of biotin–derivatized poly(L–lysine)–grafted–poly(ethylene glycol) monolayers on metal oxides.* Langmuir 17: 7313–7322
73. Harris JM, Zalipsky S (1997) *Poly(ethylene glycol): chemistry and biological applications.* American Chemical Society, Washington D.C.
74. Jenney CR, Anderson JM (1999) *Effects of surface–coupled polyethylene oxide on human macrophage adhesion and foreign body giant cell formation in vitro.* J Biomed Mater Res 44: 206–216
75. Ruiz-Taylor LA, Martin TL, Zaugg FG, Witte K, Indermule P, Nock S, Wagner P (2001) *Monolayers of derivatized poly(L–lysine)–grafted–poly(ethylene glycol) on metal oxides as a class of biomolecular interfaces.* Proc Natl Acad Sci USA 98: 852–857
76. Frederickson RM (2002) *Zyomyx, protein chips: protein chemistry comes to the surface.* Chem Biol 9: 763–765
77. Angenendt P, Glökler J, Murphy D, Lehrach H, Cahill DJ (2002) *Toward optimized antibody microarrays: a comparison of current microarray support materials.* Anal Biochem 309: 253–260
78. Vasiliskov AV, Timofeev EN, Surzhikov SA, Drobyshev AL, Shick VV, Mirzabekov AD (1999) *Fabrication of microarray of gel–immobilized compounds on a chip by co-polymerisation.* BioTechniques 27: 592–605
79. Arenkov P, Kukhtin A, Gemmell A, Voloshchuk S, Chupeeva V, Mirzabekov A (2000) *Protein microchips: use for immunoassay and enzymatic reactions.* Anal Biochem 278: 123–131
80. http://lifesciences.perkinelmer.com/areas/proteomics/chem.asp HydroGel^{TM}.
81. http://www.biocept.com/technology.html
82. Ge H (2000) *UPA, a universal protein array system for quantitative detection of protein–protein, protein–DNA, protein–RNA and protein–ligand interactions.* Nucleic Acids Research 28: e3, i–vii
83. de Wildt RMT, Mundy CR, Gorick BD, Tomlinson IM (2000) *Antibody arrays for high–throughput screening of antibody–fantigen interactions.* Nature Biotechnology 18: 989–994
84. Paweletz CP, Charboneau L, Bichsel VE, Simone NL, Chen T, Gillespie JW, Emmert-Buck MR, Roth MJ, Petricoin EF, Liotta LA (2001) *Reverse phase protein microarrays which capture disease progression show activation of pro-survival pathways at the cancer invasion front.* Oncogene 20: 1981–1989
85. Lueking A, Horn M, Eickhoff H, Büssow K, Lehrach H, Walter G (1999) *Protein microarrays for gene expression and antibody screening.* Anal Biochem 270: 103–111

86. Zhu H, Klemic JF, Chang S, Bertone P, Casamayor A, Klemic KG, Smith D, Gerstein M, Reed MA, Snyder M (2000) *Analysis of yeast protein kinases using protein chips.* Nature Genetics 26: 283–289
87. Bernard A, Michel B, Delamarche E (2001) *Micromosaic immunoassays.* Anal Chem 73: 8–12
88. Avseenko N, Morozova TY, Ataulliakhanov F, Morozov VN (2001) *Immobilization of proteins in immunochemical microarrays fabricated by elecrospray deposition.* Anal Chem 73: 6047–6052
89. Morozocv VN, Morozova TY (1999) *Electrospray deposition as a method for mass fabrication of mono- and multicomponent microarrays of biological and biologically active substances.* Anal Chem 71: 3110–3117

4

Reagent Jetting Based Deposition Technologies for Array Construction

Mitchel J. Doktycz

4.1 Introduction

Technologies utilizing arrayed biological reagents are revolutionizing bioanalytical measurements. In genomics, initial successes in gene microarray experiments for analysis of gene transcription have led to applications involving microarrays of proteins [1, 2], whole cells [3], membranes [4] and small molecules [5]. High throughput screening applications, which exploit small volume reaction mixtures, are also leveraging off of microarray technology. Deposition technologies have developed to meet the challenges inherent to these various applications and materials. Deposition technologies must be compatible with the assay requirements (e.g. reagent conservation, volume metering, array density) and bridge 'macro–scale' sample containers to microscale assay devices.

Two robust technologies are becoming conventional. Currently, most microarraying of prepared reagents is carried out using pin based, touch–off deposition techniques, which is the subject of the following chapter in this book. This technique is inherently simple and numerous variants of pin spotting are in practice or development. Another approach with gaining popularity is based on reagent jetting. Similar to ink jetting technology that is commonly used in desktop printers, reagent jetting does not require contact between the dispensing tip and surface and allows for metering of extremely small volumes of reagent. This latter technique will be overviewed herein, highlighting variants and their specific strengths.

4.2 Reagent Jetting – Technology Overview

Various approaches to reagent jetting are currently in use. These techniques borrow features and technology developed for commonly used ink jet based desktop printers [6]. These printers are typically drop–on–demand devices that use either thermal or piezo based actuation mechanisms (Fig. 4.1). The development of mass-marketed printing devices over the past several decades has facilitated the use of this technology for other purposes, including biomedical applications, solder dispensing, construction of three–dimensional ceramic structures and construction of organic-based electronic circuits [7]. Another class of reagent jetting devices is based on a high-speed solenoid valve. This latter technique is typically used for industrial applications, such as bar code printing or container labelling, and is becoming popular in liquid handling instruments.

A key strength of reagent jetting techniques is the ability to rapidly dispense extremely small volumes (picoliter level) of liquid. When compared to touch–off spotting techniques, reagent jetting is a gentle deposition technique, enabling printing on fragile substrates, and can allow for volume metering (discussed further below). General limitations of the technique are its complexity, relative to pin printing, and effective operation occurs within a pre-designed range of physical and chemical parameters. Selection of a particular reagent jetting technique depends on the intended application. Specific advantages and disadvantages of different reagent jetting techniques are discussed in the relevant sections below.

A general consideration is the relation between a dispensed volume and the resultant spot size. Using a hemispherical cap as a model for a sessile drop on a flat surface leads to the following relation between spot size (radius, r) and volume (V):

$$V = (3b + b^3)\pi r^3/6 \tag{4.1}$$

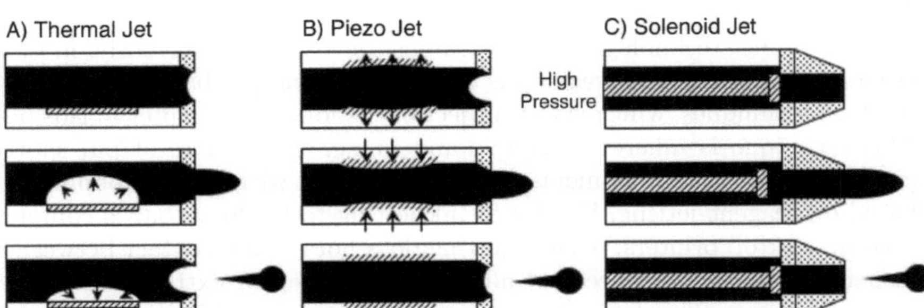

Fig. 4.1. Reagent jetting techniques: Cross sectional view of the fluid channel and nozzle, illustrating the mechanism of droplet ejection, for three different reagent dispensing techniques is shown. Thermal-based jets **(a)** operate by rapidly heating and cooling the reagent, which results in ejection of a droplet. Piezo-based techniques **(b)** employ rapid expansion and contraction of the piezo material to cause droplet ejection. Solenoid based jets **(c)** function by rapidly opening and closing a valve that controls the flow of a pressurized reagent

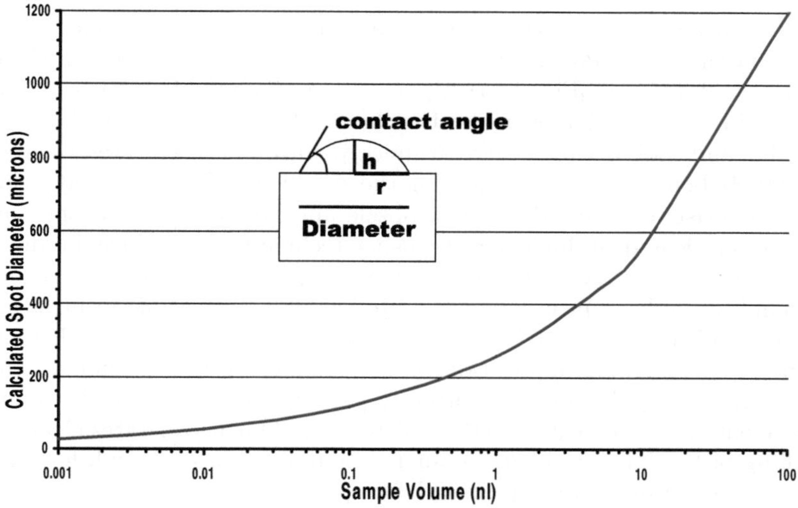

Fig. 4.2. Predicted relation between dispensed volume and spot size: The calculated spot sizes are determined based on the dispensed volume and the contact angle formed between the surface and liquid as shown in the inset drawing

where b is the ratio between the height and radius of the sessile drop. This value can be related to the contact angle that the droplet makes with the surface. These parameters are pictorially described in Fig. 4.2. Also shown in Fig. 4.2 is a graph of this relation using a contact angle of 40° (such as occurs between water and a poly–L–lysine treated glass surface). The graph shows the relation between spot diameter and dispensed volume. Approximately an order of magnitude lowering in volume is required to drop the spot diameter in half, and picoliter scale volumes are therefore required for spot diameters on the order of a few tens of microns. Such volumes are in line with reagent jet dispensing. For example, a 1200 dpi printer corresponds to spot diameters on the order of 20 μm. Alternatively, to achieve small spots, the contact angle must be increased. Although this variable is not easily changed, welled structures can be used to demarcate the deposition area. This requires careful alignment between the dispenser and the target substrate, which can be challenging when dealing with structures on the order of a few tens of microns. This relation between volume and spot size highlights one of the major challenges for employing any liquid dispensing technique for further miniaturization of arrays.

4.3 Thermal Jet Based Dispensing

Thermal jets, often referred to as bubble jets, eject droplets by superheating a small volume of liquid near the dispensing orifice (Fig. 4.1a). Typically, a resistive heating element is controlled such that the application of current causes rapid heating of the ink. This generates a vapor bubble, forcing liquid from the nozzle (middle panel, Fig. 4.1a). Upon cooling, the bubble collapses, pinching the liquid stream and allowing for the channel to refill (lower panel, Fig. 4.1a). Heating and cooling can occur very quickly, with repetition rates greater than 10,000 Hz being typical. Considering the simplicity of the required structure, which consists of a liquid channel, nozzle and heating element, thermal jets can be fabricated at high density using techniques developed in the semiconductor industry. The simple manufacturing process coupled with the high demand for desktop printers have led to low cost, disposable print heads.

A significant difference between desktop printers and those needed for high throughput screening applications is the number of 'inks' required. The few ink cartridges needed for color printing pales in comparison to the thousands of reagents used to create a cDNA microarray. This necessitates cleaning and refilling of the ink cartridges. A further complication is the ink formulation. Commercial printers are optimized for specific ink compositions and printing densities. Factors such as surface tension and viscosity must be carefully controlled. Further, these inks are often matched with the properties of the print media for optimal performance. Similar considerations are necessary for adopting thermal jet based printing for biomedical applications. To date, custom thermal jet print heads, specifically designed for microarray printing, have not been described. Nevertheless, several examples on the use of commercial printers, adapted to printing biomaterials, have been published [8–10].

To adapt a commercial printer for dispensing DNA or protein solutions, the ink cartridge must be carefully rinsed out and replaced with the biochemical in a solution of similar viscosity and surface tension as the original ink. This can be done by the addition of various reagents such as ethanol [8], glycerol [9] or a detergent such as sodium dodecyl sulfate [10]. The printed spots can be extremely small, on the order of a few tens of microns, which is consistent with the dispensed volume of a few tens of picoliters. The rapid heating, which can reach temperatures of 200–300°C, could presumably cause protein degradation which would lead to low protein activity as well as clogging of the nozzle. However, while extensive evaluation of different proteins has not been performed, the problem of protein denaturation does not appear to be significant. This is likely due to the highly localized heating which expands the liquid behind the ejected droplet. When spotting nucleic acids, the potential for denaturation may be advantageous because single stranded probes are desired.

While the use of a commercial printer for printing biomolecules takes advantage of low instrumentation costs and exploits various computer software programs for defining the printed regions, there are no simple means for changing reagents and complete recovery of unused material is not possible. Therefore, such an approach is only useful for applications where one, or a few reagents, need to be patterned.

4.4 Piezo Jet Based Dispensing

This technology operates by mechanically inducing a pressure wave into the liquid. Rapid dimensional changes of a piezoelectric material can induce this pressure pulse to eject a single droplet (Fig. 4.1b). A number of different designs are employed in desktop printers, with the piezoelectric material operating in either a push, pull, shear or squeeze tube mode [6]. The characteristics of the droplet are dependent on a number of factors including the physical and chemical characteristics of the liquid, the nozzle structure and the dimensions of the preceding fluidic chamber. Commercial desktop printers employ dozens to hundreds of individually controlled dispensers. In contrast to thermal jet printers, single channel piezo-based dispensers are commercially available for applications other than desktop printing. Instruments or components from manufacturers such as MicroFab Technologies [11], Microdrop GmbH [12], and Perkin Elmer Life Sciences [13] are commonly used for biomedical applications. Perkin Elmer's Packard BioChip Arrayer™ and SpotArray™ Enterprise are specifically designed for microarray construction.

The majority of dispensers for research applications are based on the squeeze tube design. Typically, a glass capillary is mounted inside a cylindrical piezo material. A specific voltage pulse is applied to the piezo material to create the pressure pulse. The optimal duration and amplitude of the voltage depends on the design of the device. Typically, the diameter of the dispensed droplet matches closely the diameter of the orifice. Volumes of a few picoliters are reproducibly dispensed at rates of a few thousand per second. Figure 4.3a displays an ~ 10 pl drop being dispensed from a 20 µm orifice. The image in Fig. 4.3a is actually a composite of 15 dispenses captured with a synchronized Xenon strobe lamp and illustrates the reproducibility of the technique. The spots that are formed from a single dispense are on the order of 50 µm in diameter (Fig. 4.3b). The Packard BioChip Arrayer™ uses a 75 µm nozzle and dispenses drops on the order of 300 pl. This leads to spots on the order of 200–300 µm when dispensing onto a glass slide [13]. These volumes and resultant spot sizes are consistent with the estimates displayed in Fig. 4.2.

The sub–nanoliter volumes that can be dispensed and the commercial availability of the technology are clear strengths of the piezo jetting technique. However, a complication is that the dispense nozzle must be filled with the desired reagent and a specific fluid pressure must be maintained for dispensing. Appropriate pressure in the fluid tube is necessary for preventing the reagent from dripping out the nozzle and for optimal performance of the device. One method, aspirating sample through the nozzle, requires sufficient time to stabilize the system pressure and can reduce fluid handling throughput. Further, small nozzle diameters can lead to clogging and slow aspiration rates. Alternatively, the dispenser can be dedicated to delivering a particular reagent by filling from a reservoir behind the nozzle, much like in a conventional desktop printer cartridge.

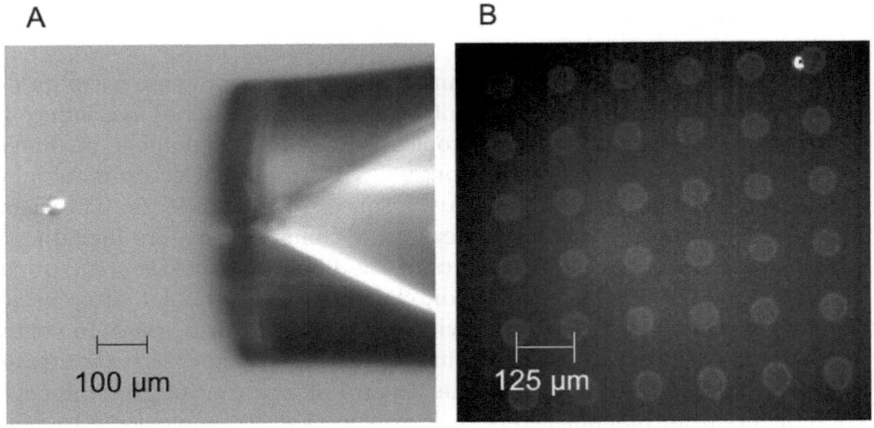

Fig. 4.3. Piezo-based reagent jetting: **(a)** composite image of a droplet ejecting from 20µm nozzle (MicroFab, Inc.). The volume droplet is on the order of 10 pl. **(b)** shows the array that results from individual dispenses

Considering the capabilities of piezo jet dispensing, many applications are under development [11]. For example, microarraying of previously prepared DNA probes is competitive with pin based deposition techniques, especially when many dispensing tips are used in parallel. Additionally, in situ construction of high density oligonucleotide microarrays appears to be a viable technique [14]. This application involves an array of piezo jets operating in a dispense mode. Each dispenser delivers a unique reagent required in the phosphoramidite-based synthesis of DNA oligomers [15]. By defining the location of individual dispenses, large arrays of long oligonucleotides (e.g. 60–mers) of designed sequence can be constructed at high density. Other applications, such as in high throughput screening of pharmaceutical compounds, have also been considered [16]. The small volumes that can be dispensed are ideal for economical evaluation of large numbers of samples.

4.5 Solenoid Jet Based Dispensing

A third commonly used reagent jetting technique is fundamentally different from the other two. The thermal- and piezo-based reagent jetting techniques function as fluid pumps. The solenoid-based technique exploits high speed valves. In operation, the valve is positioned between a pressurized fluid source and a nozzle (Fig. 4.1c). Rapid actuation of the valve causes fluid to stream from the nozzle. High speed miniaturized solenoid valves are available from the Lee Company (Westbrook, CT). These valves can open and close as often as $\sim$ 1200 Hz under pressure heads on the order of 10 psi. To operate at these rates, a voltage 'spike' ($\sim$ 40V), as short as 150 microseconds, is applied to rapidly open the valve. The valve can then be held open for longer pulses, or indefinitely, using a lower 'hold' voltage ($\sim$ 8 V).

An advantage of this approach to reagent jetting is the ability to precisely meter nanoliter–scale volumes of fluid. Typical syringe pump based liquid handling instruments operate in the microliter to milliliter range and are inappropriate for arraying or high throughput applications at smaller volumes. Conversely, the picoliter scale quantities dispensed with thermal or piezo-based techniques are too small to effectively deliver volumes in the nanoliter range. To increase or alter the delivery volume using these techniques, multiple dispenses are necessary. Even at high actuation rates, such an approach is too time consuming [16]. With solenoid-based dispensers, volumes ranging from a few nanoliters to several microliters can be rapidly delivered [17]. Flow through the valve is dependent on a number of parameters, including the applied pressure, valve opening time, fluid viscosity, and nozzle dimensions. The valve opening time is the easiest variable to control and can be modulated by simply changing the duration of the hold voltage. The linear relation between dispensed volume and valve opening time is shown in Fig. 4.4.

Several fluid dispensing devices based on solenoid valve technology have been described [17–19]. The technology is relatively simple to implement, enabling the construction of custom instruments for desired applications. Commercial systems based on solenoid valve technology are available from Carte-

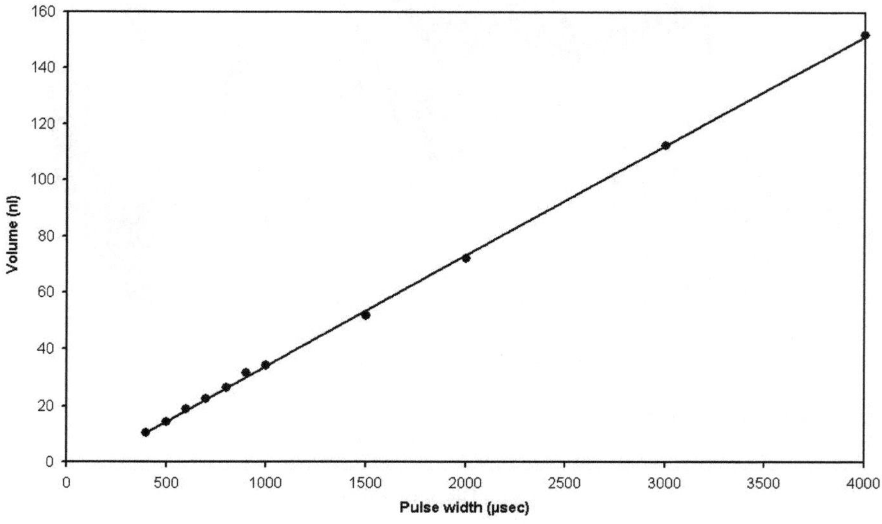

Fig. 4.4. Graph of the volume ejected as a function of valve pulse width for a solenoid-based reagent jet. A linear relation between the dispensed volume and valve opening time is observed. A pressure head of 10 psi and a nozzle of 125 µm inner diameter was used. The volumes were determined by weighing the sum of 1000 dispensing events

sian Engineering and Innovadyne Technologies. Cartesian produces a complete system, containing motion control and fluid handling. This system uses a finely controlled syringe pump for aspiration and for maintaining a desired hydraulic pressure when dispensing. Incremental steps of the syringe pump are timed relative to the solenoid valve opening to dispense reagents.

Innovadyne Technologies, Inc. manufactures ASAP^{TM} technology and is integrated into different commercial liquid handling platforms. At the heart of Innovadyne's technology is a 'hybrid valve' structure that controls the fluidic connectivity for performing different operations. An expanded view of the valve is shown in Fig. 4.5. The switching valve is a flat face configuration, similar to that found in conventional high performance liquid chromatography applications. The face of the rotor is grooved and pressed against the stator face. The stator contains fluid ports that connect to the various components via a microfluidics structure. Turning the rotor changes the fluid paths based on the design of the grooved surface. A stepper motor performs the rotation and the actuator body applies pressure to the face of the stator to prevent

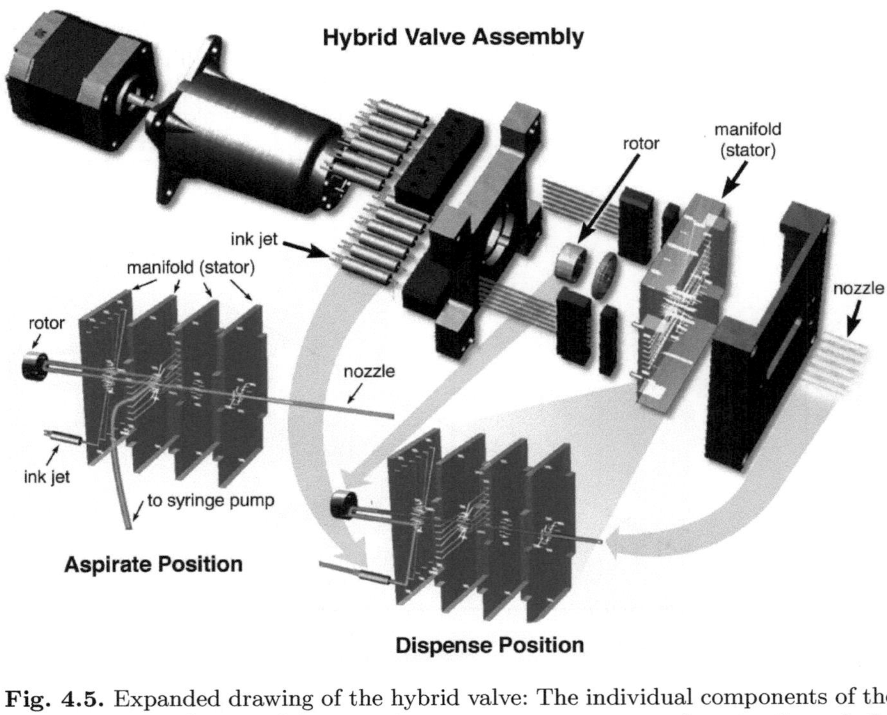

Fig. 4.5. Expanded drawing of the hybrid valve: The individual components of the hybrid valve are illustrated. In operation, a stepper motor turns the rotor through the actuator body. The position of the rotor, relative to the manifold (stator), determines the fluid pathway. The fluid pathways for the aspirate and dispense positions are shown (Reprinted with permission from Innovadyne Technologies)

leakage. With this set up, multiple fluid streams can be switched simultaneously and rapidly. Further, multiple functionalities can be integrated without interfering in the sample path. This prevents contamination of the solenoid valve that can reduce its operational lifetime. Additionally, different components such as syringe pumps, washing and purging sources, or reagent sources can be integrated depending on the application. The hybrid valve can deliver greater than 200 individual dispenses per minute and deliver volumes ranging from 50 nl to 10 μl with less than 5% coefficient of variation.

Several applications based on solenoid-based reagent jetting have been developed. Although the system can be used for microarraying, the typical lower limit on droplet volume ($\sim$ 1 nl) is too large to produce high density microarrays. Other applications exploit the technique's ability to rapidly dispense a desired volume. These applications include high throughput screening of pharmaceutical candidates or synthesis of combinatorial libraries [18]. When used as a reagent dispenser, care must be taken not to expose the valve to harsh solvents as this can lead to degradation of the valve seals. The use of solenoid-based reagent dispensing has also been described for the automated screening of protein crystallization parameters [19]. The ability to dispense on the nanoliter scale, afforded by solenoid-based reagent jets, allows significant miniaturization and higher throughput leading to significant cost savings.

References

1. Le HP 1998. *Progress and Trends in Ink–jet Printing Technology.* J of Imaging Science and Technology 42(1):49–62
2. Bateman TA, Ayers RA and Greenway RB (1999) *An Engineering Evaluation of Four Fluid Transfer Devices for Automated 384–Well High Throughput Screening.* Lab Robotics and Automation 11:250–259
3. Mueller U, Nyarsik L, Horn M, Rauth H, Przewieslik T, Saenger W, Lehrach H, Eickhoff H (2001) *Development of a technology for automation and miniaturization of protein crystallization.* Journal of Biotechnology 85, 7–14
4. Okamoto T, Suzuki T and Yamamoto M (2000) *Microarray Fabrication with Covalent Attachment of DNA Using Bubble Jet Technology.* Nature Biotechnology 18:438–441
5. Schober A, Günther R, Schwienhorst A, Döring M, and Lindmann BF (1993) *Accurate High–Speed Liquid Handling of Very Small Biological Samples.* BioTechniques 15(2):324–329
6. Roda A, Guardigli M, Russo C, Pasini P and Baraldini M (2000) *Protein Microdeposition Using A Conventional Ink–Jet Printer.* BioTechniques 28(3):492–496
7. Lahiri J, Jonas SJ, Frutos AG, Kalal P and Fang Y (2001) *Lipid Microarrays.* Biomedical Microdevices 3(2):157–164
8. de Wildt RM, Mundy CR, Gorick BD and Tomlinson IM (2000) *Antibody arrays for high–throughput screening of antibody–antigen interactions.* Nature Biotechnology 18:989–994

9. Hughes TR, M Mao M, AR Jones AR, J Burchard J, MJ Marton MJ, KW Shannon KW, SM Lefkowitz SM, M Ziman M, JM Schelter JM, Meyer MR, Kobayashi S, Davis C, Dai H, He YD, Stephaniants SB, Cavet G, Walker WL, West A, Coffey E, Shoemaker DD, Stoughton R, Blanchard AP, Friend SH, and Linsley PS (2001) *Expression profiling using microarrays fabricated by an ink–jet oligonucleotide synthesizer.* Nature Biotechnology 19:342–347
10. Hicks JS, Harker BW, Beattie KL, and Doktycz M J (2001) *Modification of an automated liquid handling system for reagent–jet, nanoliter–level dispensing.* Biotechniques, 30 (4), 878–885
11. MacBeath G and Schreiber SL (2000) *Printing Proteins as Microarrays for High–Throughput Function Determination.* Science 289:1760–1763
12. Lemmo AV, Fisher JT, Geysen HM, and Rose DJ (1997) *Characterization of an Inkjet Chemical Microdispenser for Combinatorial Library Synthesis.* Anal. Chem. 69:543–551
13. Blanchard AP, Kaiser RJ, and Hood LE (1996) *High–Density Oligonucleotide Arrays.* Biosensors & Bioelectronics 11(6/7):687–690
14. Allain LR, Askari M, Stokes DL, Vo-Dinh T (2001) *Microarray sampling–platform fabrication using bubble–jet technology for a biochip system.* Fresenius J Anal Chem 371:146–150
15. Uetz P, Giot L, Cagney G, Mansfield TA, Judson RS, Knight JR, Lockshon D, Narayan V, Srinivasan M, Pochart P, Qureshi-Emili A, Li Y, Godwin B, Conover D, Kalbfleisch T, Vijayadamodar G, Yang M, Johnston M, Fields S, Rothberg JM (2000) *A comprehensive analysis of protein–protein interactions in Saccharomyces cerevisiae.* Nature 403:623–627
16. Qureshi-Emili A and Cagney G (2000) *Large–scale functional analysis using peptide or protein arrays.* Nature Biotechnology 18:393–397
17. Calvert P (2001) *Inkjet Printing for Materials and Devices.* Chem. Mater. 13:3299–3305
18. Cooley P, Wallace D, Antohe B (2001) Proceedings, SPIE Conference on Microfludics and BioMEMS. Vol. 4560, p. 177–188, Microfluidics and BioMEMS, Carlos H Mastrangelo; Holger Becker; Ed
19. Papen R, Croker K, and Kolb A (1998) *Nanoliter Dispensing Technology.* Genetic Engineering News, 18(4)

5

Manufacturing of 2-D Arrays by Pin-printing Technologies

Uwe R. Müller and Roeland Papen

5.1 Introduction

Seldom has a simple concept had such an impact on the Life Sciences as the application of 'pin–printing' to the arraying of biological materials, creating an entirely new movement in biotechnology. While Affymetrix developed a costly high tech precision photolithography process to produce high density arrays of oligos, Schena, Davis, Brown and Shalon, then at Stanford University, used a single split pin, mounted on a home-made X–Y–Z robot, to transfer small aliquots of cDNA from a 96–well microplate onto surface modified microscope slides, thereby providing the research world for the first time access to high density microarrays [1]. What followed was a popularization of pin tool printing technology, aided by the emergence of several new companies that focused on delivering robotic instrumentation to deliver nano– and picoliter volumes of biological materials to a substrate at ever increasing density. While transferring liquids with pins, hollow needles or capillaries appears low-tech, the small amount of liquid that is being transferred and especially the need to print many different fluids without sample mixing provides a significant technical challenge. Different approaches and solutions have been developed to meet these challenges. While non-contact jetting technologies have been discussed in the previous chapter, the focus here is on a variety of pin-based techniques and procedures, as well as key elements in the printing step that are crucial for obtaining high quality arrays.

5.2 Definition of 'Contact' Pin–Printing

Contact pin–printing derives its definition from the fact that at the critical point in the process a continuity exists between the transfer device (pin), the fluid (liquid) and the receiving surface (substrate). Several important physical and chemical properties of these three elements affect their interaction and, in combination with other environmental conditions (e.g. humidity), determine the volume of the transferred liquid and the geometry of the resulting spot. Among these properties are viscosity and surface tension of the liquid, the geometry of the pin, the force of deposition and speed of retraction, and the wetting characteristics (hydrophobicity) of both the substrate and the transfer device. Contact between all surfaces with a multiplicity of different fluids also indicates the need for washing the pins between different transfers, another important parameter contributing to the reliability and quality of printing.

In difference to contact printing, ink jet-based technologies such as solenoid and piezo–electric dispensing are considered non-contact technologies, as there is never continuity between dispensing element, liquid and receiving surface. Transfer volume and spot formation are therefore determined by fewer interacting parameters, which results in somewhat better quantification and more uniform spot morphologies than achievable with contact printing, however typically at the cost of higher instrument complexity and therefore higher price. For a comparison of robotic arraying instruments, see [2,3].

In reality, even ink jet printing may be considered a contact printing technology since the drop–formation is determined by interaction of the fluid with a physical orifice [4]. The only true non-contact printing technology is therefore based on focused acoustics, where sound energy is coupled into the bottom of a container and a droplet ejected upwards by focusing acoustic energy at the meniscus. The formation of the ejected drop depends solely on the frequency, energy and duration of the tone burst and eliminates variability and limitations due to solid–liquid interactions [5].

5.3 Overview of Different Pin Technologies

While Pat Brown and others initially used only a single pin for printing an array, the need to transfer to more destinations, faster, smaller and more precisely led to many different embodiments of the transfer pin and supporting robotics. The basic pin types are reviewed below:

Solid Pins: Solid pins have excellent reproducibility for both transfer volume and spot size as long as they are adequately washed between liquid transfers. Typically only one spot can be printed and the pin needs to be re-loaded with new material after every deposition, whereby the amount of liquid loaded is proportional to the diameter of the pin. Solid pins have the lowest sample wastage ($< 15\%$) of all the pin types, have an excellent CV (coefficient of variance) for transfer volume (as low as 2%) and spot–size, and are more robust with regard to impact. The disadvantage is low throughput, and if more than one solid pin is used in a system to make up for this deficiency, the variability in spot–size and volume transferred increases proportionally as a function of pin quality (e.g. uniformity of pin diameter, surface treatment, etc). Pin performance is a function of dimensions and coating, and there are several manufacturers offering different choices [6,7].

Ring and Pin: Only one commercial arraying system (the Affymetrix GMS417; formerly Genetic Microsystems) is based on the ring and pin technique. This process involves capturing a film of sample liquid inside a small ring by dipping it into the sample solution. For sample deposition a solid pin is pushed through this ring, whereby some of the sample is carried by the flat end of the pin to the substrate surface. The continuity of the sample film is typically not disrupted by this pin movement. Thus, the ring acts as a sample reservoir allowing multiple depositions without having to return to the sample source. A CV for spot–size of $< 10\%$ (across 4 pins) has been reported [8]. The disadvantage is a higher susceptibility to environmental conditions, especially humidity, that affects both the concentration of the sample in the film as well as the stability of the film itself. This technique is also very wasteful of sample since a large dead volume is required in the source well (to cover the ring), and not all the material in the film can be transferred. For example, a typical load volume on a GMS427 ring is 1.5 µl, of which typically only 6.7 nl (4 replicates/slide $\times$ 42 slides $\times$ 40 pl/spot), or less is used, meaning that 99.5% of the sample is wasted.

Micro–Fabricated Pin Array: An extreme example of parallelization is the print plate from Corning, an etched silicon surface containing more than one thousand 100 µm posts in an array layout and matching perfectly to a funnel reservoir containing the samples to be transferred. The print plate is inserted into the mated funnel and removes a few picoliters out of each channel upon retraction, which is then deposited onto a substrate. Positionality is excellent as the spot to spot distance is not affected by robotic motion but is a feature of the print plate. The throughput and reloading is improved by moving the substrates (microscope slides) in between the reservoir (funnel) and the pinplate, keeping travel distances small and allowing quick reloading between prints. The CV's for transfer volume and spot–size are on the order of 9% (over a thousand pins), and there is little sample wastage on the pin ($< 15\%$). Disadvantages include the high set–up cost of this very specialized manufacturing equipment. This new and unique process is described in detail further in this chapter.

Dip–Pen Nanolithography™: This technology represents the smallest 'solid pin' to date and is based on atomic force microscopy. The AFM tip serves as the pen that is coated with organic molecules which are transferred via a water meniscus to the substrate surface [9–11]. This allows extreme miniaturization with spots of less than 0.5 µm in diameter. This new technology is described in more detail in Chap. 6 of this book.

Split or Quill Pin: These types of pins represent the biochemist's version of the old quill pen, basically a goose feather with a slit at the end that was used to draw up ink. For microarraying these pins are now machined with high precision to contain slits of 15–50 µm. After loading 0.1–0.5 µl they can dispense hundreds of spots through tapping on the surface to expel droplets in the nl to pl range [6, 7, 12, 13]. The exact volume is a function of the tapping force, the slit dimensions, the fluid viscosity, and other parameters. The advantage is that many spots can be printed with relative consistency without having to reload the pin. On the negative side, the tapping action of the spring loaded quills may damage delicate surface treatments or remove reactive binding groups from the surface of either the pin tip or the chip, resulting in both non-uniform deposition and variable binding efficiencies. However, these split pins can also be used in a non-contact mode to avoid these problems [14].

Stealth Pin: The Stealth contact printing technology from TeleChem uses precision pins with flat tips and defined uptake channels that act as sample reservoirs, similar to the quill pins. They are by far the most used transfer pins for arraying to date. Pins are available in a wide assortment of tip and channel sizes, allowing users to specify spot diameter and loading volume. Pins are manufactured with advanced micro–machining and polishing technologies with exceptionally tight tolerances and come in a wide assortment of sizes and reservoir capacities. The CV's for transfer volume and spot–size are on the order of $\sim 12\%$ (across twelve pins). While these pins allow multiple spots to be printed per load (> 160), they still waste a lot of sample ($> 70\%$) and require an excess amount of sample in the source well. Tips have to be pre-blotted and the transfer is sensitive to humidity and sample composition, resulting in relatively high variability in the amount of deposited material [15, 19].

Hitachi X–Cut Pin: The SPBIO^{TM} Microarray Station of Hitachi Genetic Systems uses a new pin design with an X–groove cut into the pin tip that enables it to capture larger volumes as well as control the spot morphology better. Due to the enclosing effect of the pin geometry this liquid reservoir lasts longer and evaporation has less of an effect on the concentration and spot morphology. Excellent CV's were obtained for transfer volume and spot–size (1–7%) [17]. Low sample wastage ($< 20\%$) and lower source dead volume are further advantages.

Capillary Pins: One of the early developers of capillary contact printing was former Genometrix, which used very fine capillaries connected to a microplate reservoir in order to deposit spots in the nanoliter range. While solving the reservoir problem and partially protecting the transfer liquid from evaporation, the sensitivity of the system to bubble–formation in the capillaries during loading and operation resulted in major difficulties in controlling hydrostatic pressure in each capillary line. Both non-printing events and run–outs (depositing too much sample) hampered overall reliability.

A similar technology is employed by Vysis, Inc. in their manufacture of DNA chips for the Genosensor System^{TM}. Short steel needles (25–75 µm ID) with a plastic reservoir at one end are loaded with DNA solution, and the capillary is then connected to a high precision air pressure system. Fluid is dispensed by a combination of air pressure and inertia. After a rapid down–movement of the capillary, it stops some 20–50 microns above the slide surface by which a droplet forms at the tip of the needle. Though the needle tip never touches the slide surface, the fluid droplet ($\sim$ 300 pl) makes contact and is 'ejected' by a millisecond air pulse [18]. The main advantage is that only the fluid touches the slide, which leaves its surface without any damage. In addition, the relatively large fluid reservoir allows many prints off the same needle and storage of the needle between print–runs. The main disadvantage is that only a single needle can be used, requiring accurate X–Y–Z calibration of the needle tip position after a needle change, thereby limiting the use of this system to the manufacture of relatively small arrays. Recent data on reproducibility are not available.

Micro-Machined Capillary and Quill Pins: A miniaturized version of the capillary pin is the micro-machined pin [19] that uses differences in surface tension to move the ink inside an etched channel. Spots with an average diameter of 16 ± 3 µm can be printed, which is approximately 7–fold smaller than the average spots produced by TeleChem Stealth pins. The MicroSpot pins manufactured by Oxford Laser have a slit width as small as 5 µm, but we have no data on spotting performance [20]. The *MicroSpot*™ pins from Matrix Technologies are made of tungsten and cut by laser. With a fill volume of 55 nl and a dispense volume on the order of 50 pl they can be arrayed into a 10K pin tool for dispensing of up to 100,000 spots per glass slide [13].

Massively Parallel Fiber–Optic Capillary Printing: GenoSpectra (Fremont, CA) has developed a novel high speed printing technology, termed *FiberPrint*, that is capable of depositing liquid samples onto flat surfaces in a massively parallel fashion. A fully automated *FiberPrint* system is capable of printing 10,368 uniquely addressable DNA (oligonucleotide or cDNA) probes with up to 3 repeats onto the surface of a standard microscope slide, totaling over 30,000 spots per slide. This system uses specially designed print–heads containing over 10,000 fiber optic capillaries that are bundled together to form a flat (level to within 4 µm) print–head surface (Fig. 5.1). DNA or other solutions to be deposited are stored in micro–well plates assembled in a pressure chamber. Samples are deposited in 400 pl volumes with high fidelity and spot uniformity. With an estimated throughput of 2400 slides per day the *FiberPrinter* system appears ideal for high throughput, low volume, and highly parallel deposition of liquids with CV's around 9% across 10k capillaries (Fig. 5.2). Similar to the Corning GenII System, the *FiberPrint* reduces larger source well dimensions into the smaller array–dimensions by compressing connecting capillaries into a dense print–head. (Data provided by Geno–Spectra; no references to published information are available).

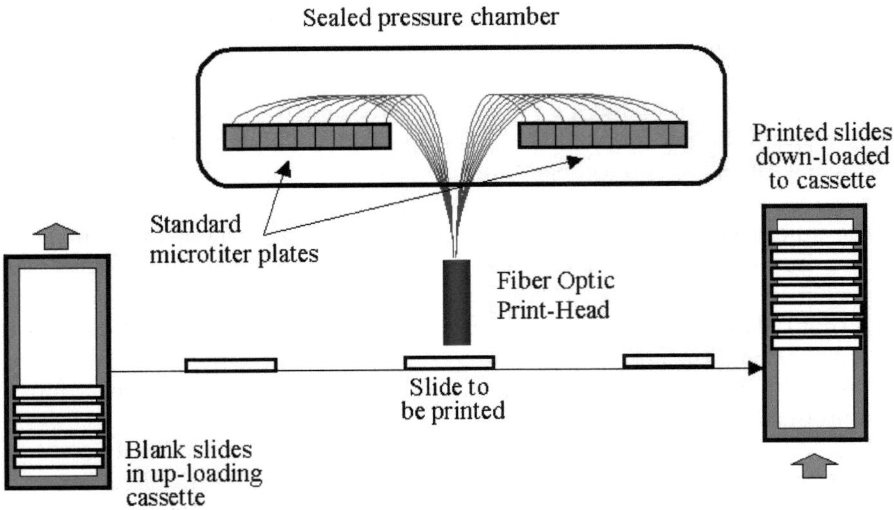

Fig. 5.1. Schematic of the *FiberPrint* system (courtesy of GenoSpectra)

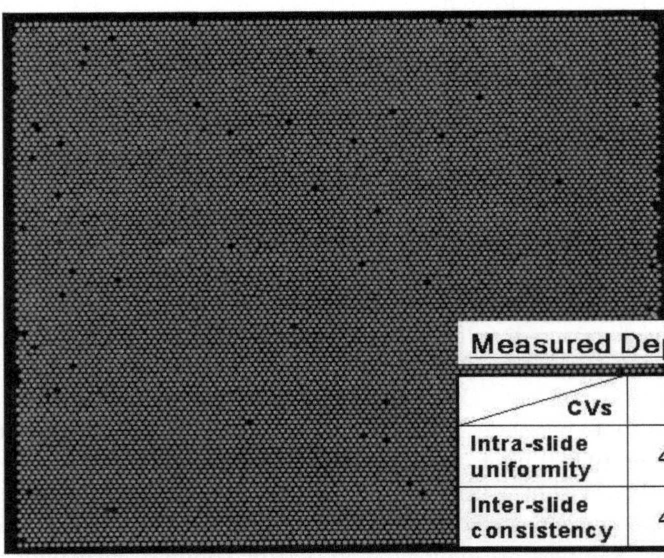

Measured Deposition Quality

CVs	Size	Intensity
Intra-slide uniformity	4.6%	8.9%
Inter-slide consistency	4.6%	14.8%

Fig. 5.2. Image and measured quality of a labelled oligo array deposited by a 10k FiberPrint printhead (courtesy of GenoSpectra)

Disposable Pins: VP–Scientific has recently introduced disposable one–time–use pre-molded pin–arrays. The pin–array is made out of polypropylene and has 96, 384 or 1536 pins that transfer between 120–135 nl per pin with a CV of 8–12% [7]. Adapters are available for integration of the disposable print head with various robotic platforms, which allows transfer of whole plates' worth of DNA samples at a time. This avoids the need for washing and therefore eliminates cross–contamination, further enhancing the integrity of the array. The adaptability to existing robotics greatly increases their usefulness for arraying.

In addition to the pins discussed above, a variety of other materials and pin designs have been explored by different groups and industrial manufacturers [19, 21].

5.4 Other System Components and Environmental Factors

While the pin is the core of a contact printing system, other system components and environmental factors influence its reliability and reproducibility. Proper monitoring, maintenance, calibration and minor adjustments of these can make the difference between optimal performance and bad arraying. Key factors include:

Pin Holders: Pins are typically held 'floating' in a pin holder, meaning that they are held in position in guiding sleeves by gravity and prevented from falling through due to a mechanical stop; yet they are free to move upwards as the pin hits the substrate. This avoids excessive wear of the delicate pin tip and minimizes damage to the substrate surface. Critical in this arrangement of the guiding sleeves are tight tolerances to minimize angular deflections of the pins, but enough space to prevent bonding between the tip and the holder, which can result in tips no longer reaching the substrate. Telechem Stealth pins also feature a pin 'collar' that prevents rotation, providing near frictionless printing. Material choices are also important here to prevent static build–up on the tips and print head, which can dramatically lower print quality.

Multiple Pins: Machine tooling of pins is a delicate procedure, and given the small dimensions of the pin tip, it is often necessary to match pins in the same grid to minimize performance variations from pin to pin. Maintaining near perfect parallelism between the print head and the plane of the substrate becomes increasingly difficult with increasing number of pins and increasing size of the print head. Adjusting this planarity by mechanical means in addition to pin selection is essential to good printing and affects the longevity of the pins. Typical configurations for print heads are 4 pins in a 2×2 or 2×6 format with 9 mm center to center spacing for 96–well source plates, and a

4×4, 4×8 or 4×12 format with 4.5 mm center to center spacing for 384–well plates. These dimensions are mostly dictated by the fact that the usable surface area on a glass microscope slide is limited to approximately 22×60 mm (excluding label and edges to fix a hyb–chamber). With a single transfer pin it is possible to maintain the relative position of the samples after printing on the array the same as in the source plates. For multiple pin configurations this is not possible due to the fixed format of the pins and the dimensional difference between source plate and array, requiring sample tracking software for dealing with large numbers of samples.

Environmental Control: As soon as the pin, tip or capillary is loaded with sample, a race against time starts since evaporation at the liquid interface will change both the volume on the pin tip and the concentration of the biomolecules in that volume. Solid pins are most vulnerable to evaporation, but even for quill pins the amount of liquid available for deposition will eventually be reduced by evaporation, acting as a counter force in the substrate–liquid–pin interaction, and slowly reducing the volume deposited in each spot. Application notes by MiraiBio [17] clearly show the effect of evaporation on the X–cut solid pins. But even capillary pins, while protecting the transfer liquid better, are subject to evaporation. Typically a 70 micron orifice capillary will concentrate an analyte at the bottom of the tip by about 10% per second for the bottom half nanoliter. This often results in what is called "the first drop effect", whereby the first spot may result in a higher signal intensity than subsequent spots. Evaporation control is also important for the source plates, as lengthy exposure to typical laboratory environments may concentrate the DNA solutions in the microplates and create variability between different samples. Furthermore, the rate at which the deposited fluid dries on the substrate surface affects spot morphology. Therefore, most manufacturing quality arraying robots are equipped with some type of enclosure to maintain a consistent humidity level, ideally between 55–70%. For high density arrays it is also advisable to keep the source plates cooled to minimize evaporation over the term of the printing run. Additional cooling or heating of the substrate may be required for printing of protein arrays to either minimize the risk of protein denaturation or to enhance surface reactivity.

Due to the micron dimensions in which spot sizes are measured, it is clear that dust particles, lint and other airborne debris can have a detrimental effect on the array quality either by clogging up the capillary channels in quill or capillary pins or by 'smudge' deposition, thereby disrupting surface tension and affecting spot morphology. In addition, any dust particles that stick to the slide surface and are not removed by the hybridization or washing process will typically affect the imaging, since such particles scatter light and tend to also fluoresce across the visible spectrum. Deionization of the air as well as selection of anti-static materials in the system can be very helpful in avoiding that slides become dust traps.

Positional Robotic Control: The limitation on array density is primarily determined by spot diameter and secondarily by the positional reproducibility and accuracy of the robotic XYZ stage. Typical spot sizes for pin based printers are between 100 and 200 µm in diameter, though significantly smaller spots can be made with Nanoplotters (see Chap. 6). While even relatively low cost stages provide positional accuracy and reproducibility in the range of tens of microns, manufacture of high density arrays and prolonged production runs may require more precise stages with single digit micrometer precision and accuracy as well as positional feedback to compensate for system errors. Precise control of speed, acceleration and positional accuracy in the Z–axis are also critical to contact arraying, as described below in the printing process section.

Washing System: Essential for consistent print performance and low carry–over is a good washing system. While some commercial arraying systems rely on a simple water rinse to clean the tips between different samples, other systems use additional ultrasonic cleaning or pressurized–jet streams of water that are directed at the tips for a more efficient rinse. A combination of both approaches as well as procedures that use specific cleansing and soaking fluids have been reported [15, 22].

5.5 Pin Printing Process

5.5.1 Dynamics of Spot Formation

As mentioned above the elements interacting in spot formation and spot morphology are the geometry and surface properties of the pin and the substrate, as well as the viscosity, composition and resulting surface tension of the sample. Some of these issues have already been addressed in Chaps. 2 and 3 of this book, and for the more intricate physics involved we refer the reader to the literature [23–26]. Our focus here is on the key issues in the mechanics of the process.

In the first step the pin is dipped into the ink reservoir, whereby the penetration depth and time of the pin in the fluid is important. Over–immersion of the pin will result in loading too much sample, causing the deposition of too large a drop for single transfer solid pins, and even for the first depositions of quill pins. It can also lead to unwanted carry–over in subsequent cycles. Obviously, immersion times have to be sufficiently long to fill up the capillaries in the case of quill style pins. Before starting the print, it is advisable to blot the loaded pins for a specific number of 'pre-print' spots on a sacrificial substrate to condition the pins and eliminate 'first spot effects'.

In the second step the pin is contacted with the substrate surface for delivery of the ink. Several parameters influence the amount of liquid that is deposited and the resulting spot size. First, the force (speed) with which the tip impacts the substrate is particularly critical for capillary and quill type pins, requiring good control of deceleration in the Z–axis. Upon contact, the pin diameter and the topography of the tip and substrate determine the type of gap that is formed, the capillary forces that are generated, how much liquid is squeezed between pin and substrate, and how much is pushed beyond the perimeter of the tip. For an aqueous sample the capillary forces will hold the liquid between the pin tip and the substrate surface in an area slightly larger than the pin diameter. How much of this fluid remains on the substrate depends on the dwell time and Z–retract speed of the pin, the surface tension and viscosity of the fluid, and the contact angles at the liquid substrate interface. For very large contact angles the deposited fluid may first chaotically recede while evaporation diminishes the drop volume before it is pinned down. If the pin is retracted too fast, satellite spots may be created as the liquid thread between retracting tip and substrate breaks unevenly. These satellites can cause contamination of other spots in the array (Fig. 5.3) [26].

Assuming that all environmental, surface and mechanical parameters can be maintained consistently, the amount of biomolecule solution that is transferred becomes a function of the ink composition and biomolecule concentration, pointing to the need for uniform concentration and fill levels in the source plates. After deposition, the final spot morphology is mostly depen-

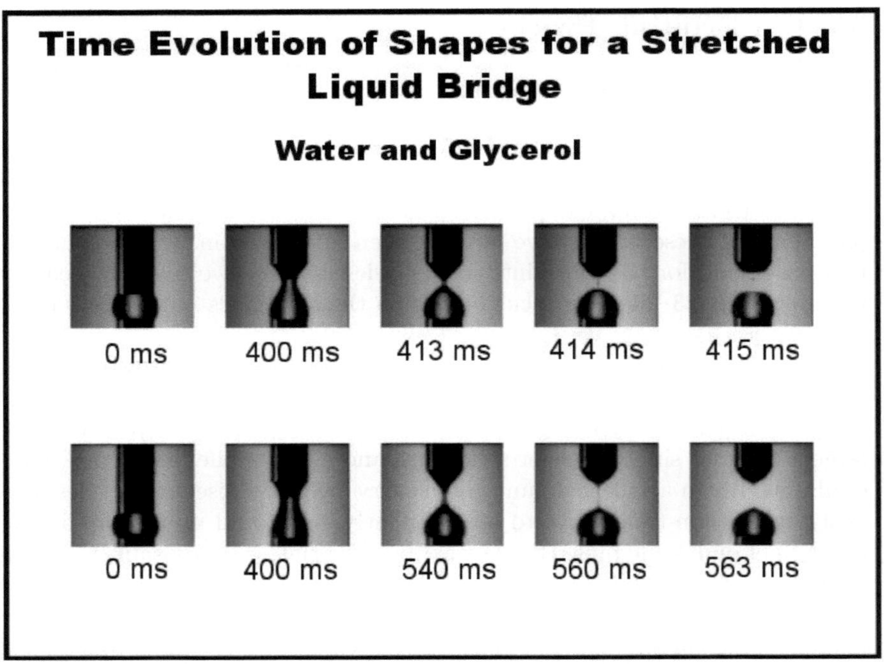

Fig. 5.3. Spot formation and liquid column break-up on a substrate (courtesy of Prof. Osman Basaran, Purdue University)

dant on the rate of evaporation, contact wetting angles and composition of the fluid [24, 25].

5.5.2 Importance of the Substrate and its Surface

From a pin–printing perspective the ideal substrate surface is flat and its coating is uniform. The most prevalent substrate for spotting DNA is still the microscope slide coated with either an amino–silane or aldehyde group. The process of attaching organosilanes with various functional groups to glass has been known for over 30 years and can be easily duplicated in any lab. Producing such surfaces in high volume and with reproducible and stable contact angles, however, remains an art, a fact that is reflected in the high 'value add' that the surface coating brings to a basic glass slide. Other than the contact angle, the most relevant surface characteristic for the fluid deposition process is the thickness and pore–size of its coating. In fact, recent developments have increased the amount of material that can be transferred to the surface by introducing a 3–dimensional nature, providing more surface area and even giving the surface the wicking effect of a membrane [27–29]. This in itself may have a significant effect on the variability in spot size and drop volume, but is especially critical for contact deposition, since the impact of the pin may damage or alter the surface and its wicking characteristics.

The detailed physico–chemical properties of the substrate surface and how that impacts the amount and the mechanism by which the biological material is bound in the arraying process has been the subject of two of the preceding chapters and will not be further discussed here. Also note that the optical characteristics of the glass and its coating are equally important for the detection process, since most current assay formats rely on optical read-out (see Chaps. 8 and 11).

5.5.3 Software and Data–Tracking

Software control is a critical component of contact printing systems. They provide the user with an interface to manage the operation and to fine–tune critical variables and system parameters such as array spacing, number of pre-print spots, Z–motion control, dwell time, wash and dry sequences. More expanded configurations also include sample tracking software that allows the source well coordinates to be related to its spot location within the printed array on the destination substrate. Integration of sample tracking software with a data management system enables a scientist to rapidly design array experiments as well as de-convolute and link experimental results back to the printing process for optimization. A large number of software packages, either system specific or generic, are available from different vendors, and have integrated quality control features that can monitor the printing process and alert the user to deviations from operational specifications.

5.6 Example of a High Throughput Pin–Printing System for Manufacturing of 2D Arrays – the Corning GENII System

A remarkable new technology for high speed printing of high density arrays was recently developed by Corning, Inc. Recognizing the opportunities in this field for a company with high quality engineering and manufacturing expertise, researchers at the Corning research facilities in Avon, France and in Corning, NY modified an existing proprietary technology that was originally developed for printing of colored dots onto the back of TV screens with 6 sigma reproducibility. The basic components and operating principle of this technology are shown in Fig. 5.4. Relying on an extrusion technology developed for producing catalytic converters for the automotive industry, a ceramic preform is fabricated, consisting of a honeycomb like structure that contains approximately 2000 circular channels of $\sim$ 1 mm diameter. This structure is then locally reheated and redrawn (b), whereby the integrety of all channels is maintained. A precursor printhead (c) is cut from the conical section, and, after polishing both ends, the internal channel surfaces as well as the end surfaces are treated. The final print–head (d) has a funnel-like structure where the channels at the narrow bottom end have an internal diameter of less than 200 μm. For each printhead, a unique pin–plate (f) is etched from silicon to contain an array of $\sim$ 100 μm diameter pins (g) to fit the bottom end of the print–head. Pin–plate and printhead are then assembled into a mechanical device that can move the pin–plate into or out of the channels with high precision. A computer controlled robot station is then employed to load approximately 6 μl of DNA solution from pre-formatted microplates into each channel (typically only the center 1100 to 1200 channels are used). Capillary forces move each fluid to the bottom end of the printhead and maintain them near the end surface, where they can come in contact with the inserted pin–plate. Once loaded, the completely assembled print–head is mounted onto a manufacturing bench that provides for the precise movement of a glass slide (h) between the print–head and the pin–plate, when the latter is in the down position. In a synchronised motion the pin–plate moves up to pick up a few picoliters of fluid from each channel with the tip of each pin, retracts to allow for a slide to move into position, and then moves up again to make contact with the slide. The completely assembled GENII manufacturing system holds an array of 10 printheads with a continuous path for the glass slides for the printing of up to 10 subarrays per slide, i.e. over 10,000 spots per slide in about 1 minute.

In addition to high speed, this system has the advantage of high reproducibility, since thousands of slides can be made in a single print–run without having to reload any DNA solutions. Comparative studies at Corning between a robotic quill–type pin printing system and the GenII system have shown that judged by the quality of hybridization data, the GenII system delivered equal or better arrays. When combined with a quality slide surface, such as

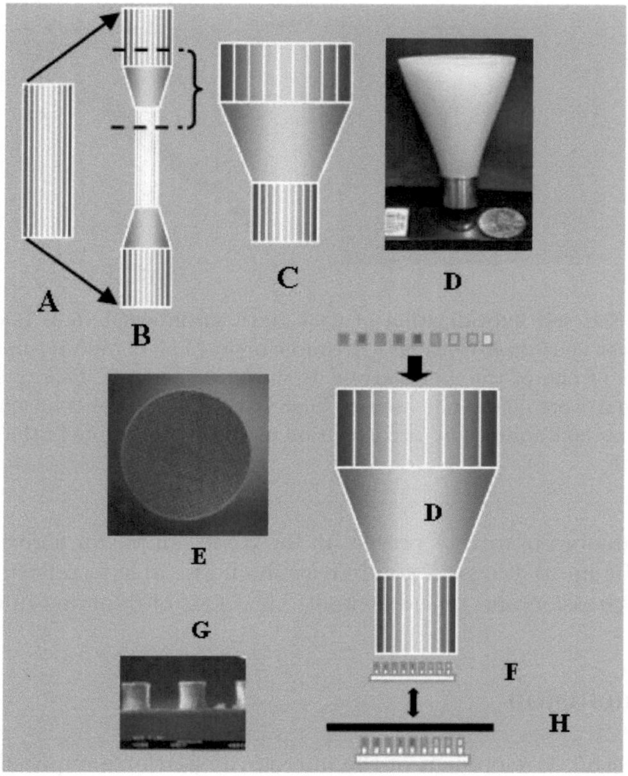

Fig. 5.4. Corning's GENII High Speed Array Printing System. **(a)** Extruded preform, **(b)** sintered and redrawn preform, **(c)** Print head cut from redrawn preform, **(d)** finished printhead, **(e)** top-view of printhead revealing honey comb structure, **(f)** pin–plate, **(g)** scanning EM of pinplate, **(h)** glass slide (Images Courtesy of Corning, Inc.)

the Corning GAPS^{TM} slides [30], and a high sensitivity assay system (see Chap. 11) the arrays manufactured by the GENII system produced 3 logs of dynamic range and CV's of $< 9\%$ for the same spots between multiple slides that were sampled from different manufacturing runs.

As discussed in Chap. 11, a quality control for array performance typically includes a so called self–self hybridization, whereby RNA from the same source is labelled separately with a green (Cy3) and a red fluorophor (Cy5) by reverse transcription, and the resulting cDNAs are mixed and hybridized to the array. The results of such a test with RNA extracted from breast cancer cells is shown in Fig. 5.5. The image reveals spots of similar color composition but varying intensity. This is expected and is quantitatively demonstrated in the graph. The composite color of each spot should be the same, since the red/green ratio for each spot should be similar. The total intensity however, should correlate

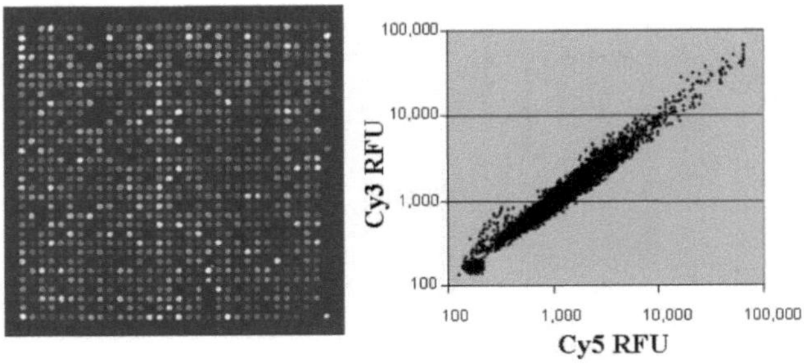

Fig. 5.5. Self–self hybridization of Cy3 and Cy5 labelled total RNA from MDA breast cancer cells on a Corning 4K cancer array. The composite image combining both colors of one of the 4 sub-arrays is shown on the left. Empty spots are from channels that were either not used or filled with a DNA free solution to control for channel cross-contamination. A correlation analysis of all spots in the array is shown on the right

with the amount of mRNA present in the RNA sample for a given gene, which can vary by up to 4 logs. A good array should be able to reflect this variation in gene expression and reveal at least 2.5–3 logs of dynamic range.

5.7 Conclusion

The successful development of the microarray platform required a merging of the latest technologies in chemistry and biology with those from physics and engineering. At the basis was the classical robot equipped with novel pin–tools to enable the significant growth and popularity of this new technology. Whether DNA, protein, lipids, whole cells or small molecules, the pin–printer has provided a platform from which new miniaturized assay chemistries, surface treatments and detection systems could be developed. While definitely not the most economic mode of printing large quantities of high density arrays, it is still one of the most accessible technologies to researchers all over the world to perform array-based experiments with relatively low capital investment.

Believing that an efficient printing system is the key to success in high density arraying, as many as twenty different companies emerged early on to develop and commercialize array–printing instrumentation, varying from manual tools and desktop spotters all the way to sophisticated clean room sized industrial printing presses. Given the typical cost of several hundred U.S. dollars for a commercial expression microarray, for example, there appears to be a significant 'home–brew' arraying market. Yet, the recent wave of consolidations, buy–outs and even business closures of printer (as well as array) manufacturers may suggest otherwise. Given that the microarray field is still growing, it will be interesting to see whether the array printer will follow the path of the thermocycler with a place in every molecular biology lab, or that of the DNA synthesizer, a tool that has largely disappeared from the average biochemistry lab, since home–brew oligo synthesis is no longer cost-effective.

References

1. Schena M, Shalon D, Davis RW, Brown PO. Quantitative monitoring of gene expression patterns with a complementary DNA microarray. Science 1995; 270:467–70
2. Bowtell DD, Sambrook JFE. *DNA Microarrays:A Molecular Cloning Manual*, New York: Cold Spring Harbor Laboratory Press, 2002; http://www.nature.-com/ng/journal/v32/n4s/extref/ng1030–S4.pdf
3. Leung YF, Pang CP, Browne K. *Guide to microarray hardware – a researcher perspective.* 2001; http://www.images.technologynetworks.net/resources/-comptab.asp
4. Rose D, Lemmo T. *Challenges in implementing high–density formats for high throughput screening.* Laboratory Automation News. 1997; 2:12–19
5. Ellson R. *Transfer of Low Nanoliter Volumes between Microwell Plates Using Focused Acoustics–Automation Considerations.* JALA 2003; 8:34–39
6. Genetix. http://www.genetix.com/productpages/consumables/slides/aQuPins.-htm
7. V&P Scientific I. 2003; http://www.vp–scientific.com/index.htm
8. Genetic MicroSystems. GMS 417 Arrayer User Guide. 1999;A–1
9. Demers LM, Ginger DS, Park SJ, Li Z, Chung SW, Mirkin CA. *Direct patterning of modified oligonucleotides on metals and insulators by dip–pen nanolithography.* Science 2002; 296:1836–8
10. Hong S, Mirkin CA. *A nanoplotter with both parallel and serial writing capabilitites.* Science 2000; 288:1808–11
11. Lee KB, Park SJ, Mirkin CA, Smith JC, Mrksich M. *Protein nanoarrays generated by dip–pen nanolithography.* Science 2002; 295:1702–5
12. Majer Precision Engineering. DNA Array Pins. http://www.majerprecision.-com/pins.htm
13. Matrix. http://www.matrixtechcorp.com/tech/AT–pdf/MicroSpot_Tech.html
14. Zeng J, Deshpande M, Kan H–C, Gilbert JR. *A dynamic spotting method for split–pin based microarrays.* Technical Proceedings of Micro Total Analysis Systems, Micro–TAS 2001, Monterey, CA 2001
15. Rose D. *Microfluidic Technologies and Instrumentation for Printing DNA Microarrays.* Microarray Biochip Technology. Eaton Publishing, 2000:35
16. TeleChem International I. Stealth Micro Spotting Pins and Printheads. 2003; http://arrayit.com/Products/Printing/Stealth/stealth.html
17. MiraiBio. Spotting Pin Evaporation 1, Application Sheet 3. 2003; http://www.miraibio.com/pdf/tech/appnotes/AppSht3_Evaporation1_pub.pdf
18. Müller UR, Daywalt M, Che D, VanBriesen A. *Dispensing system for DNA microarrays.* The American Journal of Human Genetics 63, A237: 88
19. Tsai J, Kim CJ. *A silicon micromachined pin for contact droplet printing.* Proceedings of the Second Joint EMBS/BMES Conference, Houston, TX, USA. 2002

20. Oxford Lasers. High Precision CVL Micro–Cutting. http://www.oxfordlasers.-com/Industrial/uchi/CSE4BiomedicalPin.htm
21. George RA, Woolley JP, Spellman PT. *Ceramic capillaries for use in microarray fabrication.* Genome Res 2001; 11:1780–3
22. Holloway AJ, van Laar RK, Tothill RW, Bowtell DD. *Options available – from start to finish – for obtaining data from DNA microarrays II.* Nat Genet 2002; 32 Suppl:481–9
23. Chen AU, Notz PK, Basaran OA. *Computational and experimental analysis of pinch–off and scaling.* Phys Rev Lett 2002; 88:174501
24. Deegan RD. *Pattern formation in drying drops.* Phys Rev E Stat Phys Plasmas Fluids Relat Interdiscip Topics 2000; 61:475–85
25. Deegan RD, Bakajin O, Dupont TF, Huber G, Nagel SR, Witten TA. *Contact line deposits in an evaporating drop.* Phys Rev E Stat Phys Plasmas Fluids Relat Interdiscip Topics 2000; 62:756–65
26. Zhang X, Padgett R, Basaran OA. *Nonlinear deformation and breakup of stretching liquid bridges.* J. Fluid Mech. 1996; 329:207–245
27. Beattie KL, Beattie WG, Meng L, Turner SL, Coral-Vazquez R, Smith DD et al. *Advances in genosensor research.* Clin Chem 1995; 41:700–6
28. Lee PH, Sawan SP, Modrusan Z, Arnold LJ Jr, Reynolds MA. *An efficient binding chemistry for glass polynucleotide microarrays.* Bioconjug Chem 2002; 13:97–103
29. Mirzabekov A, Kolchinsky A. *Emerging array–based technologies in proteomics.* Curr Opin Chem Biol 2002; 6:70–5
30. Corning I. GAPS Coated Slides. 2003; http://catalog.corning.com/Lifesciences

6

Nanoarrays

Dan V. Nicolau, Linnette Demers, and David S. Ginger

6.1 Introduction

The field of microarray technology progressed, in the most general terms, along three directions: increase of the number of tests (biomolecules or cells) on the same chip; increase of the number of tested biomolecules on the same unit area (i.e. density); and increase in the sophistication of the biochips, with many alternative designs being proposed. The first two trends walk in the steps of the evolution of microchips proper, i.e. larger chips and higher density on the chip, but the last similarity (i.e. 'smarter' design) should be analyzed in more detail. While semiconductor technology imposed very early in its history a 'champion' device, i.e. bipolar and later CMOS transistor, microarray technology does not have yet a 'champion'. It follows that microarray technology is still to reach its maturity, with all the benefits (e.g. effervescent innovation) and drawbacks (e.g. difficult standardization) that arise from this still-emergent stage.

However, seen from another angle, microarray technology is much closer to a 'technology crisis'. It has been argued for decades, and proven wrong every time, that semiconductor technology will come to a halt due to the inability of lithography to print smaller features at the pace asked by the unforgiving Moore's Law [1], i.e. halving of the printed critical size on the chip every 18 months. Apart from the apparently endless capacity of microlithography to push the resolution limits, fundamentally the crisis has been always far away. Microelectronics and – nowadays – nanoelectronics 'operate' with electrons (which are much smaller than 1 nm), while the most advanced lithography is asked to print features of many tens of nanometers. Even if we consider the quantum effects, the present lithography can print features that are at least ten times larger than the critical technological barrier. On the other hand, individual DNAs and proteins, the smallest 'building elements' of microarrays, are several to several tens of nanometers in size. Consequently advanced lithography is already capable of printing features that are on the same order of magnitude, if not smaller than, the 'modules' to be printed! Furthermore, for cell and tissue arrays, patterning resolution is a non-issue.

Then, why do we need nanoarrays? Firstly, the decrease of the 'feature size' does increase the capability of the microarrays via the decrease of the cost (less volume of analyte required per probed biomolecule) and via the efficiency and reliability (increase of the number of the probed biomolecules and/or tests on the same chip). This amplification of capability regarding the decrease of the sample volume and hence associated costs, which is further explored in a later section, is presented in Fig. 6.1. But the development of nanoarrays also allows much more, that is the fabrication of arrays with different complexity and functionality. For instance, the probing of single biomolecules, which is conceivably possible with nanoarray technologies, will address the present potential problem of probing the bioactivity of biomolecules collectively in 'lumps', rather than individually as it happens in actual natural biomolecular recognition.

Secondly, the capability of addressing single biomolecules allows the conceptualization of totally different micro/nanoarrays. For instance instead of 'passive', 'one–use' arrays, where biomolecules are probed (similar to a 'read' function) once, one can think of 'active', 'multiple–use' arrays, where biomolecules perform repetitive functions, e.g. 'computation'. Another layer of complexity can be added if we progress from 'static' arrays, where the biomolecules reside and are probed on one location, to 'dynamic' arrays with biomolecules that move either laterally or circularly in order to perform repetitive tasks e.g. sensing, power generation and again computation.

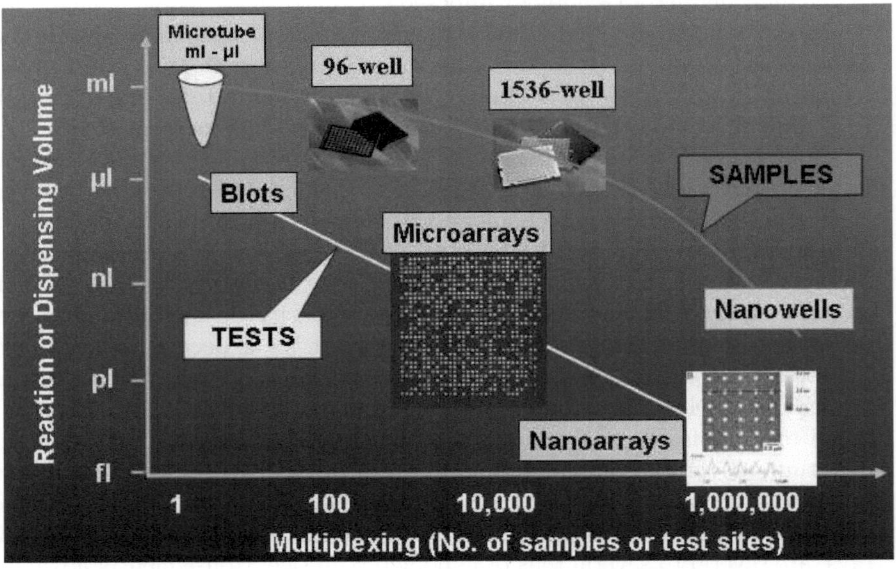

Fig. 6.1. Evolution of the sample volume versus multiplexing density (courtesy of Uwe Muller)

6.2 Passive Nano–scale Arrays

Typically, robotically-spotted microarrays contain spots of 100 microns, with up to 10,000 different cDNA sites on a chip. In situ synthesis, using 20 μm^2 spots is currently capable of producing up to 400,000 distinct oligonucleotides on a chip [2]. A reduction in feature size from 20–200 μm to microns or sub–microns would vastly increase the amount of genetic information that could be screened simultaneously under certain conditions on one chip. This type of scaling, with appropriate readout system, could enable SNP analysis via tiling arrays of the entire human genome on a single 2×2 cm array [3]. In general, achieving such high resolution with directly patterned oligonucleotide probes would enable the study of binding and detection in arrays that are up to 10,000 times more complex, in the same area, than is presently possible. A decrease in feature size will also lead to assays wherein a fixed number of targets are screened with correspondingly smaller requirements of sample volume. Importantly, patterning at this scale will not only require, but greatly facilitate the development of high throughput, high resolution screening tools.

In principle, there are two major strategies for the fabrication of nanoarrays, which are common to the micro/nanolithography for both microchips and microarrays. First, one can alter the properties of an area e.g. with light, creating different chemical functionalities or hydrophobicities locally. Subsequently, this 'island' is used for further fabrication, e.g. immobilization of the target biomolecule. It has been shown (discussion in Chap. 3) that this fabrication strategy has certain fundamental limitations in terms of the achievable resolution. Second, one can deposit locally the chemical species (e.g. target biomolecules in solution) by mechanical means directly on the surface with e.g. a nano–sized 'pencil' – a strategy similar to several 'new generation lithographies' [4]. The deposition by mechanical means can also be performed in a non-contact manner using technologies developed on the back of ink jet printer technology.

Current methods for preparing microarrays vary with the specific application, and include contact and non-contact methods of spotting oligonucleotides or cDNA, or a combination of photolithography and in situ synthesis for oligonucleotides. However, without major investment in high end projections systems, conventional lithography techniques cannot fabricate features in the 150–200 nm range, due to the diffraction of light. Extreme UV lithography and other next–generation photolithography strategies may offer the required resolution, but at ever increasing mask and fabrication facility costs, and operating under increasingly harsh conditions that may not be compatible with biological materials. Specifically, it will eventually become economically prohibitive to scale down microarray spots with conventional photolithography. As a comparison, the estimated cost of conventional microelectronics fabrication facilities will reach 200 billion dollars by 2015 [5]. Thus, there has been a significant effort on the behalf of the research and industrial communities to develop strategies to replace conventional robotic spotting and photolithographic methods for generating sub–100 nm biological nanoarrays. For instance, microcontact printing, developed at Harvard, is a direct–printing method that uses photolithographically generated masters to generate elastomer stamps which can be 'inked' with molecules and used to directly transfer the molecules in the form of a pattern to a substrate [2]. This technique is useful for forming large area patterns of organic or biological materials in a massively parallel fashion with pattern resolutions approaching 100 nm. However, this parallel technique is limited in its capabilities for generating multiple, chemically diverse, high resolution patterns in alignment on a surface.

The patterning strategies for biological arrays that rely on direct deposition avoid the indirect, resist or optical mask-based approaches. For instance, inkjet or other dispensing technologies capable of depositing nanoliter sized droplets of material are now employed to form array spots on the order of hundreds of microns. Advanced technologies of this type, such as that of Picoliter Inc. that uses acoustic droplet ejection technology, are capable of delivering picoliter volumes in a non-contact fashion, yielding spot sizes on the order of tens of microns. Still, true nanoscale patterning demands deposition volumes several orders of magnitude smaller than what is currently possible. In addition to the challenge of direct nanoscale delivery of biological molecules, ultra-precise nanoscale lateral positioning technologies must be developed and exploited, screening approaches for nanoscale bio-assays must be considered, as well as methods for increasing throughput and reliability for printing large numbers of distinct biological species. Recently, a number of compelling examples have emerged from the scanning probe microscopy community that address some or potentially all of these challenges.

6.2.1 Fabrication of Nanoarrays with sub–100 nm Resolution

Combinatorial Nano–surfaces Fabricated via Micro–ablation

Biomolecules, in particular proteins, strongly interact with the surfaces they are immobilized on. Nano–structures would have both the ability to probe large biomolecules individually, because they have comparable dimensions with the probed biomolecules, and also to make this probing largely parallel because nanostructures are amenable to large area densities. In general microfabrication is incapable of producing nanostructures, but recently [6] laser micro–ablation has been used for the fabrication of structures that are micron–sized laterally but nano–sized vertically. The micro–wells are fabricated via the localized laser ablation of a protein-blocked thin (tens of nm) metal (e.g. gold) layer deposited on a transparent polymeric (e.g. PMMA) film. The micro–ablation of gold induces local chemical and physical changes in the top surface of the polymer as well as a higher specific surface, which cooperate to achieve a higher and more reproducible surface concentration of proteins in micro–wells. The fabrication method can use a sequence of local ablation and 'flood' coverage with protein solution, or the ablation of the whole micro–assay followed by the 'spatially-addressable' deposition of different protein solutions with a pico–liter pipette (Fig. 6.2). It was observed that the micro–assays comprising line-shaped micro–structures offer a higher reproducibility and the opportunity to encode the information (e.g. type of protein, concentration) through a combination of vertical lines in a 'bar code', 'informationally-addressable' mode and not in a spatially-addressable mode like in the classical microarrays. It has been found that the 'combinatorial' nature of the inner surface of the channels (Fig. 6.3) allows for the increased adsorption of molecularly different proteins, from 3 to 10 times more than the adsorption on similar flat surfaces, with a higher amplification of smaller, globular proteins.

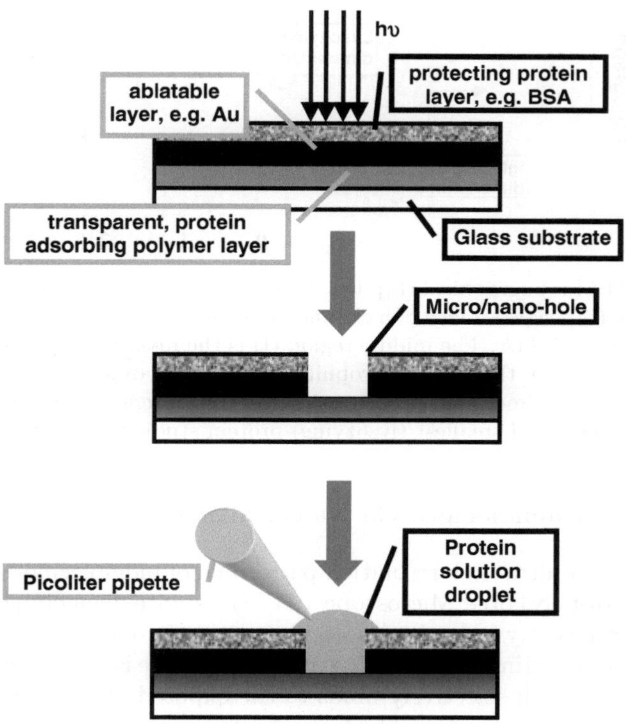

Fig. 6.2. Procedure for the fabrication of microwells and deposition of protein solution droplets

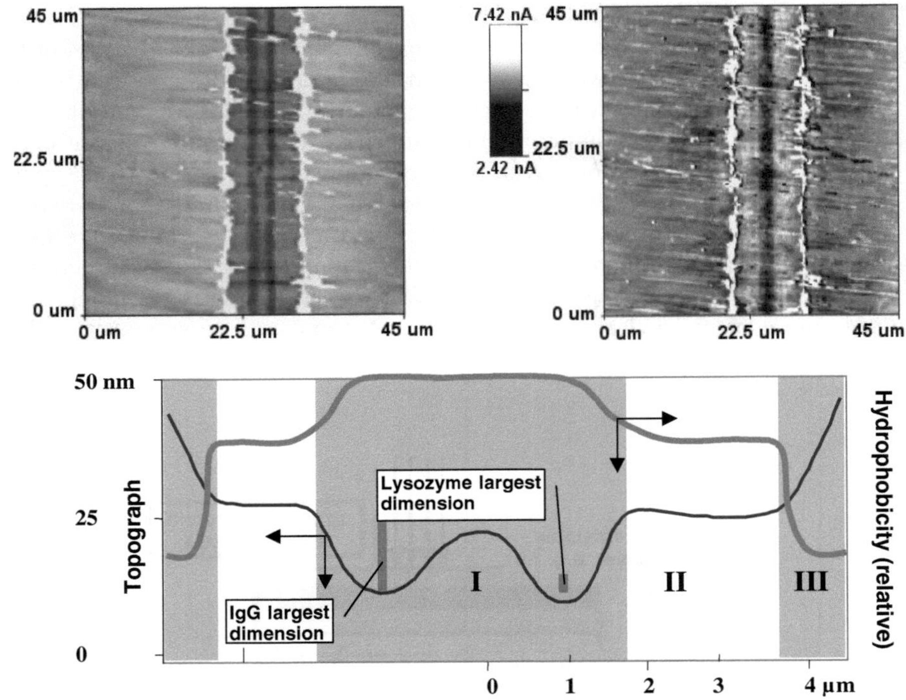

Fig. 6.3. AFM topographical (top left, brighter areas indicate elevated regions) and lateral force (top right) image of a channel fabricated via the ablation of a 30 nm Au layer on top of PMMA. The middle region (I) is the most hydrophobic, whereas the outer region (III) is the most hydrophilic. The lateral dimensions are much larger than the vertical dimensions. The vertical bars represent the largest dimension of the largest (IgG) and smallest (lysozyme) protein studied, respectively

Patterning Biomolecules via Nanografting

One method of ultra-high resolution patterning of biomolecules uses the probe tip of an Atomic Force Microscope (AFM) as an extremely sharp stylus to literally scrape away a swath of molecules on a resist-coated surface. In this approach, 'Nanografting', the freshly patterned surface is flooded with a patterning molecule which selectively binds to the exposed areas. Variations on this approach have been developed by Gang-Yu Liu's group and others as relatively facile methods for fabrication of one molecule thick patterns of biomolecules such as oligonucleotides and proteins with line patterns less than 100 nm, and down to 10 nm on select substrates [7–9]. In a typical experiment, alkanethiol resists are deposited as self-assembled monolayers on ultra flat gold surfaces. A solution of oligonucleotides modified by alkanethiol tags bathes the substrate during patterning such that the molecules assemble into the gaps made by the path of the AFM tip in the resist layer (Fig. 6.4). A similar strategy has been employed for patterning protein molecules with exposed cysteine residues. In its present form, nanografting is a serial technique, thus inherently slow and limited in its application to miniaturization of bioarrays. In addition, the basic requirement of a monolayer resist limits the choice of substrate to those that can be well passivated with resists which are in turn easily removed. Furthermore, the etching/backfilling process may not be amenable to a parallel process because of the difficulty of selectively filling in etched features with different molecules (e.g. different DNA sequences) on the sub–100 nm scale. Finally, the method is essentially a negative tone lithography, inappropriate for multiple patterning, which is required for an array-like application. Direct–printing nanolithography techniques such as those described below will be useful for overcoming such limitations. However, nanografting is a potentially useful technique for specialized applications that require high

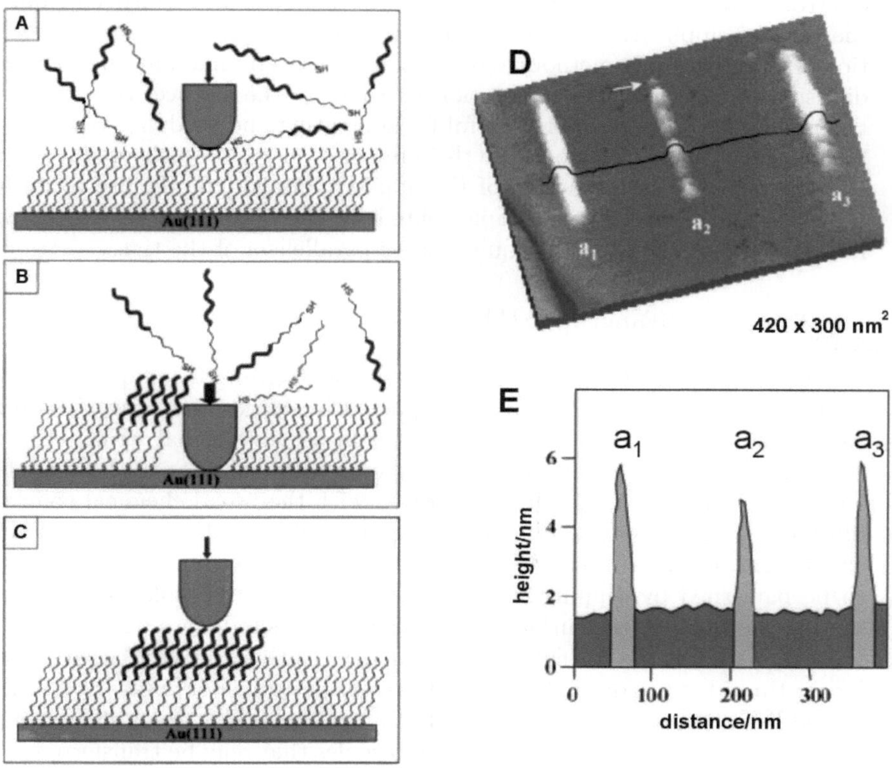

Fig. 6.4. (A)–(C) Schematic representation of general nanografting method for patterning biomolecules on gold surfaces. **(D)** AFM image of 3 DNA lines on generated on Au via nanografting, and **(E)**, a line scan through the features in (d) showing height of individual DNA molecules in the pattern (Reprinted with permission from [8]. Copyright 1997 American Chemical Society Publications)

resolution patterns of a single type of molecule, for instance to examine effects of nanoscale confinement of oligonucleotide or protein molecules, investigate new readout methods for miniaturized bioanalysis, and for preliminary research in the area of bioelectronic circuits.

Direct Nanopipet Deposition

The ability to generate multicomponent arrays of biomolecules requires development of techniques for directly depositing materials on surfaces. In one example of efforts in this direction, Klenerman et al. used a modified version of Scanning Probe Microscopy called scanning ion–conductance microscopy to directly deposit biomolecules such as biotinylated DNA onto streptavidin-coated glass surfaces and protein G onto positively charged glass surfaces [10]. In these experiments, nanopipets with inner diameters of 100–150 nm in an electrolyte solution as reservoirs for charged biomolecules which flow out of the tip with application of the appropriate bias (Fig. 6.5). The spatial resolution of the patterning methodology is limited to several microns due to lateral diffusion of the molecules in solution en route to the surface. However, this technique may be particularly useful for generating and studying gradients of biomolecules on a surface because the number of molecules delivered from the tip per unit time is a function of the applied voltage [10]. To use to its full potential and in order to be implemented for the fabrication of nanoarrays, however, the method would require major parallelism of the tips.

Dip Pen Nanolithography™

Recently a new SPM-based direct–write nanopatterning technique, 'Dip Pen Nanolithography™' (DPN™) was reported by Mirkin and coworkers from Northwestern University [11–15]. Based upon a conventional AFM, DPN™ combines resolutions comparable and in some cases superior to those of electron beam lithography (15 nm linewidths) with the broad chemical compatibility obtained by operating under ambient conditions. In a typical DPN™ experiment, a conventional AFM probe tip is coated with a molecule or 'ink' to be patterned by dipping the tip in a solution of the molecules. By contacting the tip with the surface molecules are deposited via a water meniscus that condenses at the tip–substrate contact. With this diffusion-based process longer tip–substrate dwell times lead to larger pattern spot areas [11,15]. Due to its direct deposition nature, the DPN™ process has been shown to be very general, both with respect to the molecules that may be transferred from the AFM tip to a surface (small organic surfactants, charged macromolecules such as conjugated polymers and proteins, sol–gel forming materials, and even nanoparticles) [11,16–19] and the substrate (metals, e.g. gold; insulators, e.g. silicon oxide; and semiconductors, e.g. GaAs). The main requirement for transport is that there is some interaction, covalent or physical between the ink and the surface. For instance, alkanethiols form a coordination bond with a

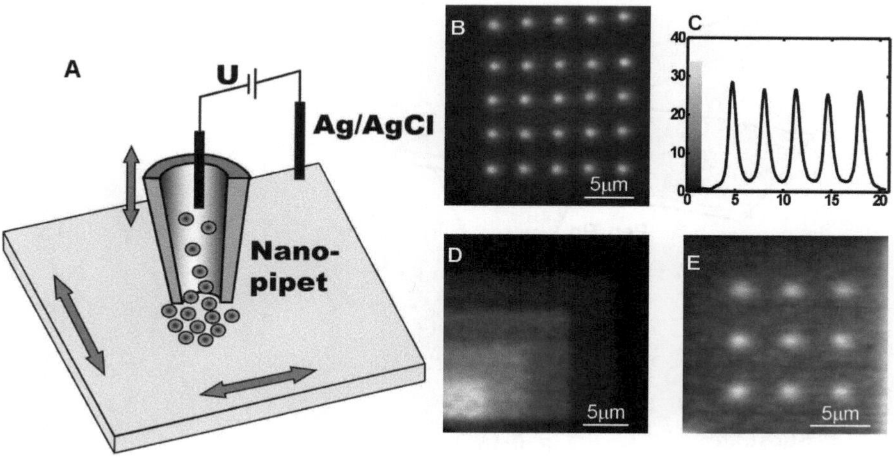

Fig. 6.5. **(A)** Schematic of nanopipet strategy for deposition of biomolecules, **(B)** Fluorescence microscopy of biotin-modified DNA on streptavidin-coated glass. **(C)** Line scan showing spot profile of bottom row in B. **(D)** DNA patterns with increasing surface concentrations on glass, and **(E)** Fluorescence micrograph of protein G on a positively-charged glass surface (Reprinted with permission from [10]. Copyright 2002 Academic Press Inc Elsevier Science)

gold surface [20, 21]. While there are techniques that can be used to produce extremely fine structures on a surface (such as electron–beam or focused ion beam lithography), the challenge for the fabrication of nanoarrays lies in generating complex patterns composed of different materials, placed in precise locations relative to each other [22]. With DPN^{TM}, one can exploit the ability to write and read high resolution chemical patterns with the same tool. Thus, multiple chemical or biological patterns can be generated using DPN^{TM} with precise ($\sim$ 5 nm) alignment registration. Among patterning techniques that can operate at sub–micron and sub–100 nm dimensions, such as e–beam lithography or contact stamping methods, DPN^{TM} is the only technology that can directly deposit biological molecules under ambient conditions with ultrahigh precision and registration. Moreover, these molecules can be deposited in either ambient or inert environments without exposing them to ionizing UV or electron–beam radiation. Also, several different kinds of molecules can be deposited without exposing the patterned molecules to harsh solvents or chemical etchants, and without risking cross–contamination. The desired chemistry is carried out exactly, and only, where it is desired (Fig. 6.6).

Preliminary experiments suggest that DNA patterning via DPN^{TM} is not only possible, but can be highly controllable in terms of pattern size/shape, and that the immobilized DNA is functional and accessible to specific binding of labelled targets [18]. Initial studies of direct transfer of DNA from an AFM tip to both metal and insulator substrates identified several key com-

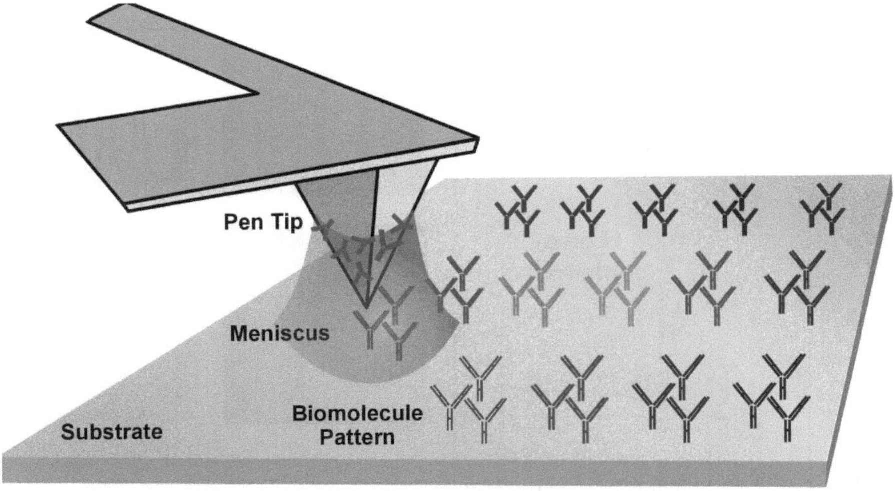

Fig. 6.6. Schematic of the DPN^{TM} process for direct deposition of biological molecules

ponents which modulate DNA patterning, including precise control of the ambient humidity and careful functionalization and inking of the AFM tips. In addition to tip–coating and humidity, a judicious choice of ink–substrate combination can facilitate the DPN^{TM} process. For example, hexanethiol-modified oligonucleotides were used to directly pattern gold substrates with features ranging from 50 nm to several micrometers in size. For nanoarrays on oxidized silicon wafers or glass surfaces, acrylamide modified oligonucleotides are deposited directly via DPN^{TM} onto activated (mercapto–propyl–trimethoxylsilane, MPTMS) substrates where subsequent reaction (under ambient temperature, 45% relative humidity) forms a covalent link to the surface. Similar chemistry has been developed for glass and quartz substrates. Non-specific binding of target oligonucleotide was minimized by passivating the unpatterned regions of the substrate by reaction with buffered acrylic acid monomer at pH 10. The feature size of individual DNA spots is controllable over a range of several orders of magnitude via the tip–surface dwell time, as observed with other DPN^{TM} systems [11]. For example, 100 nm spots can be deposited in times less than 1 second. Moreover, the rate of patterning is controllable by tuning the relative humidity of the patterning chamber. For example, the diameter of a spot created by holding the AFM tip for 10 seconds changes from less than 50 nm to 1 μm with a relative humidity increase from 30% to 80%. The selectively and function of patterned oligonucleotides was verified by hybridization of complementary fluorophore-labelled DNA or oligonucleotide derivatised gold nanoparticle probes of different sizes. For example, a 2-component DNA pattern consisting of micron scale features was characterized first by epi–fluorescence microscopy of bound fluorophore-tagged complements, then by AFM topography measurements of two different sizes of DNA-modified gold nanoparticles (Fig. 6.7). Importantly, only fluorescence corresponding to the complementary target and the patterned area was detected, and the AFM topography images show that the gold particles react only with the correct oligonucleotide spot. In these preliminary experiments, spot shape, size, and emission intensity is extremely uniform, within individual features, and from spot to spot. With this technique, DNA spots with diameters as small as 50 nm were prepared, i.e. 10,000 times smaller (in terms of area density) than those in conventional microarrays. With the resolution demonstrated herein, arrays with $\sim$ 100,000 oligonucleotide spots can be generated in an area the size of a typical AFM scanner (100 $\times$ 100 μm), making it possible to investigate scanned probe methods of microarray readout.

The DPN™ technique has recently been extended to deposition of proteins. In particular there have been reports of direct patterning of thiolated collagen and collagen-like peptides onto gold surfaces [17], human chorionic gonadotropin antibody onto 3 glycidoxy–propyl–trimethoxysilane modified glass surfaces [23], as well as a number of immunoproteins, enzymes, and viruses [24–26]. Significant effort has been directed towards the characterization of the resulting protein nanostructures with regard to their activity. Although DPN™ is a gentle lithographic technique, surface interactions and covalent or non-covalent attachment chemistry could potentially serve to denature some classes of proteins. Researchers in the Mirkin group at Northwestern University have begun studying the complex issues involved in preserving the biological activity of immunoproteins such as IgG during a DPN™ experiment [24]. The use of additives such as glycerin to the protein patterning solution was found to enhance patterning by diminishing the negative effects of drying the deposited proteins. For instance, Lim et al. used the DPN™ technique to deposit human IgG and rabbit IgG nanostructures on oxidized silicon surfaces through covalent attachment to carbonyl groups on the surface (Fig. 6.8a). The activity and identity of the immobilized proteins was confirmed by binding fluorescently-tagged antibodies specific for the two different nanopatterns [24]. The resulting two-color fluorescence images revealed specific biological activity and predicted cross–reactivity for the two patterns (Fig. 6.8b and c).

6.2.2 Strategies for Increased Throughput for Ultra-High Density Nanoarrays

In order to generate biological nanoarrays with significant improvements in complexity over those prepared by standard photolithographic or robotic spotting methods with adequate throughput, it is critical to develop nanopatterning technologies that operate in a massively parallel fashion. The common tools for generating microarrays deposit or assemble in situ hundreds of thousands of different probe features using photolithographic masks, or spot biomolecules directly using four or many pin configurations. The most signifi-

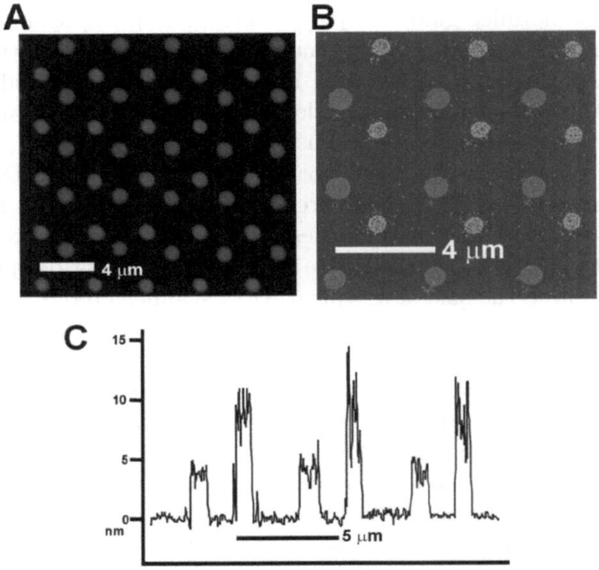

Fig. 6.7. Epi–fluorescent (**A**) and AFM topography (**B**) and (**C**) images of two different sizes of DNA-modified gold nanoparticles

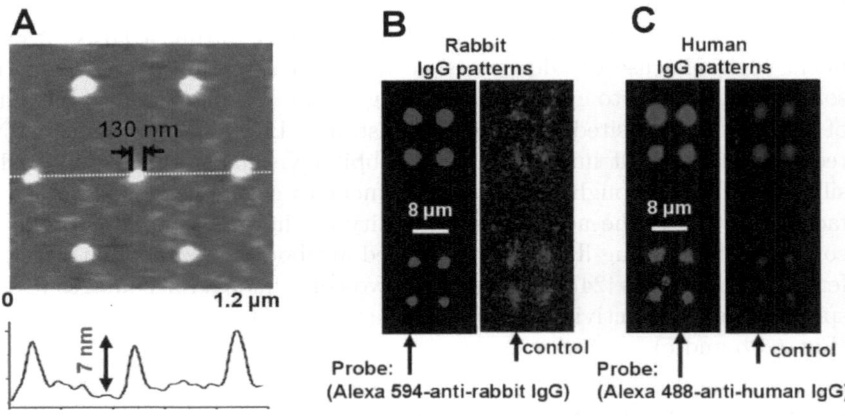

Fig. 6.8. Protein nanostructures deposited by DPN^{TM}. (**A**) Tapping mode . AFM image of high resolution IgG pattern on silicon oxide (line scan shows height of individual molecules in the pattern. Epi–fluorescence of (**B**) Rabbit IgG DPN^{TM} pattern with Alexa 594-labelled anti-rabbit IgG probe, (**C**) Human IgG pattern with Alexa 488-labelled anti-human IgG probe. Note the faint spots indicating cross–reactivity for (B) and (C) (Reprinted with permission from [24]. Copyright 2003 Wiley-V C H Verlag GMBH)

cant barrier to using the scanned probe nanolithography techniques described up to this point for arraying applications stems from the serial or 'single pen' nature of the techniques. Recognizing this limitation, several important advances have been made by researchers at IBM [27] and also at Stanford University [28–30] in the direction of parallel scanning probe methods. In particular, researchers at IBM have fabricated devices wherein 32 × 32 arrays of individually addressable and actuated cantilevers have been etched into a chip in an area of 3 mm × 3 mm [27]. Individual tips on this device are used to read and write 'bits' in a 100 × 100 micron area of a polymer film via thermally induced nanoindentation for ultra-high density data storage applications. In addition, Quate and co-workers at Stanford have developed a number of 1–D and 2–D probe arrays for both imaging and lithographic applications [28–30]. These results indicate that the MEMS technology expertise is available for designing and fabricating pen arrays suitable for deposition rather than indentation. The next challenge is to interface these engineering advances with direct write lithography methods such as DPN™. In order to increase the throughput and area accessible to scanning-probe techniques, several groups around the world are pursing the development of parallel-probe cantilever arrays. A number of academic groups, as well as NanoInk Inc. are implementing similar MEMS based parallel-probe strategies designed specifically with the constraints of DPN™ applications in mind.

The simplest implementation of parallel-pen DPN™ is thus a passive probe array. In this case, the pens are not actuated independently but are simultaneously brought into contact with the surface and scanned together, allowing the duplication of a single pattern a number of times equal to the number of probes in the array. An example of preliminary efforts in this direction was reported by Hong and coworkers [12]. More recently, the microfabrication facility at NanoInk Inc. has produced cantilever arrays composed of more than 1.2 million pens on a single 4 inch wafer (Fig. 6.9a). Since active feedback is applied to only a single cantilever in the array, and the others are allowed to track the topography passively, specific constraints on the registration between the array and the surface, as well as the flexibility of the cantilevers must be met [31]. This ongoing work could eventually produce nanoarrays of more than 300 billion spots on a 4 inch wafer (50 nm diameter spots separated by 100 nm).

Independent control of each probe tip is another strategy with a different set of applications. Individual tip actuation can be accomplished using piezoelectric, capacitive, or thermoelectric actuation. In the first generation of active parallel-probe DPN™ arrays, researchers have used thermoelectric actuation: resistive heating of a multilayer cantilever results in differential expansion of the components, leading to bending of the probe (Fig. 6.9b). Using this approach, a range of complex patterns can be generated at high speed because the contact between each tip and the writing surface is independently controlled. For instance, a 10–pen array can be used to write the numerals 0–9 simultaneously. The final challenge of complete MEMS integration of DPN™

technology is the automation of tip coating and ink delivery. For certain applications it appears that custom microfluidic systems will ultimately be used to control the inking of individual cantilevers in a parallel probe–array. The realization of such systems will depend on the development and adaptation of a number of technologies. Large-scale integration of microfluidic technologies is still challenging, and arrays of 1000 individually addressable wells represent the current state of the art [32]. Thus, to meet the inking needs of parallel probe arrays (with an ultimate goal of being able to deliver a different ink to each probe in a large pen array), arrays of addressable ink wells must also be developed. Indeed it is clear that ultra-high density nanoarrays will re-

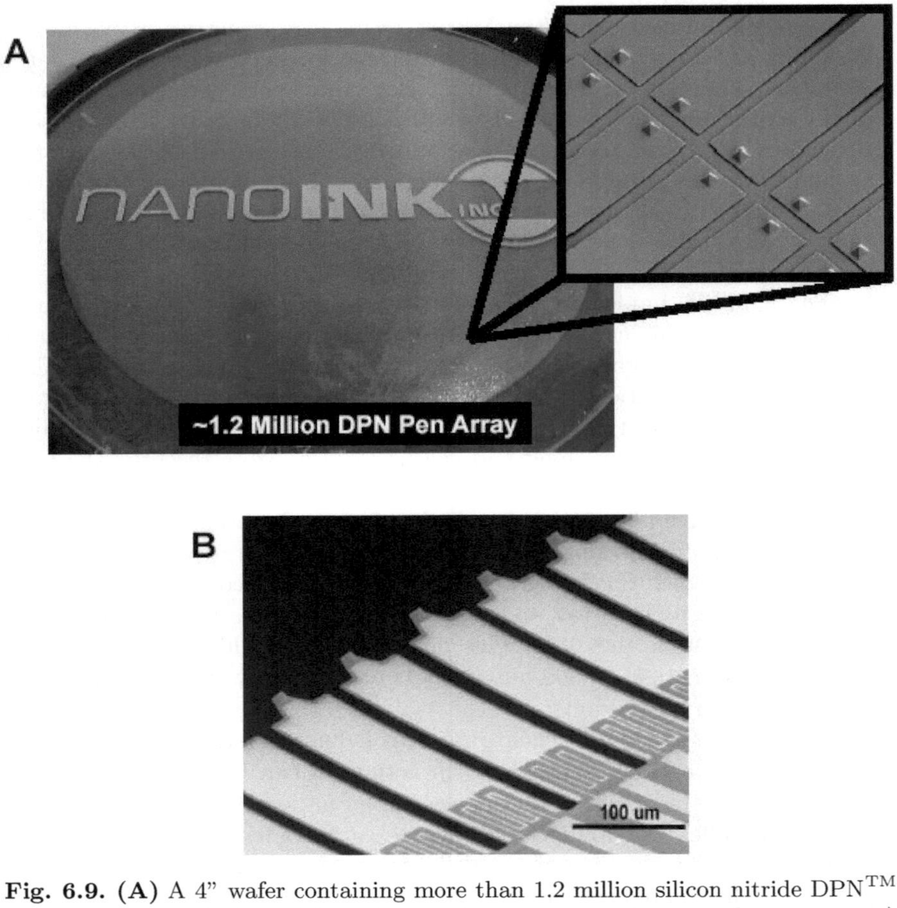

Fig. 6.9. (A) A 4" wafer containing more than 1.2 million silicon nitride DPN^{TM} pens (inset is an SEM at 500× showing individual cantilevers and writing tips.) (Courtesy of NanoInk, Inc.) **(B)** SEM of active DPN^{TM} probes equipped with thermoelectric actuation technology (courtesy of Chang Liu, University of Urbana–Champaign)

quire incredibly complex sample handling. The synthesis and purification of large numbers of oligonucleotides for instance is daunting and may become prohibitive as the number of distinct probe features increases. An alternative strategy for fabrication of such high density nanoarrays may eventually exploit an in situ synthesis approach, whereby monomers are delivered sequentially by probe tips, building the probe molecules at each feature from the chip up in a strategy similar to that currently used by Affymetrix.

6.2.3 Strategies for Nanoarray Detection and Analysis

New technologies for generating nanoarrays with sub–100 nm sized features offer an opportunity for investigation and development of new detection methodologies that can operate below the diffraction limit of light.

While present-day detection methods may be inadequate for screening such high density arrays, miniaturization on the scale accessible with DPN™ will allow the development of screening methods that are suitable for such nanoscale structures. There are many scientific opportunities in this regard: when a feature composed of receptors is miniaturized to the scale of the biological analytes or their attached labels, almost every mechanical, electrical and chemical property of the receptor feature is changed upon reaction with the analyte. These properties, including size, shape, electrical conductivity, and hydrophilicity, can all be monitored in situ with an AFM or with on-chip electronic circuitry. In the long term, it may even be possible to direct the attachment of proteins and virus particles in specific orientations to study reactivity as a function of structural configuration. One promising strategy for detecting analyte binding to nanoarrays is the use of labelled nanoparticle probes. Nanoparticles can be prepared from a host of different materials in different sizes and shapes and can be functionalized with biological recognition molecules such as antibodies or oligonucleotides [33–37]. Some of the particles have been shown to bind specifically to surface-immobilized receptors or complementary nucleic acids where they are detected using optical or electrical readout. This strategy is proving to be a particularly useful method of identifying and possibly quantifying binding in microarray assays due to the striking properties of the nanoparticles [18, 25]. For example, in addition to height change measurements after particle binding, there are already examples of electrical detection of DNA targets using DNA-modified gold nanoparticles between microelectrodes [36], as well as reports of detection strategies that make use of the strong resonant scattering [38, 39], optical absorbance [40], or fluorescent properties of certain metal or inorganic nanoparticles [41]. In general such strategies are amenable to spatially-resolved characterization of nanoarrays on surfaces through the wide variety of tools accessed by scanning probe microscopy, from topography, to friction, magnetic force, and even near field scanning optical configurations. This approach has been recently used in conjunction with topographical AFM to detect the selective binding of different sized DNA-modified nanoparticles to two component nanoarrays as in Fig. 6.7b. [18,25,42].

Furthermore, there is also the opportunity for label-free screening of nanoarrays using SPM imaging. For instance, Fig. 6.10 shows how such a DPN^{TM}-fabricated 2-component protein array could potentially be used as a label free protein screening tool. Arrays of rabbit IgG and lysozyme proteins are deposited onto a gold substrate via DPN^{TM}. Reacting the nanoarrays with rabbit anti-IgG leads to a 1:1 binding of the antibody–antigen IgG pairs, resulting in a doubling of the feature height of the IgG rows, which is observed in the micrographs. Perhaps more significantly, the control array (lysozyme in alternating rows) shows no change in feature height, indicating that the coupling retains its specificity. Although the orientation of the surface-bound IgG protein is not controlled in this experiment, a sufficient fraction of the population apparently provides the solution-borne antibodies with access to the relevant binding domains of the protein [25].

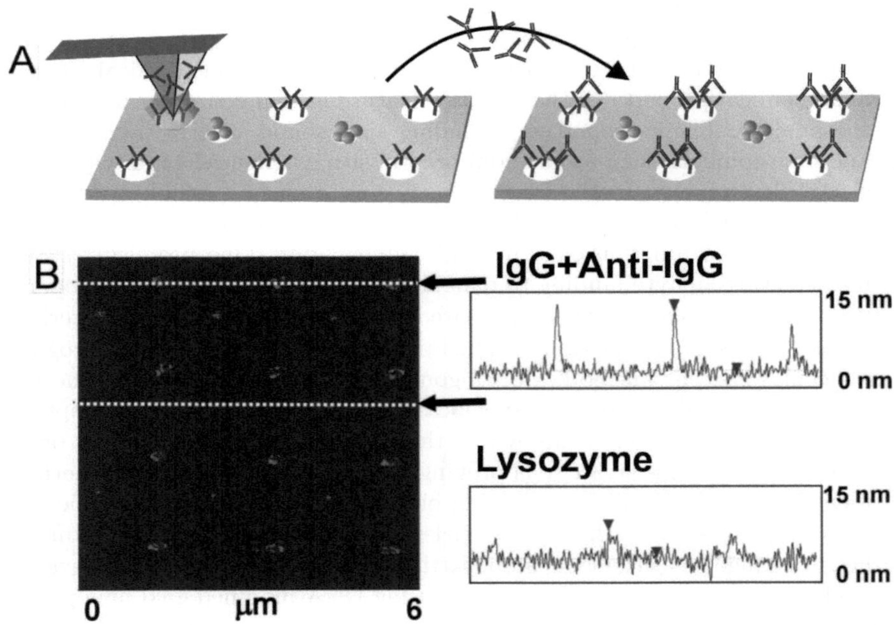

Fig. 6.10. Label-less protein detection using Tapping Mode AFM. Alternating rows of rabbit IgG and lysozyme nanostructures were exposed to anti-rabbit IgG. **(A)**,**(B)**: Line scans of IgG row and lysozyme row after anti-rabbit IgG treatment showing selective height increase at the IgG and no change at the lysozyme sites (Reprinted with permission from [25]. Copyright 2003 Academic Press Inc Elsevier Science)

Clearly, the practicality of screening nanoarrays using scanned probe technology is currently limited by the slow imaging speed. A single 10 × 10 micron AFM image can take up to 20 minutes to acquire and conventional AFM at its fastest only acquires several frames a minute. However, SPM technology is advancing to address this particular issue. For example, Infinitesima's novel technology combines a resonant scanned probe system with near-field optical detection to produce images in ten milliseconds or less, nearly video rate. With all the other advantages of conventional SPM it is ideally suited to following dynamic processes in situ and in almost any environmental conditions.

The path towards miniaturization will not proceed without requiring obstacles to be overcome along the way. However, it is widely thought that the potential rewards clearly justify the effort. One potential difficulty comes from cross–reactivity and non-specific binding of analyte or other species to the array spots. Although non-specific binding is a problem for any surface-based assay, it is likely to become more problematic as screening goes nanoscale: on a nanoscale receptor spot it would be possible for a few non-specifically-bound particles to completely overwhelm the intended signal. Fortunately, going nanoscale offers new possibilities to alleviate the non specific binding. On one hand, for a small sacrifice in information density, redundancy and 'error checking' could be built into any array. On the other hand, controlling (and screening) the chemical environment with nanoscale precision could offer the opportunity both to reduce the frequency of non-specific binding events, and to more readily identify them when they do occur. Finally, with direct techniques such as DPN^{TM}, cross–contamination of the patterned array features is entirely eliminated.

6.3 Computational Nanoarrays

The function of 'classical' bio–arrays, be they micro– or nano–, is to provide information regarding the biomolecular recognition through the docking of probe biomolecules on target biomolecules (or cells) spatially encoded on the surface of the array. But molecular recognition may be just the first of a concatenation of stages that represent a process of computation, in which case the last configuration of the microarray represents the 'solution'. Although these functional arrays are not nanoarrays in the sense of lateral or vertical resolution as described in the previous section, they perform their function truly at the nano–level.

DNA computing is a new method of physical computing which is based on the molecular recognition of complementary biomolecules (DNA) and the massive parallelism that can be achieved through cycles of DNA synthesis, PCR, ligation, electrophoresis and use of restricting enzymes. This new computing method appears to be particularly suited to problems that cannot be solved by even the most advance traditional electronic computers that operate sequentially. Traditionally these are called NP problems, referring to the exponential (i.e. Nondeterministic Polynomial) time required to reach a solution for a linear increase of the size of the problem.

Adleman [43] was the first to describe a DNA-based method which solves the Hamilton path problem (e.g. finding an airline path that passes several cities optimally visiting each just once) in polynomial–time. The trade–off in Adleman's experiment was to use a large number of 'computers' (i.e. DNA molecules) which perform operations in a massively parallel manner against time (number and type of physical procedural steps). The nodes and the pair between nodes were encoded in DNA strands, which self–assemble in all possible arrangements following Watson–Crick complementarity. Some of these dsDNA may contain possible solutions, which can be selected, amplified and detected using classical molecular biology techniques.

After the initial proof of concept carried out by Adleman [43], DNA computing received a lot of attention due to its potential for problem-solving efficiency, data storage capacity, energy efficient computation and new mathematical outlook on computation. Essentially, the basic operations of the DNA computing are: *Amplify; Merge; Detect; Sequence-separate; Length-separate;* and *Position-separate*. Using this basic mathematical apparatus, many algorithms have been proposed to solve specific problems using DNA computing, among others, the satisfiability problem [44], the maximal clique problem [45], the graph coloring problem [46], with many other (e.g. breaking the Data Encryption Scheme, Travelling Salesman Problem, decide graph connectivity, 'knapsack' problem) being possible.

The critical factor on which the success of DNA computing in solution–phase depends is the capacity to achieve very small error rates for various biochemical operations. Because the grand idea behind DNA computing is to perform massively parallel operations, it follows that an efficient computation relies on an as complete as possible search of the possible solutions space (DNA strands). Classically, this can be achieved by a high ratio of DNA strands available per number of candidate solutions. As the complexity of the problem (expressed in terms of the dimensions of the input) increases, this ratio decreases for a given initial amount of DNA, i.e. the average number of strands encoding one candidate solution becomes smaller. This places demands on the maximum acceptable error rate or, equivalently, on the minimum amount of DNA needed. Thus, DNA computing as defined suffers from a 'scalability' problem. This has prompted the search for means to better control the error rates in DNA computing operations (e.g. PCR, hybridization). One avenue for improving experimental control during DNA computing experiments is to immobilize the DNA strands on a surface before manipulation.

Microarray technology helped move the concept of manipulation of DNA molecules for DNA computing from solution-based to surface-based processes. For instance, Smith et al. [47] proposed a new surface-based DNA computation (Fig. 6.11). Firstly, ssDNA molecules that correspond to 'all' possible solutions to a problem ('make' function) are synthesized and covalently immobilized ('attach' function) on a surface. Then, subsets of the surface-bound

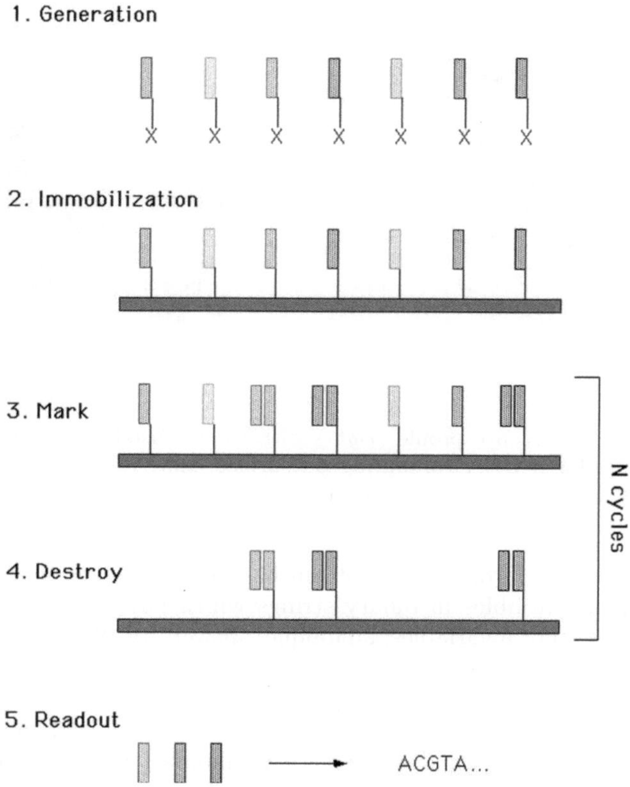

Fig. 6.11. Schematic of DNA computation at surfaces (Reprinted with permission from [47]. Copyright 1998 Mary Ann Liebert Inc Publishers)

combinatorial ssDNA library are recognized by hybridization to their complements ('mark' operation), making these parts double stranded. An enzyme (e.g. exonuclease) destroys the non-hybridized oligonucleotides ('destroy' function). Finally, the previously hybridized oligonucleotides are regenerated ('unmark' operation). All strands that do not represent the solution are removed via the repetition of the 'mark', 'destroy' and 'unmark' operations, leaving only the 'solution' bound on the surface. Finally, the solution is read through sequence of decoupling from the surface, PCR and further hybridization to a designed microarray (Fig. 6.12). Frutos et al [48] developed the method further, proposing the use of enzymatic ligation reactions of DNA 'words' on surfaces for DNA computing.

This method of computation has been used by Liu et al. [49] for solving a simple case of the 3–SAT problem, which is considered to be the hardest of all NP problems. The solution of the 3–SAT problem has to satisfy a set of logical clauses, each composed of three true or false variables, connected

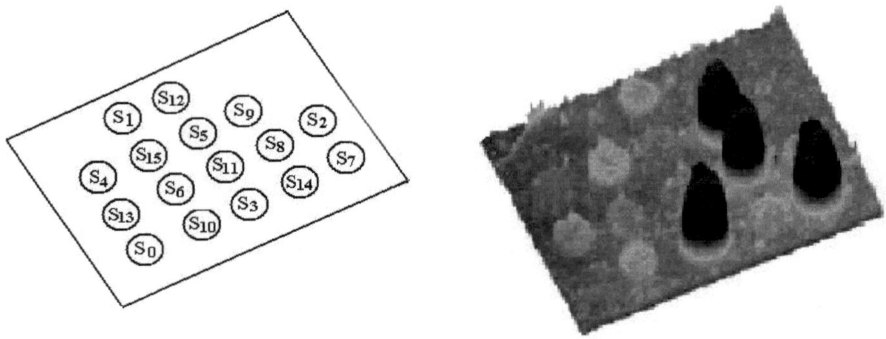

Fig. 6.12. Fluorescence profile (*right*) with the surface-bound oligonucleotide locations (*left*) for a DNA computing on surface chip (Reprinted with permission from [47]. Copyright 1998 Mary Ann Liebert Inc Publishers)

by 'or' logical operators. The problem has been solved in a reasonable time by coding the variables in binary strings which have been in turn coded in ssDNA strings. For n variables, 2n unique 'answer' (or 'Watson') strands exist, e.g. TGCGG = 001, complemented by unique 'Crick' strands. The solution is accepted if it satisfies all the logical clauses of a 3–SAT formula. If ssDNA strands representing all candidate solutions are immobilized on a gold surface, the addition of Crick strands will create a combination of ss– and dsDNA. The non-solution ssDNA, which do not satisfy the first clause encoded in the added Crick strands, are destroyed by enzymes leaving still–possible solutions locked in the dsDNA strands, which are subsequently melted – and the process starts again for the next clause. The last remaining strand is the solution which is decoded in a microarray format. The synthesis of DNA strands aside, the computation proceeds in 3k + 1 steps for the exploration of all 2n possibilities (k is the number of clauses). This procedure is much more efficient than the best conventional computer algorithm [50], which scales as 1.33^n (n = number of variables). To put things in perspective, a 3–SAT problem with 30 clauses and 50 variables would be solved classically in about 1.6 million steps, but the method described above would solve it in 91 steps [51].

From a mathematical point of view, surface-based DNA computing is a competitor to solution–phase DNA computing. It is known [51] that surface-based DNA chemistry supports general circuit computation on many inputs in parallel efficiently and that the number of parallel operations needed to decide the satisfiability of a Boolean circuit is proportional to the size of the circuit. Both solution phase and surface-based DNA computation present advantages and disadvantages. Surface-based DNA computing is more molecularly–efficient, because less strands are lost at each step and subsequently, there are less pressures on the needed initial representation redundancy, due to the immobilization of the oligos at the surface. Other advantages include ease of purification and the ability to use more advanced biochemical techniques, in particular those developed for microarrays. However, these gains come at the price of a massively reduced *physical* density (from 3D storage to 2D storage). Additionally, the number of operations per second is limited by the slower enzyme kinetics and lower hybridization efficiency. Finally, the surface-based method does not eliminate scaling problems since discrimination of single-base mismatches becomes more difficult as the strand length increases and the operations are not error-free. The most serious of these limitations is the loss of information density. One must either increase the surface area (e.g. by using microbeads instead of a planar surface) or attempt to employ a local three–dimensional surface chemistry.

6.4 Dynamic Nanoarrays

Another characteristic of the 'classical' micro/nanoarrays is their single-use. Once their function, be that simple molecular recognition or biomolecular computation, is fulfilled and the information is passed further to appropriate information processing systems, the product –the microarray– becomes obsolete and therefore micro/nanoarrays are essentially single-use devices (with the notable exception of Nanogen's approach derived from biosensors). More advanced devices would be designed to use molecular recognition *for*, rather than *being*, their function, which would be then continuous rather than one–off. These future devices, which would operate in a highly parallel arrangement, possibly in a microarray format, would comprise moving elements that are propelled by external means, or preferably self-propelled. The first option, i.e. external powered dynamic devices, has been launched by microfluidics and manipulation of magnetic beads. However, it is the self-propelled dynamic devices that offer the highest expectations of technological revolutions. Fortunately, Nature offers several working models of molecular motors, many tested in primitive hybrid dynamic nano–devices.

Protein molecular motors, which work either as a pair in tandem, i.e. linear motors, or single, i.e. rotary motors, transform chemical energy, through the hydrolysis of adenosin–triphosphate (ATP), into mechanical energy or movement. Molecular motors, which are ubiquitous proteins, are responsible for biological functions as diverse as cell movement and division, transport of vesicles and muscle function.

Two experimental techniques, motility assays and single molecule visualization, manipulation and measurement, resulted in important advances in the understanding and quantification of the functions of molecular motors. Motility assays, which were pioneered some 15 years ago, are essentially primitive nano–devices operating in a 'distributed' microarray format, which allow the probing of the functions of molecular motors in a 'black box' manner. On the other hand, single molecule techniques allow the measurement of fundamental parameters, e.g. forces, and are therefore useful for the design of future nanodevices based on molecular motors.

Rotary Motors

Protein molecular motors perform their function through either rotary or linear motion. Although it has been demonstrated that actin filaments also perform a rotary motion along their axis when sliding atop of myosin functionalized surfaces [52], there are two motors that operate in a truly rotary mode, i.e. the bacterial flagellum motor and the ATP synthase enzyme. The latter appears to be the smallest (approximately 12 nm, [53]), the most efficient (generating some 100 pN nm with almost 100% efficiency [52]), and the quickest (unloaded rotational velocity of approximately 17 r.p.s, [54]) rotary motor. All of these advantages make this system quite attractive for its use in hybrid nanodevices. ATP synthase is a large enzyme, which synthesizes ATP in the mitochondria. Similar enzymes can be found in other organisms, e.g. plant chloroplasts and bacterial cell membranes, with the latter being specifically appropriate for robust hybrid nanodevices. The structure of the protein comprises the actual engine (F_1 unit) mounted on a 'pedestal' (F_0 unit) as in Fig. 6.13 [55, 56].

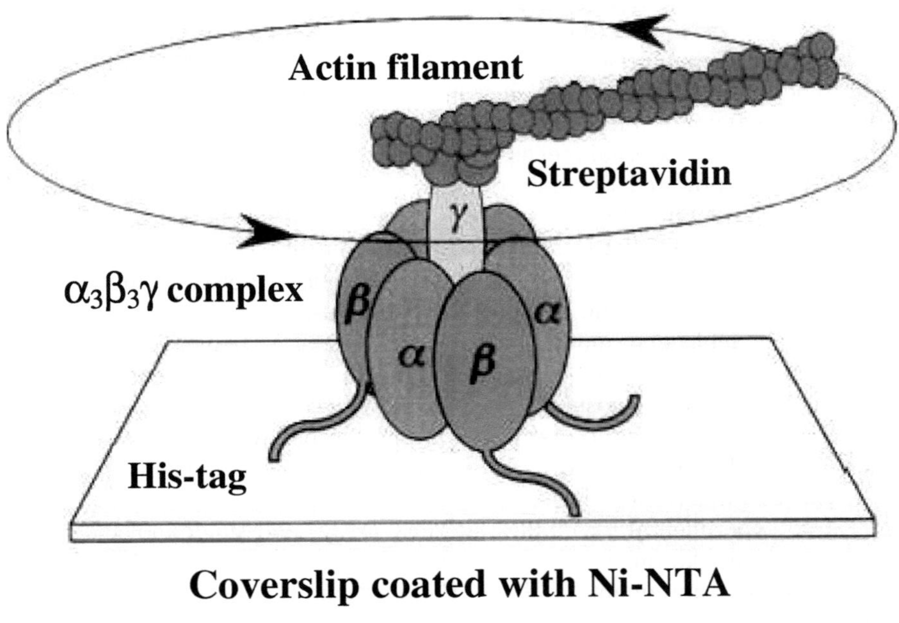

Fig. 6.13. Architecture of the F_1 ATPase rotary motor anchored on a surface at the non-working end [56]

Montemagno and co-workers' crucial work [57] provided the proof of principle for the building of a hybrid nanodevice based on a rotary motor. Their hybrid nanodevice powered by a rotary molecular motor consisted of three major elements: (i) a microarray of a nano-sized nickel posts, fabricated by e–beam lithography; (ii) a thermostable form of Ni-selective F_1–ATPase which selectively attach on the Ni nano–posts; and (iii) Ni nanopropellers (Ni rods) with functionalized surfaces that allow specific attachment of the lever of the motor. The design, the fabrication concept and the microarray organization of the hybrid nanodevice are presented in Fig. 6.14. Despite the low fabrication yield (only 5 out of 400 propellers rotated) no backward steps have been observed, possibly due to the high ATP concentration. Also the device showed a 2.5 hours long endurance cycle. Subsequent work [58, 59] discussed the many engineering issues produced by the difficult interfacing between inorganic nano-engineered objects and very delicate proteins.

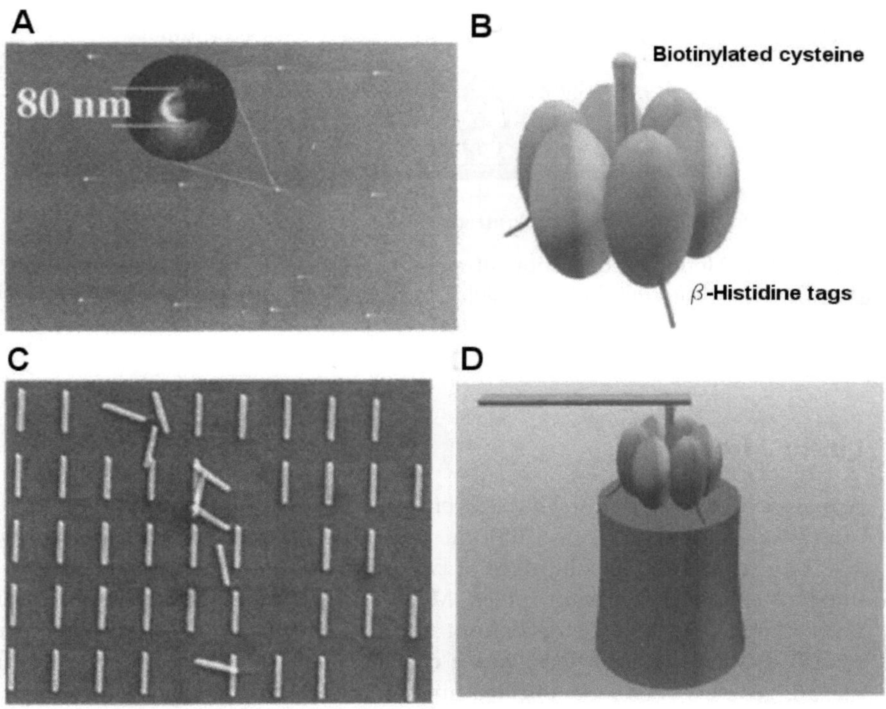

Fig. 6.14. Hybrid dynamic device in a microarray architecture. **(A)** Top view of the pole; **(B)** Molecular engineering of the rotary motor for anchoring on the surface and attachment of the Ni nanorod; **(C)** Top view of an array of Ni rods mounted on rotary motors; **(D)** Side view of a rotary motor mounted on a pole (Reprinted with permission from [57]. Copyright 2000 American Association for the Advancement of Science)

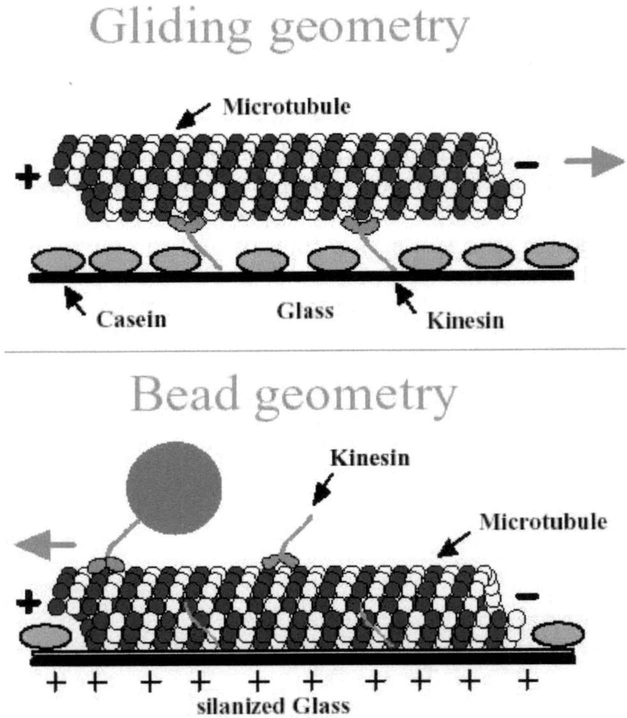

Fig. 6.15. Modes of operation of motility assays for linear molecular motors (Reprinted with permission from [60]. Copyright 2001 Academic Press Inc Elsevier Science)

Linear Motors

Apparently, linear motors have been studied more intensively than rotary ones because of the directed motion which can be used for transport of cargos. Linear motors are comprised of two protiens operating in tandem, i.e. the filament (F–actin or microtubules, MTs) and the motor (myosin, or kinesin, respectively). As mentioned before, motility assays are effectively primitive hybrid dynamic nanodevices, which can have two generic architectures: (i) a gliding geometry with the surface functionalized with the motor protein and the filament/MT sliding atop, possibly carrying a fluorescent tag; or (ii) an inverted, or bead, geometry with the filaments/MTs immobilized on the surface and the motor proteins, which are immobilized on cargo beads, 'walking' on tracks. Fig. 6.15 [60] presents the two motility assay geometries for the kinesin/MT system. Motility assays, which have been proposed for almost 2 decades for both actin–myosin [61] and kinesin–MT [62] systems, are still the technological paradigm of hybrid dynamic nano–devices based on molecular motors due to their extreme ease of use and low cost. More advanced devices, however, will need to incorporate 'smarter' features.

The success of the future dynamic nanodevices based on linear molecular motors will depend on successful resolution of several technological problems: (i) confinement of the movement of motile elements exclusively on fabricated paths; (ii) enforcement of unidirectional polarity of the movement; (iii) endurance of the devices; and (iv) appropriate applications and designs. The first technological barrier has been addressed in many studies in the last decade, in both motility assay architectures. The gliding motility assay architecture has been used more extensively due to easier fabrication, e.g. movement of actin or MT on motor-functionalized tracks [63–66] and channels [67–69]. The bead architecture has more operational potential because the unipolarity of the movement is naturally achieved through the built-in directionality information in the filaments/MTs. Fig. 6.16 presents a color encoded trajectory of actin filaments with movement confined in myosin-functionalized channels [67]. However, because the filaments/MTs have to be unipolarly aligned – another difficult technological problem, the bead architecture is less successful, with the notable exception of a very early study at NRL [70]. The more difficult problem of unidirectional movement has been also recently addressed through the use of strong electric fields outside the flow cell [71] which enforce the movement of actin filaments in a preferential direction, and the use of arrow shaped channels [72] to favor the movement of MTs in one direction due to their relative rigidity. The third technological hurdle (device endurance) is very much modulated by the stability of the motor proteins, which are reputedly prone to denaturation following minute changes in carefully optimized buffer media. Many operational aspects of hypothetical biomolecular dynamic devices based on linear molecular motors have been recently comprehensively reviewed [60].

Possible Applications of Hybrid Dynamic Nano–Devices in a Micro/Nanoarray Format

Whatever their future use, hybrid nano–devices based on rotary or linear molecular motors are likely to operate in largely parallel manner, with many individual 'clusters' of nano–devices organized in a microarray architecture. While the devices based on rotary motors have been already presented in a simple and explicit microarray format, the devices based on linear motors would raise interesting design issues. The possible applications for future dynamic nanodevices will use their natural functions, i.e. power generation and cargo transport, as their primary purposes or for different purpose, e.g. sensing and computation.

Power generation would be the most obvious application with both rotary and linear motors being in principle capable of inducing electric currents if a complex between the motile element and a metallic micro-sized object (a rod or bead) is moving in the vicinity of an electric current capturing device (e.g. a metallic wire loop). Many elements of the devices proposed by Montemagno's team prompt to this application (but not exclusively) with the rotary motor based devices already being organized in micro– and nanoarray formats (Fig. 6.13). An engineering study [73] tested the hypothesis of obtaining a reasonable electric current from an array of nano–electric generators, i.e. beads moved by actin–myosin system in microfabricated structures. It was found that, in an ideal situation, although the generated electric field is approximately a few tens of pV per bead, a purposefully designed array would amplify this to several nV – enough for micro-level local needs, e.g. powering electronic circuits.

A number of *devices performing mechanical functions* have been recently proposed by a group at University of Washington. A light-powered nano–assembly line uses molecular shuttles, which exploit UV-induced release of caged ATP combined with enzymatic ATP degradation, and which carry cargos along engineered paths [74]. Also a forcemeter capable of measuring the strength of biological receptor/ligand pairs i.e. pN forces, has been demonstrated [75]. The device is assembled from nanoscale building blocks, using a cantilevered MT as a beam of known stiffness, loaded by a second MT transported by kinesin.

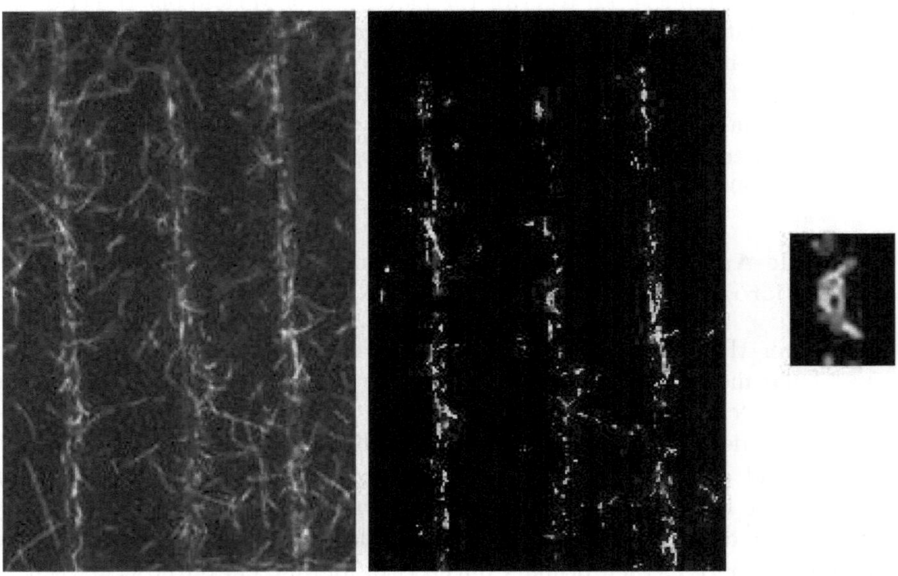

Fig. 6.16. Confinement of the movement of actin filaments in channels (left) and color-coded trajectories of actin filaments (right: red and purple – start and end of sequence). (Adapted from [67]. Copyright 2002 Kluwer Academic Publishers)

Imaging devices are another possible application. Vogel's group proposed an imaging device based on the kinesin–MT system [76]. Information about surface properties such as topography is obtained by repeated acquisition of an optical signal from a large number of microscopic, self-propelled probes moving on random paths across a surface. Nicolau et al. [65] observed that the fluorescence of rhodamine-labelled actin filaments decreases when the filaments pass across hydrophilic (myosin-poor) surfaces. This system can be then used for the readout of encoded surface properties with nanometer precision.

Molecular motors based devices can be also used, in principle, for *biosensing applications*. If biomolecular recognition can induce a dramatic change in the movement characteristics of motile elements, e.g. motor functionalized beads, or antibody decorated filaments, then a very sensitive biosensing device is available. The detection of the movement characteristics can be detected by a giant magneto resistance (GMR) detector and integrated on a chip if the beads are magnetic. The sensitivity of such a device is also its drawback, especially in the context of the sensitivity of protein molecular motors to minute changes in environmental conditions. However, the major benefit of such biosensing molecular motors-based devices lies in the motility of the 'molecular sensor' which allows for improved process kinetics by adding a moving component to the otherwise diffusion-limited tangent–probe binding process. This is especially important for detection of sensors aimed at highly toxic or pathogenic agents, where speed of detection is critical.

Finally, molecular motors based devices can be used for *computation*, in a similar, algorithmically–speaking, fashion as DNA computing. It has been recently proposed [77] that motile elements can explore in a highly parallel manner graphs that encode a mathematical problem. The most intuitive example would be to solve the travelling salesman problem in a maze that represents at a small scale the air paths in Adleman's experiment.

6.5 Conclusion

We hope we have shown in this chapter that nanoarray technology opens many new fields for microarray industry in many ways. In the immediate to short term, static nanoarrays open the possibility of probing biomolecular recognition on an enormous scale and also at the single molecule level. DNA computation microarrays are possibly the best technological avenue for DNA computing, which in itself is a tremendous development. New possibilities, unforeseen at the moment, would be opened by the development of dynamic nanodevices working in a microarray architecture.

References

1. Dammel, R. *Diazonaphthoquinone–based Resists*; SPIE Tutorial Texts; SPIE: Bellingham, WA, 1993; Vol. 11, Chapter 1
2. Pirrung, M. C. Angew. Chem. Int. Ed. 2002, 41, 1276–1289

3. Lander, E. S. Nature Genet. 1999, 21, 3–4
4. Xia, Y.; Rogers, J. A.; Paul, K. E.; Whitesides, G. M. Chem. Rev. 1999, 99, 1823–1848
5. Geppert, L. IEEE Spectrum 1996, 33–38
6. Ivanova, E., Wright, J.P., Pham, D.K., Filipponi, L., Viezolli, A., Nicolau, D.V. Langmuir, 18, 9539–9546, 2002
7. Liu, G.-Y.; Xu, S.; Qian, Y. Acc. Chem. Res. 2000, 33, 457–466
8. Xu, S.; Liu, G. Langmuir 1997, 13, 127–129
9. Shoer, J. K.; Crooks, R. M. Langmuir 1997, 13
10. Bruckbauer, A.; Ying, L.; Rothery, A. M.; Zhou, D.; Shevchuk, A. I.; Abell, C.; Korchev, Y. E.; Klenerman, D. J. Am. Chem. Soc. 2002, 124, 8810–8811
11. Piner, R. D.; Zhu, J.; Xu, F.; Hong, S.; Mirkin, C. A. Science 1999, 283, 661–663
12. Hong, S. H.; Mirkin, C. A. Science 2000, 288, 1808–1811
13. Hong, S. H.; Zhu, J.; Mirkin, C. A. Langmuir 1999, 15, 7897–7900
14. Hong, S. H.; Zhu, J.; Mirkin, C. A. Science 1999, 286, 523–525
15. Piner, R. P.; Hong, S.; Mirkin, C. A. Langmuir 1999, 15, 5457–5460
16. Ivanisevic, A.; Mirkin, C. A. J. Am. Chem. Soc. 2001, 123, 7887–7889
17. Wilson, D. L.; Martin, R.; Hong, S.; Cronin-Golomb, M.; Mirkin, C. A.; Kaplan, D. L. Proc. Natl. Acad. Sci. USA 2002, 98, 13660–13664
18. Demers, L. M.; Ginger, D. S.; Park, S.-J.; Li, Z.; Chung, S.-W.; Mirkin, C. A. Science 2002, 298, 1836–1838
19. Su, M.; Liu, X.; Li, S.-Y.; Dravid, V. P.; Mirkin, C. A. J. Am. Chem. Soc. 2002, 124, 1560–1561
20. Troughton, E. B.; Bain, C. D.; Whitesides, G. M.; Nuzzo, R. G.; Allara, D. L.; Porter, M. D. Langmuir 1988, 4, 365–385
21. Bain, C. D.; Whitesides, G. M. J. Am. Chem. Soc. 1989, 111, 7164–7175
22. Pique, A.; Chrisey, D. B., Eds. *Direct-write technologies for rapid prototyping applications: sensors, electronics, and integrated power sources*; Academic Press: San Diego, CA, 2002
23. Noy, A.; Miller, A. E.; Klare, J. E.; Weeks, B. L.; Woods, B. W.; DeYoreo, J. J. Nano Letters 2002, 2, 109–112
24. Lim, J. H.; Ginger, D. S.; Lee, K. B.; Heo, J.; Nam, J. M.; Mirkin, C. A. Angew. Chem. Int. Ed. Engl. 2003, 42, 2309–2312
25. Lee, K. B.; Lim, J. H.; Mirkin, C. A. J. Am. Chem. Soc. 2003, 125, 5588–5589
26. Lee, K.-B.; Park, S. J.; Mirkin, C. A. Science 2002, 295, 1702–1705
27. Vettiger, P.; Cross, G.; Despont, M.; Drechsler, U.; Durig, U.; Gotsmann, B.; Haberle, W.; Lantz, M. A.; Rothuizen, H. E.; Stutz, R.; Binnig, G. K. IEEE Transactions on Nanotechnology 2002, 1, 39–55
28. Chow, E. M.; Yaralioglu, G. G.; Quate, C. F.; Kenny, T. W. Appl. Phys. Lett. 2002, 80, 664–666
29. Sulchek, T.; Grow, R. J.; Yaralioglu, G. G.; Minne, S. C.; Quate, C. F.; Manalis, S. R.; Kiraz, A.; Aydine, A.; Atalar, A. Appl. Phys. Lett. 2001, 78, 1787–1789
30. Degertekin, F. L.; Hadimioglu, B.; Sulchek, T.; Quate, C. F. Appl. Phys. Lett. 2001, 78, 1628–1630
31. Zhang, M.; Bullen, D.; Chung, S.-W.; Hong, S.; Ryu, K. S.; Fan, Z.; Mirkin, C. A.; Liu, C. Nanotechnology 2002, 13, 212–217
32. Thorsen, T.; Maerkl, S. J.; Quake, S. R. Science 2002, 298, 580–584
33. Storhoff, J. J.; Mucic, R. C.; Mirkin, C. A. J. Cluster Sci. 1997, 8, 179–216
34. Storhoff, J. J.; Elghanian, R.; Mucic, R. C.; Mirkin, C. A.; Letsinger, R. L. J. Am. Chem. Soc. 1998, 120, 1959–1964

35. Niemeyer, C. M.; Burger, W.; Peplies, J. Angew. Chem. Int. Ed. 1998, 37, 2265–2268
36. Niemeyer, C. M. Angew. Chem. Int. Ed. Engl. 2001, 40, 4128–4158
37. Park, S. J.; Taton, T. A.; Mirkin, C. A. Science 2002, 295, 1503–1506
38. Haes, A. J.; Van Duyne, R. P. J. Am. Chem. Soc. 2002, 124, 10596–10604
39. Taton, T. A.; Lu, G.; Mirkin, C. A. J. Am. Chem. Soc. 2001, 123, 5164–5165
40. Taton, T. A.; Mirkin, C. A.; Letsinger, R. L. Science 2000, 289, 1757
41. Parak, W. J.; Boudreau, R.; Le Gros, M.; Gerion, D.; Zanchet, D.; Micheel, C. M.; Williams, S. C.; Alivisatos, A. P.; Larabell, C. Adv. Mater. 2002, 14, 882–885
42. Demers, L. M.; Park, S. J.; Taton, T. A.; Li, Z.; Mirkin, C. A. Angew. Chem. Int. Ed. Engl. 2001, 40, 3071–3073
43. Adleman, L. M. *Molecular computation of solutions to combinatorial problems.* Science 1994, 266, 11 November
44. Lipton, R. J. *DNA solution of hard computational problem.* 1995, Science 268, 28 April
45. Ouyang, Q.; Kaplan, P. D.; Liu, S.; Libchaber, A. *DNA solution of the maximal clique problem.* Science 1997, 278, 17 October
46. Liu, Y., Xu, J. Pan, L., Wang, S. *DNA Solution of a graph coloring problem.* J. Chem. Inf. Comput. Sci. 2002, 42, 524–528
47. Smith L. M, Corn R. M, Condon A. E, Lagally M. G, Frutos A. G, Liu Q, Thiel A. J. *A surface-based approach to DNA computation.* J Comput Biol. 1998 Summer, 5(2):255–67
48. Frutos, A. G., Smith, L. M., and Corn, R. M. (1998) *Enzymatic ligation reactions of DNA "words" on surfaces for DNA computing.* J. Am. Chem. Soc. 120, 10277–10282. York, Vol. 1
49. Liu, Q, Wang, L., Frutos, A. G., Condon, A.E. Corn, R. M. Smith, L. M. *DNA computing on surfaces.* Nature, Vol. 403, 13 Jan 2000 175–178
50. Schöning, U. in Proc. 40th Ann. IEEE Conf. Found. Comp. Sci. (FOCS) 410–414 (IEEE Comp. Sci., Los Alamitos, California, 1999)
51. Ogihara M. Ray, A. *DNA computing on a chip.* Nature Vol 403, 143–144, 2000
52. Kinosita Jr, K., Yashida, R., Noji, H., Ishiwata, S., Yoshida, M. Cell 93, 21 (1998)
53. Noji, H., Yasuda, R., Yoshida, M., Kinosita Jr, K., *Direct observation of the rotation of F1ATPase.* Nature, 386, 299–302, 1997
54. Bachand, G.D. Montemagno, C. D. *Constructing Organic/Inorganic NEMS Devices Powered by Biomolecular Motors.* Biomedical Microdevices 2:3, 179–184, 2000
55. Yoshida M, Muneyuki E, Hisabori T. *ATP synthase–a marvelous rotary engine of the cell.* Nat Rev Mol Cell Biol. 2001 Sep;2(9):669–77
56. Noji H, Yasuda R, Yoshida M, Kinosita K Jr. *Direct observation of the rotation of F1-ATPase.* Nature, 1997, 386 :299–302.
57. Soong, R.K., Bachand, G.D., Neves, H.P., Olkhovets, A.G., Craighead, H.G., Montemagno, C.D. *Powering a nanodevice with a biomolecular motor.* Science, 290, 1555–, 2000
58. Bachand, G.D. Montemagno, C. D. *Constructing Organic/Inorganic NEMS Devices Powered by Biomolecular Motors.* Biomedical Microdevices 2:3, 179–184, 2000
59. Soong, R.K., Neves, H.P., Schmidt, J.J., Bachand, G.D., Montemagno, C.D. *Engineering Issues in the fabrication of a hybrid nano–propeller system powered by F1–ATPase*, Biomedical Microdevices, 3:1, 71–73, 2001

60. Hess H, Vogel V. *Molecular shuttles based on motor proteins: active transport in synthetic environments.* J Biotechnol. 2001 Nov; 82(1):67–85
61. Yanagida, T., Nakase, M., Nishiyama, K. & Oosawa, F. (1984). *Direct observation of motion of single F–actin filaments in the presence of myosin.* Nature 307, 58–60
62. Spudich, J. A., Kron, S. J. & Sheetz, M. P. (1985) Nature, 315, 584–586
63. Suzuki, H., Oiwa, K., Yamada, A., Sakakibara, H., H. Nakayama and Mashiko S. 1995. *Linear arrangement of motor protein on a mechanically deposited fluoropolymer thin film.* Jpn. J. Appl. Phys. 34:3937–3941
64. Suzuki, H., Yamada, A., Oiwa K., H. Nakayama and S. Mashiko. 1997. *Control of actin moving trajectory by patterned poly(methylmethacrylate) tracks.* Biophys. J. 72:1997–2001
65. Nicolau, D.V., Suzuki, H., Mashiko, S., Taguchi, T., Yoshikawa, S. *Movement of actin filaments on microlithographically–functionalized myosin tracks.* Biophysical Journal, 77 (2), 99044–99065, 1999
66. Dennis, J. R., Howard, J. & Vogel, V. (1999). *Molecular shuttles: directing the motion of microtubules on nanoscale kinesin tracks.* Nanotechnology 10, 232–236
67. Mahanivong, C., Wright, J. P., Kekic, M., Pham, D. K., dos Remedios, C., Nicolau, D.V. *Manipulation of the Motility of Protein Molecular Motors on Microfabricated Substrates.* Biomedical Microdevices 4(2): 111–116; 2002
68. Clemmens, J.; Hess, H.; Howard, J.; Vogel, V. *Analysis of Microtubule Guidance in Open Microfabricated Channels Coated with the Motor Protein Kinesin.* Langmuir; 2003; 19(5); 1738–1744
69. Bunk, R., Klinth , J., Rosengren , J., Nicholls, I., Tagerud, S., Omling, P., Mansson, A., Montelius, L. *Towards a 'nano–traffic' system powered by molecular motors.* Microelectronic Engineering 67–68 (2003) 899–904
70. Turner, D. C., Chang, C., Fang K., S. L. Brandow and D. B. Murphy. 1995. *Selective adhesion of functional microtubules to patterned silane surfaces.* Biophys. J. 69:2782–2789
71. Riveline, D., Ott, A., Julicher, F., Winkelmann, D. A., Cardoso, O., Lacapere, J. J., Magnusdottir, S., Viovy, J. L., Gorre-Talini, L. & Prost, J. (1998). *Acting on actin: the electric motility assay.* Eur Biophys J 27, 403–8
72. Hiratsuka, Y., Tada, T., Oiwa, K., Kanayama, T., Uyeda, T.Q. P. *Controlling the Direction of Kinesin–Driven Microtubule Movements along Microlithographic Tracks.* Biophysical Journal, 81 2001 1555–1561
73. Fulga, F., Myhra, S., Nicolau, Jr. D. V., D.V. Nicolau, *Interrogation of the dynamics of magnetic microbeads on the meso–scale via electromagnetic detection.* Smart Materials & Structures, 11 (5) 722–727, 2002
74. Hess, H., Clemmens, J., Qin, D., Howard, J. & Vogel, V. (2001). *Light–Controlled Molecular Shuttles Made from Motor Proteins Carrying Cargo on Engineered Surfaces.* Nano Letters 1, 235–239
75. Hess, H.; Howard, J.; Vogel, V. *A Piconewton Forcemeter Assembled from Microtubules and Kinesins.* Nano Lett. 2002; 2(10); 1113–1115
76. Hess, H.; Clemmens, J.; Howard, J.; Vogel, V. *Surface Imaging by Self–Propelled Nanoscale Probes.* Nano Lett. 2002; 2(2); 113–116
77. Nicolau, D. V., Jr.; Nicolau, D. V. *Computing with the Actin–Myosin molecular motor system. In Biomedical Applications of Micro– and Nanoengineering.* Nicolau, Dan V. (Ed.) SPIE Proc. 4937, 219–225, 2002

7

The Use of Microfluidic Techniques in Microarray Applications

Piotr Grodzinski, Robin H. Liu, Ralf Lenigk, and Yingjie Liu

7.1 Introduction

The area of hybridization arrays enjoyed unprecedented growth in the last decade [1,2]. These arrays, allowing for a highly parallel analysis of a multitude of single-stranded DNA fragments, found use in many different areas, ranging from microscale sequencing and cDNA expression microarrays for analysis of gene expression [3, 4] to drug discovery and development [5] and single nucleotide polymorphism (SNP) analysis [6].

Conventional DNA microarray chips are still hampered, however, by numerous imperfections. They usually use sizable sample volumes of $\sim$ 200 µl, which prohibits evolution towards further chip miniaturization. Current on-chip hybridization assays take several hours to be completed, since the majority of them rely solely on diffusion to control the reaction kinetics. Finally, most of the available array chips are not equipped with on-chip sample preparation provision, therefore requiring elaborate robot-based sample preparation using traditional bench techniques. Slow reaction kinetics and lack of integrated sample preparation prohibits further penetration of the microarray technology into diagnostic applications.

The recent, rapid developments in chip micro-fabrication technologies and microfluidics provide potential for elevating many of current deficiencies of microarray techniques. Microfluidic chips (also called "lab chips") contain interconnected fluidic microchannel networks, reaction chambers, mixers, and valves, and can carry out conventional biochemical measurements with increased speed and reliability [7]. They have the capacity to improve reaction kinetics with the use of target stirring or mixing techniques, thereby allowing expansion to high throughput analysis. Also, with the incorporation of micro–Total Analysis Systems (µTAS) on the chip, they have the potential to integrate front-end sample preparation with back-end hybridization detection stages.

In this chapter, we will discuss chip technologies developed at Motorola Labs and address the use of microfluidics in conjunction with microarray hybridization detection techniques. We will cover three general areas pertinent to 1) multi-sample analysis in 'biochannel' devices, 2) improvement of reaction kinetics using acoustic microstreaming target mixing and target oscillation in the biochannel, and 3) integration of on-chip PCR amplification to bring sample preparation and hybridization detection into a single chip.

7.2 Biochannel Hybridization Arrays

Conventional DNA hybridization assays rely solely on the diffusion of target to surface-bound probes. This diffusion limitation of the reaction leads to hybridization times on the order of 3 to 12 hours, depending on the size and concentration of the target and on the hybridization conditions. While amplification of genetic material has become faster with the development of rapid micro-system PCR cycling methods [8–10], detection is still hampered by the slow process of DNA hybridization. It has been recognized that mixing is important to achieve maximum rates of hybridization [11] and various methods have been devised to accelerate this process. They include electronic enhancement of DNA hybridization [12, 13], dynamic DNA hybridization using paramagnetic beads [14, 15], rotation of the whole device [16], and the use of a micro porous three–dimensional biochip with the hybridization solution being pumped continuously through it [17].

While ultra-high density arrays are powerful tools for expression analysis studies, highly parallel low or medium density arrays will be useful in many other applications such as clinical diagnostics and pharmacogenomic applications based on genotyping and SNP scoring. Therefore, the ability to perform massively parallel assays with only a few micro–liters of sample/reagent per assay would provide substantial time and cost savings, and hence is highly desirable. The 'biochannel' approach presented here addresses these points: it enables the simultaneous analysis of a multitude of samples at a time, requires only small sample volumes, improves hybridization kinetics, and provides ease of integration with other micro-fluidic device components. Fig. 7.1 depicts the 'working space' for biochannel devices, plotted as number of samples versus number of targets analyzed within one chip. Biochannel structures offer distinct advantages for analysis of a large number of different samples in the array environment, with a low to medium density of detection probes.

Two different sets of chips have been prepared: 1) plastic, multi-channel arrays for multi-sample analysis with optical detection schemes [18–20] and 2) hybrid arrays for single sample analysis used for studies of reaction kinetics with electrochemical detection schemes [20].

7.2.1 Biochannel Devices with Optical Detection

The first generation of multi-channel arrays was built using microfabricated PDMS networks containing channels which were ~ 200 µm wide, ~ 50 µm

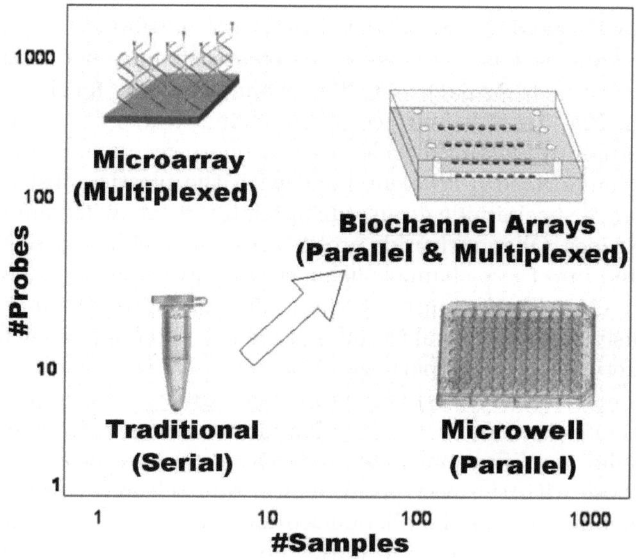

Fig. 7.1. Operational space for biochannel devices

deep, and few centimeters long [18,19]. These channel networks, fabricated using a molding process [21, 22], were then aligned and bonded to CodeLink^{TM} glass-based microarray slides (developed by Motorola Life Sciences, currently part of Amersham Biosciences operation) to form a closed channel array. The flat glass slide was coated for the immobilization of oligonucleotides (SurModics, Eden Prairie, MN) and spotted with DNA oligonucleotide probes (100 μm diameter). A selective oxygen plasma surface treatment and bonding/alignment technique was developed to obtain a robust but reversible bonding between the PDMS and microarray glass chip. Although this fabrication approach was easy to implement, its yield was low due to channel–to–channel cross–talk. Accurate alignment of the chip and the channel network were also difficult.

Fig. 7.2. Evolution of the fabrication process for biochannel devices, **(a)** oligo probes are spotted on the flat surface and overlaid with PDMS channel network, **(b)** oligo probes are spotted into the channels directly and overlaid with flat cover piece

In order to avoid the above deficiencies, we have modified the fabrication procedure and used hot embossing to create channel networks in polycarbonate (1 × 3 inch format) first. The evolution of the fabrication method is depicted in Fig. 7.2. The surface of the channels was functionalized for the immobilization of oligonucleotides with a photo-reactive bi-functional linker molecule that formed a covalent bond with the plastic substrate, the other end carrying a succinimide group which readily reacts with amino-terminated oligonucleotides. Oligonucleotide probes (Operon Technologies, Alameda, CA) were spotted into these channels using contact printing ('Spotbot', Telechem, Sunnyvale, CA). To demonstrate the ability of the biochannel device to simultaneously analyze several samples at once, a detection assay for surrogates (due to safety concerns) of pathogenic bacteria strains (*E. coli*, *S. epidermidis*, *E. faecalis* and *S. salivarius*) was performed (Table 7.1). Unique sequences for identification of the organisms were found and primer sets were developed to allow specific amplification. Several probes were evaluated for each amplicon and those with the best performance were selected (results not shown). After immobilization of the oligonucleotide probes, the channels were sealed using tape into which inlet and outlet ports had been cut using a computer controlled CO_2 laser tool (Universal Laser Systems, Scottsdale, AZ).

Table 7.1. Nosocomial etiologic agent surrogate genetic targets

Agent	Surrogate	Strain	Genetic Targets	Relevant Characteristics	Amplicon size
Staphylococcus aureus	Staphylococcus epidermidis	ATCC 14990	ArgABC	AA uptake	371
Enterococcus faecalis	Enterococcus faecalis	ATCC 19433	DnaE	DNA replication	195
Streptococcus Group B	Streptococcus salivarius	ATCC 9758	Dal	D-Ala Ligase	293
Escherichia coli	Escherichia coli K12	DH5α(pBS)	bla	Amp^R	627

To generate the samples, an aliquot of 10,000 bacteria cells was asymmetrically amplified using a ratio of 1:100 of forward to fluorescent (Texas Red®) reverse primer. The PCR mixture contained 0.005 μM forward and 0.5 μM reverse primer, 400 μM dNTP, 80 mM KCl, 16 mM Tris–HCL (pH 8.3), 2.5 mM $MgCl_2$, and 0.05 U/μl *Taq* polymerase. Cycling parameters were: 35 cycles (94°C for 60 seconds, 55°C for 60 seconds, 72°C for 60 seconds), ending with 72°C for 6 minutes to extend all unfinished DNA strands. One PCR amplification product was introduced in each channel, and after a washing step the tape cover was removed and the bottom of the channels scanned in a commercial laser scanner (GeneScan 4000, Axon, Union City, CA).

The results of the hybridization in biochannel devices with single-stranded amplicon are shown in Fig. 7.3, which presents the fluorescent scanner image of two adjacent channels that have been hybridized with samples obtained by PCR amplification of *E. coli* and *E. faecalis* samples. Probes for each of the pathogenic bacteria surrogate strains were printed in duplicate, with all channels being treated identically. The mainly single-stranded amplification product of the PCR was hybridized directly to the channel network, without the addition of hybridization buffer. The salinity was only one–tenth of conventional hybridization solutions, which have a salinity of at least 0.5 M. Although slow hybridization would be expected under these low-salt conditions, the salinity was sufficient to produce intense, specific hybridization signals in only 30 minutes.

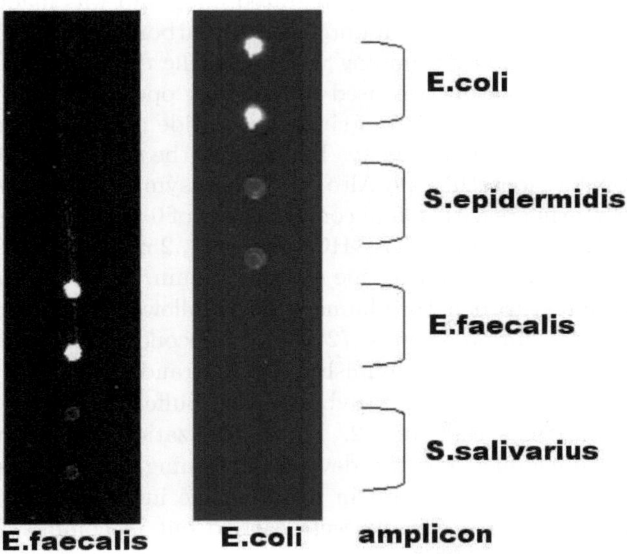

Fig. 7.3. Fluorescent scanner image of two biochannel device channels after hybridization. The left channel has been hybridized to amplicon obtained with a PCR from *E. faecalis*, the right channel with *E. coli* amplicon

7.2.2 Biochannel Devices for Electrochemical Detection – Reaction Kinetics Studies

In order to evaluate the kinetics of hybridization in the biochannel devices, we chose to use electrochemistry-based single-nucleotide polymorphism (SNP) detection arrays (eSensor^{TM}) from Motorola Life Sciences [24]. The use of a homogenous assay allowed for continuous measurement of DNA hybridization.

The channel network made of double-sided tape was placed over eSensor^{TM} array chips. To accelerate hybridization, a pump was integrated into the device. The pump consisted of a thin-film heater evaporated onto the plastic cover of the chip, in contact with an air–pocket. Cyclical heating and cooling of this air volume resulted in pumping of the hybridization solution inside the channel to overcome the diffusion-limited reaction.

ESensor^{TM} chips for the experiments were provided by Motorola (Motorola Life Sciences, Pasadena, CA). The chips had 16 electrodes, with electrodes 1–4 and 13–16 containing identical probes and the remaining electrodes containing negative controls. The channels were made in 200 μm thick double-sided adhesive tape with a Teflon® core (Fralock, Canoga Park, CA). The tape was patterned by a computer-controlled CO_2 laser tool. The channel was covered by a 500 μm thick polycarbonate slide, into which inlet and outlet holes had been drilled. The heater for the integrated air pump was made by vacuum–evaporation of a platinum–film onto the polycarbonate piece. Contact to the heater coil was made by clamping wires onto the metal film. A conventional low-voltage power supply was used to manually operate the heater.

An assay for the detection of single-nucleotide polymorphisms in HFE–H gene was used as the model assay. To generate the samples, 100 ng of human genomic DNA (Clontech, Palo Alto, CA) was asymmetrically amplified using a set of three primers with a final concentration of 0.5 μM each primer, 400 μM dNTP, 50 mM KCl, 10 mM Tris–HCL (pH 8.3), 2 mM $MgCl_2$, 0.05 U/μl *Taq* polymerase, and 100 μg/ml bovine serum albumin. Cycling parameters were: 95°C (3 minutes) to denature human DNA, followed by 40 cycles (94°C for 45 seconds, 58°C for 55 seconds, 72°C for 60 seconds), and ending with 72°C for 6 minutes to extend all unfinished DNA strands. The PCR–product was mixed with signaling probes in hybridization buffer (Motorola Life Sciences, Pasadena, CA) in a ratio of 1:2. The hybridization cocktail was manually filled into the channel. For the devices containing an integrated pump, the pump was switched on and off in regular time intervals of 3 minutes. The diffusion-controlled experiments were carried out in commercial eSensor^{TM} cartridges (Motorola Life Sciences, Pasadena, CA) with an internal volume of 65 μl. All hybridizations were performed at room temperature, with devices placed horizontally. The signals were read using a Hydra®600 instrument (Motorola Life Sciences, Pasadena, CA) using eSensor^{TM} software (Motorola Life Sciences, Pasadena, CA); the AC voltammetry technique to gather the electrochemical signal is described in more detail elsewhere [24].

Figure 7.4 shows fabricated eSensor^{TM} biochannel devices inside electrical connectors. Figure 7.4a depicts the device used for diffusion controlled experiments, and Fig. 7.4b shows the device with an integrated electrical heating coil, consisting of a metal–film evaporated onto the area of polycarbonate cover in contact with an air–pocket inside the channel. The total channel volume is 25 μl. When voltage is applied to the heater, the air pocket expands, pushing the hybridization solution through the channel and into a reservoir. Care must be taken to prevent the liquid from being pushed too far, which

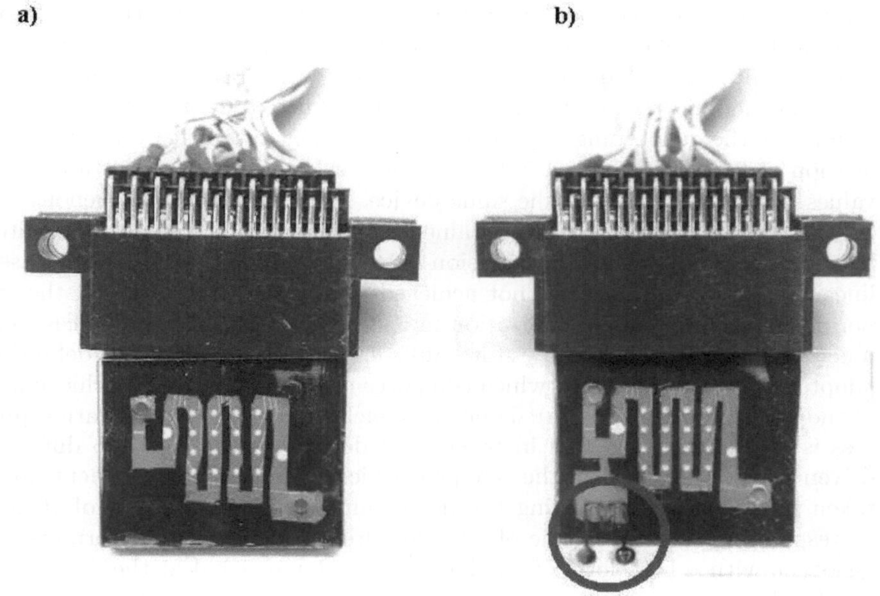

Fig. 7.4. ESensor^{TM} chip covered with biochannel microfluidic channels **(a)** and biochannel with integrated air-pump **(b)** to allow for oscillation of the hybridization mixture

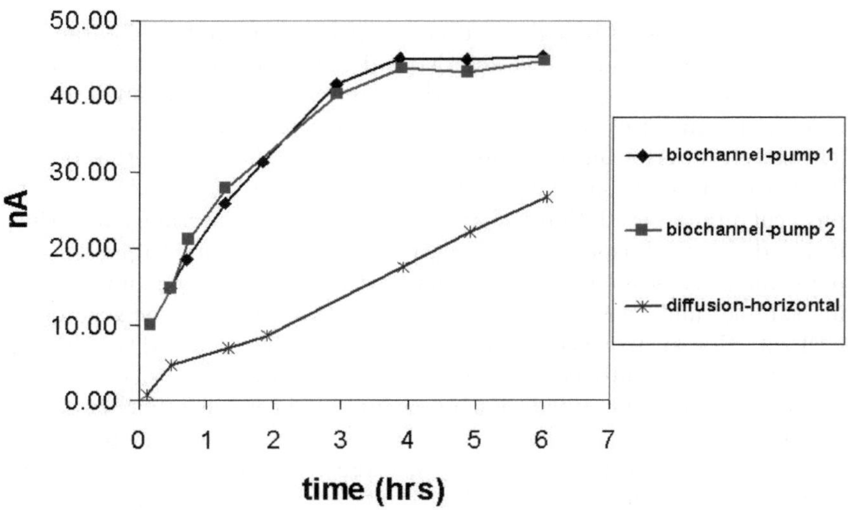

Fig. 7.5. Comparison of hybridization kinetics in biochannel with integrated pump and diffusion-controlled hybridization chamber

would expose the electrodes to air. By repeatedly switching the power on and off, fluid oscillation can be achieved. Due to the slow actuation process of the pump, the chosen oscillation frequency was 0.167 Hz for one expansion/contraction cycle, corresponding to a mass–flow rate of 0.4 μl / sec. The results for the genotyping experiment using target obtained by PCR amplification of human genomic DNA are shown in Fig. 7.5. All values are mean values from 4 electrodes in the same device. Because of the homogenous nature of the assay, results were obtained at different time points to monitor hybridization kinetics. In the diffusion-controlled device, the signal increased linearly, and equilibrium was not achieved within the time–frame of the experiment. The rates of hybridization in the pumped biochannel devices were much higher, reaching steady–state after 4 hours. Using the rate definition adopted from reference [24], which compares the time required to achieve half of the saturation (maximum) signal, we conclude that the hybridization process is accelerated $\sim$ 6–fold in biochannel devices as compared to diffusion driven chips. Moreover, in the pumped devices, the first measurement point taken immediately after filling the device already shows a signal of 10 nA, corresponding to a S/N value of over 20, already sufficient to determine the genotype with a high level of confidence. This large signal at the first time–point is likely to be due to the passing of target molecules in close proximity to the surface-bound probe molecules during the loading process, with subsequent rapid hybridization.

7.2.3 Simulation of Hybridization Biochannel Reactors

Hybridization assays in a given reactor depend on a number of parameters related to probe and target characteristics (length, concentration, binding rates, surface immobilization characteristics), and parameters related to physical reactor design (size, shape, probe patch locations, sample motion, diffusion lengths etc). Assessing the effects of these different parameters on the hybridization rates using experiments can be a challenging task. With proper physical inputs, simulations can provide very detailed information on the physical and chemical aspects of a given reactor, allowing one to predict reaction performance, assess effects of different physical strategies (e.g. mixing, oscillating sample) and allow pre-fabrication optimization of a given reactor design. CFD–ACE+, an advanced multiphysics solver [25], was used to perform coupled flow and chemistry simulation of hybridization reactors. The detailed set of equations and simulation procedures is given in reference [20]. Here, we present only the results relevant to assessing a relationship between the target oscillation within the channel, target concentration, and reaction kinetics. These biochannel simulation results are compared with those obtained for bulk, static (diffusion-controlled) hybridization reactors.

In Fig. 7.6, the normalized surface target concentration histories are plotted for each of the individual reactors. Figures 7.6a, 7.6b and 7.6c show the hybridization behavior for 10 nM, 1 nM and 0.1 nM sample target concentrations, respectively. A comparison of Fig. 7.6a to Fig. 7.6c reveals that increasing the sample target concentration speeds up the surface reaction in all of the configurations. At all sample concentrations, the static biochannel device exhibits the slowest hybridization rates. This result is expected for two reasons. First, the transport of sample targets to the probes is controlled by diffusion. Second, and more importantly, the channel configuration limits the amount of targets that are easily accessible to each of the probe sites. The bulk reactor, which is also a diffusion-dominated device, shows a better performance, while the device with the oscillating sample provides for the best hybridization performance. This result is as expected, since convection provides a faster means of transporting the targets to the surface. One of the findings from these simulations is that at the highest concentrations, the bulk reactor hybridization rates approach those for the oscillating biochannel device. This indicates that at higher target concentrations, the overall rates of hybridization may be governed more by chemical kinetics at the surface than by the efficiency of transport of target species to the surface.

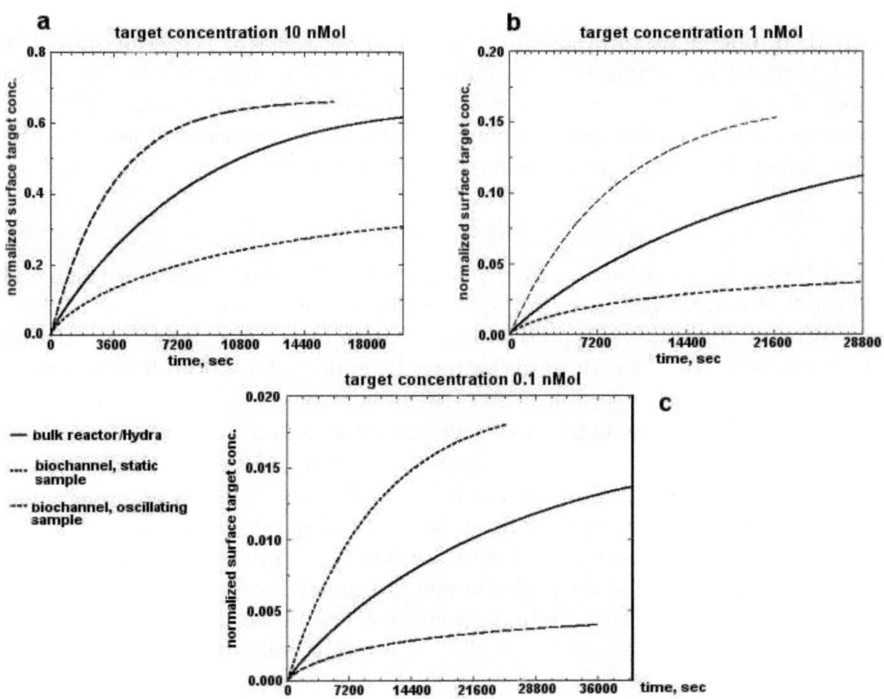

Fig. 7.6. Time-variation of the surface-bound target for three different types of hybridization reactors, **(a)** at 10 nMol target concentration, **(b)** at 1 nMol target concentration, **(c)** at 0.1 nMol target concentration

7.3 Chips with Cavitation Microstreaming Mixers – Kinetics Studies

The biochannel oscillation technique discussed in the previous section was successfully applied to the improvement of hybridization kinetics. This technique is limited, however, to low- and medium-density 1–dimensional arrays. We have also developed a more general mixing technique which can be used on 2–dimensional arrays of any size. This technique relies on the principle of cavitation microstreaming [26] and has many advantages over most existing techniques used for hybridization enhancement, including simple apparatus, ease of implementation, low power consumption ($\sim$ 2 mW), and low cost.

The mixing enhancement was tested using dye experiments, and the technique was subsequently used to enhance DNA hybridization in both optical detection-based and electrochemical detection-based DNA microarray chips [27, 28].

7.3.1 Theory of Cavitation Microstreaming

An air bubble in a liquid medium can act as an actuator (i.e., the bubble surface behaves like a vibrating membrane) when it is energized by an acoustic field. The behavior of a bubble in a sound field is determined largely by its resonance characteristics. For frequencies in the range considered here ($\sim$ kHz), the radius of a bubble at resonant frequency f is given by:

$$2\pi af = \sqrt{3\gamma P_0/\rho} \tag{7.1}$$

where a is the bubble radius, γ is the ratio of specific heats for the gas, $\mathrm{P_0}$ is the hydrostatic pressure and ρ is the density of the liquid.

When a bubble undergoes vibration within a sound field, the frictional forces generated at the air/liquid interface induce a bulk fluid flow around the air bubble, called cavitation microstreaming or acoustic microstreaming [26]. It was found that cavitation microstreaming is orderly at low driving amplitudes when the insonation frequency drives the bubbles at their resonance frequency for pulsation and when the bubbles are situated on solid boundaries. Bubble-induced streaming is strongly dependent on frequency for a given bubble radius, and on bubble radius for a given frequency. Acoustic microstreaming arising around a single bubble excited close to its resonance frequency produces strong liquid circulation flow in the liquid chamber. This liquid circulation flow can be used to effectively enhance mixing beyond the diffusion-limited process.

Although cavitation microstreaming has been studied since the 1950s [29, 30], we have not found any report on the use of this phenomenon to enhance micromixing. One challenge here is to precisely control the size of the air bubbles. In this work, we have developed an air bubble trapping design using micromachined air pockets for mixing enhancement.

7.3.2 Proof–of–concept Chips for Mixing Experiments

Practical embodiment of the chip capable of inducing acoustic microstreaming within the cavity is depicted in Fig. 7.7. The chamber is constructed by sealing a conventional DNA microarray glass chip with a polycarbonate cover layer using a double-sided adhesive tape (3 M, St. Paul, MN). The adhesive tape, with thickness of 200 μm, serves as a spacing gasket to define the shape and dimension (16 × 16 mm) of the chamber. The cover layer has a desired number of air pockets distributed uniformly above the chamber with a pitch of 2 mm. The air pockets (500 μm in depth and 500 μm in diameter) were machined using a Prolight milling machine (Light Machines, Manchester, NH) and were used to trap air bubbles in the reaction solution. A piezoelectric (PZT) disk (15 mm diameter, APC Inc., Mackeyville, PA) was bonded on the external surface of the cover layer using a super glue (Duro^{TM} Loctite Corp., Avon, Ohio).

In order to evaluate mixing efficiency, control experiments were performed using a colored dye. The chamber contents were irradiated with the sound generated by the PZT disk driven by a HP functional generator (Hewlett–Packard Co., Palo Alto, CA). Visual observations were made from above using a stereoscope. One–half of the chamber was filled with DI water and the other half with a red dye solution (a mixture of phenolphthalein and sodium hydroxide solution, both from Aldrich Chemical Co., Milwaukee, WI) in order to visualize motion of fluid elements in the chamber. The frequency employed was 5 kHz (square wave) with a peak–to–peak amplitude (V_{pp}) of 40 V.

The fluidic dye experiments showed that sonic irradiation caused little motion of the liquid if air bubbles were excluded from the chamber. However, with air bubbles that have a resonant frequency matching the insonation frequency induced by the PZT transducer, a gross liquid motion was seen to take place

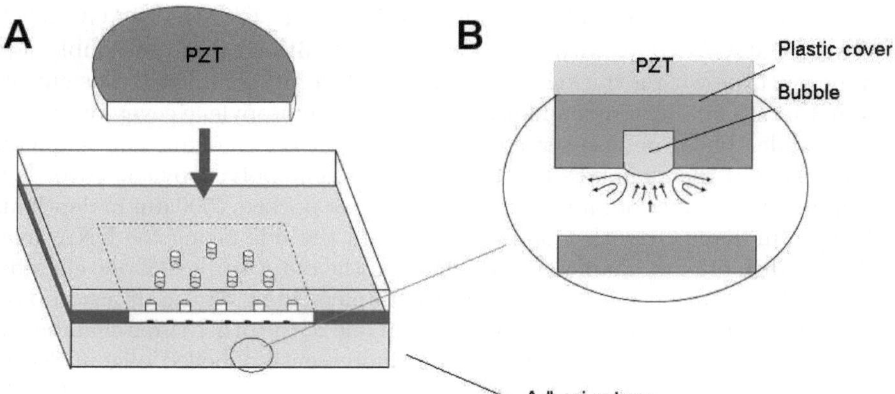

Fig. 7.7. Schematic showing a chip realization of cavitation microstreaming phenomenon, **(A)** overview; **(B)** sideview

around individual bubbles. Since the top pockets were uniformly distributed above the chamber, the resulting cavitation microstreaming dominated the mixing in the whole chamber (16 × 16 × 0.2 mm). Complete mixing was achieved across the whole chamber within 6 seconds, while diffusion-based mixing (i.e., without acoustic mixing) in the same chamber took approximately 8 hours to complete (considering diffusion in lateral direction). Dye experiments were also performed to investigate the relationship between mixing rate and acoustic parameters. It was found that the use of square waves resulted in faster mixing then the use of sinusoidal waves at the same V_{pp}. Lower voltage amplitudes also resulted in less mixing enhancement. The most effective mixing enhancement was provided by pulsation of a desired number of air bubbles having a size and resonant frequency selected in accordance with the insonation frequency induced by the PZT transducer (7.1). A more detailed discussion of these dye mixing experiments can be found in reference [28].

7.3.3 High density DNA Microarray Hybridization

High density DNA microarray hybridization experiments were performed to evaluate the effect of mixing enhancement on hybridization efficiency and uniformity as compared to conventional diffusion-based hybridization. A fluorescent detection-based microarray biochip consisting of a high density array of oligonucleotide probes dispensed on a 1 × 3 inch pre-treated glass slide (developed by Motorola Life Sciences, currently part of Amersham Biosciences operation) was used. Two different oligonucleotide probes (NEO and YJEK, both obtained from Operon Technologies Inc., Alameda, CA) and a positive control were arranged in a uniform pattern across the entire slide. Both NEO and YJEK are Cy3-labelled bacterial oligonucleotides. The sequence of the NEO probe is GCGTTGGCTACCCGTGATATTGCTGAAGAG with a 5′ amine. The sequence of the YJEK probe is TTTGTAGATTAGCACTGGAACTGGCACCGC with a 5′ amine. A 1 × 3 inch piece of double-sided adhesive tape with a thickness of 0.25 mm (3 M, St. Paul, MN) was cut into four 15 × 12 mm windows and used to bond a polypropylene cover layer to the glass slide. The tape also served as a spacing gasket to define the shape and dimension of the chambers on the glass slide. The polypropylene cover layer contained a number of uniformly distributed air pockets (500 μm in depth and 500 μm in diameter with a pitch of 2 mm) on the side facing the DNA array. A PZT disk (15 mm diameter) was glued on the outer surface of one chamber, in which cavitation microstreaming was implemented. Static diffusion-based hybridization was performed as a control in one of the other three chambers on the same chip. During hybridization, a fluorescently-labelled oligonucleotide target solution (45 μL) containing 50% formamide (Sigma Chemical Co., St. Louis, MI) and 10 nM Cy3 labelled NEO- and YJEK-specific targets (Operon Technologies Inc., Alameda, CA) was loaded into each detection chamber. The PZT transducer was driven at 5 kHz (sinusoidal sound wave) and 10 V_{pp}. The device was kept in a temperature-controlled chamber at 37 °C. Hybridization was carried out for 2 hours, after which the polypropylene layer was removed from the array glass slide, which was subsequently washed with TNT solution (TRIS/Sodium Chloride/Tween, from Sigma Chemical Co., St. Louis, MI) for 30 minutes at 42°C and rinsed three times with water. The glass slide was then scanned using a microarray scanner (Axon Instruments, Inc., Union City, CA).

Cavitation microstreaming was implemented in one of the four chambers (each 15 × 12 × 0.25 mm) on a fluorescent detection based microarray biochip consisting of a high density array of two types of oligonucleotide probes (NEO and YJEK) and a positive control. The continuous repetition of the two probe oligonucleotides in a uniform pattern across the entire slide allowed for signal comparisons across the entire array area. This is critical in understanding the signal homogeneity. The resulting fluorescent scanning images are shown in Fig. 7.8. Fluorescent intensity data for the mixing-enhanced array and the static hybridization array (diffusion-based) were analyzed. As shown in Figs. 7.9a and b, the average signal intensity of the mixing array is five times greater than that of the static hybridization array, and signal uniformity (co-variance) is also greatly improved by implementation of cavitation microstreaming. These results indicate that hybridization reactions in oligonucleotide array formats can generally be affected by the level of mixing of the target ligand. Efficient and effective micromixing maximizes delivery of the sample targets to the array surface, and thus significantly improves hybridization efficiency and uniformity.

7.3.4 Hybridization Kinetics Study

An assay for single nucleotide polymorphisms (SNP) associated with hematochromatosis (HFE–H) was performed in an eSensor^{TM} device (Motorola Life Sciences, Pasadena, CA) equipped for induction of cavitation microstreaming. The use of the eSensor^{TM} device allowed for continuous measurement of DNA hybridization signals during the reaction due to the homogenous nature of the assay, thus allowing hybridization kinetics to be studied [20]. Each device consisted of a plastic cover layer assembled with a printed circuit board (PCB) chip with 16 detection electrodes. Four electrodes contained identical oligonucleotide probes for HFE–H gene while the remaining electrodes contained other probes and negative controls. The plastic cover layer contained a 4 × 4 array of air pockets (500 µm in depth and 500 µm in diameter) facing the DNA probes on the PCB substrate. A PZT disk was glued on the outer surface of the cover layer to induce cavitation microstreaming during the hybridization.

Target solution preparation and measurement protocols were the same as those described in section 7.2.2. The hybridization cocktail was loaded into the eSensor chip with an internal volume of 65 µL. Hybridization was performed at 35°C. During the hybridization process, the PZT was driven at 5 kHz and 10 V_{pp} (square sound wave). The signals were read using a

Hydra®600 instrument (Motorola Life Sciences, Pasadena, CA). For comparison purposes, the same hybridization reaction was also performed in a conventional diffusion-based eSensor™ chip using the same amplicon mixture. Hybridization kinetics as a function of acoustic amplitude (V_{pp}) were also studied using amplitudes of 5 V_{pp} and 40 V_{pp}, as compared to 10 V_{pp}, while maintaining the same frequency of 5 kHz.

Kinetic data from the genotyping experiments using target DNA obtained by PCR amplification of human genomic DNA were collected by monitoring the electrochemical signal as a function of time. Figure 7.10 summarizes the hybridization kinetics results for a mixing-enhanced device and a diffusion-based device under the same assay conditions. The results show that in the static (diffusion-based) device, the hybridization signal evolved slowly and increased linearly. Saturation of the hybridization signal was not achieved within the time frame of the experiment. Moreover, the standard deviation associ-

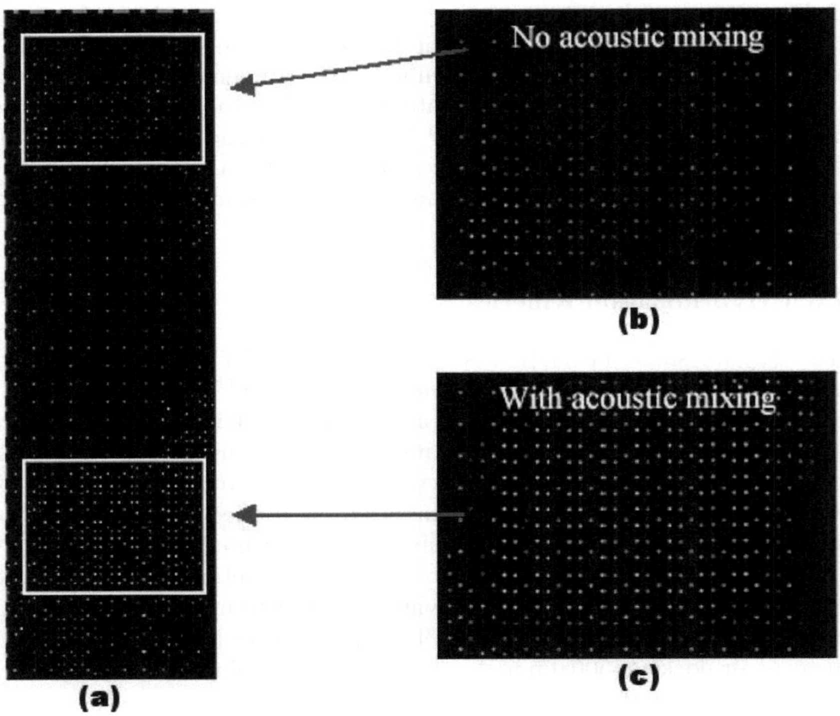

Fig. 7.8. **(a)** Fluorescent image of a 4-chamber, high density array, biochip after a 2-hour hybridization reaction. One chamber (15 × 12 × 0.25 mm) undergoes static hybridization **(b)**, while hybridization in another chamber (15 × 12 × 0.25 mm) is aided with cavitation microstreaming **(c)**

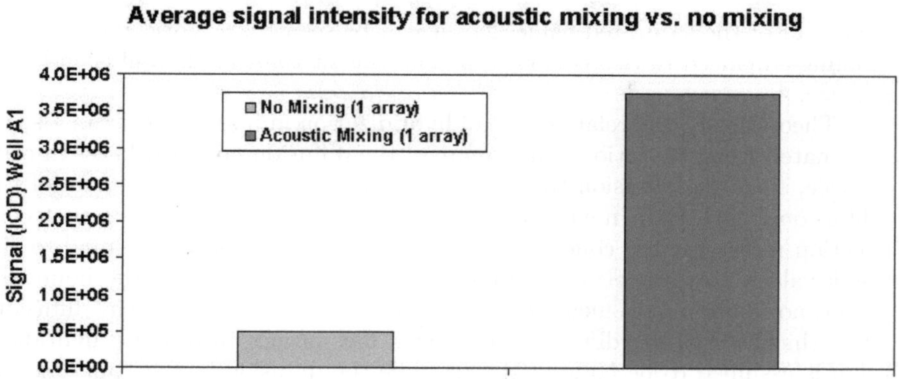

(a)

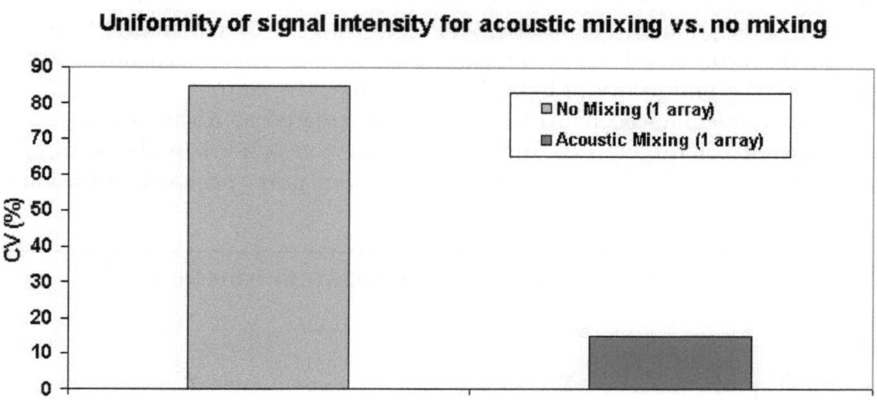

(b)

Fig. 7.9. Numerical data gathered from fluorescent images shown in Fig. 7.8. **(a)** averaged fluorescent intensity of probes NEO in the mixing-enhanced array vs. static hybridization array, **(b)** uniformity of signal intensity of probes NEO for the mixing-enhanced array vs. static hybridization array (note: 0% CV = fully uniform)

ated with each data point indicates that the static hybridization results in relatively large electrode–to–electrode variation. For the hybridization assay coupled with cavitation microstreaming, the signal increased more rapidly, and showed a much more uniform distribution (small standard deviations) compared to the pure diffusion-based device. After 40 minutes of hybridization, the sample in the mixing-enhanced device reached a saturated current value. It took approximately 6 hours for the static sample to reach the saturated level (data not shown). If the relative rates of hybridization in the two devices are calculated as the ratio of the time it takes for the signal to reach one–half of the saturated value [24], it can be seen that hybridization in the mixing-enhanced device occurs $\sim$ 5 times faster than in the diffusion-based device.

Theoretically, the relative rate of hybridization in both devices can also be estimated from the ratio of the square of the diffusion layer thickness in each device, since the diffusion time constant is proportional to the square of the diffusion length [31]. In such a diffusion layer model, it is assumed that convection maintains the concentrations of all species uniform and equal to the bulk values up to a certain distance from the surface. Within the boundary layer, no solution movement occurs and mass transfer takes place by diffusion. The thickness of the diffusion layer on a flat surface in a static hybridization is assumed to be 50 μm [24, 28], while the diffusion layer on an acoustic microstreaming-enhanced surface is estimated to be 20 μm using the steady oscillation model [32]. This theoretical ratio of 6.25 is in close proximity to the observed relative rate of hybridization. Moreover, the standard deviation of each data point in the mixing-enhanced assay is much smaller than that in the diffusion-based assay, showing that cavitation microstreaming greatly enhances the uniformity of hybridization across the chip.

A mixing-enhanced hybridization can be treated as a three-step process: 1) transport of targets (via diffusion and/or convection) in the solution to the diffusion (stagnant) boundary layer; 2) transport (primarily diffusion) of

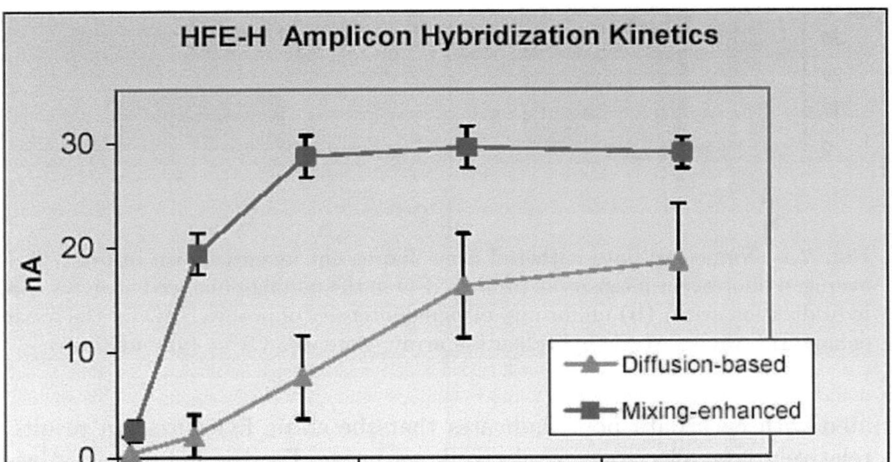

Fig. 7.10. Hybridization kinetics study performed on static hybridization vs. mixing-enhanced hybridization (square sound wave, 5 kHz and 10 V_{pp}) using eSensor^{TM} chips. Each data point is the mean value obtained from four electrodes with identical DNA capture probes in the same device

target within the diffusion boundary layer to the probes on the chip surface; and 3) reaction of target with probes on the surface. Since the last step is a chemical process of association and dissociation at the surface on which extended research has been reported [33, 34], we have focused on the first two steps. Both fluidic and hybridization experiments have demonstrated that cavitation microstreaming not only provides rapid lateral mass transport of fluidic elements, but also enhances the vertical mass transport of target DNA in the solution. The combination of rapid lateral and vertical fluid movements results in rapid transport of targets in solution to the diffusion boundary layer and thus allows for continuous replenishment of fresh DNA targets around probes that have been depleted of complementary targets. As a result, the hybridization rate is increased. Moreover, the rapid fluid movement associated with cavitation microstreaming in a shallow hybridization chamber reduces the thickness of the diffusion boundary layer by 2.5–fold. Targets are therefore in closer proximity to the immobilized probes on the chip surface, resulting in faster hybridization due to shorter diffusion lengths.

The rapid lateral fluidic movement, as observed in the fluidic dye experiments, also ensures a homogenous mixture of targets and sufficient fluid exchange across the large surface area of the chip, thus allowing for uniform hybridization signals to be achieved. Uniformity of the hybridization signal is critical, especially for high density microarrays and/or for detection of low-abundance targets. Lack of lateral convection can lead to non-homogeneous array performance and hybridization differences that are independent of differences in target concentration. Although the enhancement of hybridization rates using acoustic microstreaming is not as significant as that in the biochannel [20], flow–through [17], and electronic DNA [12, 13] devices, cavitation microstreaming has distinct advantages over the above methods, due to the rapid lateral mass transport that can be achieved, resulting in significantly enhanced uniformity of hybridization. Moreover, cavitation microstreaming requires a very simple mixing apparatus, and thus can easily be incorporated into most existing biochip devices.

7.4 Integrated Microfluidic Reactors for DNA Amplification and Hybridization

The use of microfabrication technologies has created the potential to integrate biological sample preparation with DNA analysis in a single Lab–on–a–Chip device [35, 36]. The prospective goal is to fully integrate sample collection and pretreatment with the DNA extraction, amplification, and detection into a single microfluidic platform. The ability to perform all of the steps of the biological assay, in a single self-contained microchip, promises significant advantages in terms of speed, cost, sample/reagent consumption, contamination reduction, efficiency and automation [37, 38].

In recent years, developments in Lab–on–a–Chip technologies have been substantial. Previously, integrated micro devices with reagent mixing, enzymatic reactions, and DNA sizing by electrophoresis were demonstrated [39]. The integration of micro PCR with microchip capillary electrophoresis (CE) has also been developed [40, 41]. The devices reported by Burns et al. [42] were capable of metering aqueous reagents, mixing, amplification, enzymatic digestion, electrophoretic separation, and detection with no external lenses, heaters, or mechanical pumps. Other integrated devices, demonstrated by Sosnowski et al. [43], utilized electrical forces to accomplish such functions as cell separation, sample transport, hybridization acceleration, and denaturation. In another report [35], integrated monolithic genetic assay devices have been fabricated in polycarbonate to carry out serial and parallel multistep molecular operations, including nucleic acid hybridization. Recently, Taylor et al. [44] reported on devices capable of carrying out automated sample preparation followed by real time PCR detection of pathogens. Similarly, Wilding, Kricka and Fortina [45] have developed a prototype of an integrated semi-disposable microchip analyzer. The system, which is currently under further testing, is capable of cell separation and isolation, PCR amplification, and amplicate detection.

The overall performance of an integrated device does not depend only on that of its individual functional units, but also on that of the functional integration. As a result, microvalves have become critical components for the further development of Lab–on–a–Chip technology. Some very ingenious microvalves have been designed and built as alternatives to silicon based microvalves [46, 47]. Electrokinetic valves have been successfully used for sample injection in microchip CE, on-chip fluid mixing, and dilution [36, 48]. Hydrophobic passive valves have been implemented in microfabricated centrifugal microfluidic systems [49]. Systems containing on–off valves and switching valves have been built in elastomeric materials by soft lithography [50]. Polymer monoliths containing grafted thermally responsive polymers have been thermally controlled to block or allow flow in micrometer size structures [51]. Various designs of hydrogel valves, which operate on the principle of hydrogel volume change with external stimuli, have enabled the fabrication of an organic microfluidic system [52]. Because of the unique valving requirements (high pressure, biocompatibility, and device complexity) for the integration of PCR and hybridization functionality, none of these valves could be implemented into our monolithic integrated devices.

7.4.1 Integrated Chip Design and Fabrication

Here [53], we discuss plastic, disposable devices capable of carrying out PCR amplification, hybridization, and hybridization wash assays. These microfludic devices were fabricated into polycarbonate plastic using CO_2 laser machining. Reagent transport through the device was provided by syringe pumps, which were docked onto the device. Peltier thermal electrical devices powered the heating and cooling functionality of the device. Oligonucleotide probes were deposited inside plastic hybridization channels using surface attachment chemistry and spotting techniques previously discussed in section 7.2.1. Novel Pluronics phase change valves accomplished the integration of such functional units as PCR amplification, hybridization and hybridization wash on the same device. An air permeable hydrophobic membrane valve was implemented into the device to allow for the flow of solution into the sealed hybridization chamber. All of the reagents needed for the assay were loaded into the device before the assay. Genomic DNA from the bacteria *Escherichia coli* K–12 (*E. coli*) and *Enterococcus faecalis* (*E. faecalis*) were used to amplify the *E. coli* K–12 MG1655 gene (221 bp) and the *E. faecalis* DNAE gene (195 bp) by single or multiplex asymmetrical PCR (A–PCR) reactions. The single strand amplicons were hybridized to the detection probes inside the hybridization channel. The performance of each individual functional unit and that of the integrated system were tested.

7.4.2 Pluronics Phase Change Valves

Microvalves are critical to the successful integration of PCR amplification with DNA hybridization assays. Suitable microvalves have to meet a number of requirements. First, the valves must be able to withhold the pressure generated during the PCR reaction, caused by degassing and air expansion at elevated temperature. If the valve fails, the PCR sample will be pushed out of the PCR chamber, resulting in failed PCR reaction. The amount of pressure required to prevent degassing has been estimated by Chiou et. al. [54] to be about 3.1 psi. The evaluation was performed using solubility data for air in water and Henry's law. The presence of an air gap between the PCR chamber and the valves will cause additional internal pressure build–up. Heating of this air gap will generate an additional 3.7 psi pressure at 94°C (using the ideal gas law), therefore the valve must be able to withhold at least 6.8 psi total pressure to ensure the successful confinement of the PCR sample during thermal cycling. Second, because valves will be in direct contact with PCR solution, the valve material must not inhibit the PCR reaction. Third, the valve needs to be easily opened after the PCR reaction to allow PCR solution to flow into the hybridization channel.

Pluronics F127, a commercially available surfactant, is composed of uncharged $(EO)_{106}(PO)_{70}(EO)_{106}$ triblock copolymers. Solutions of Pluronics within a concentration range of 18–30% are low viscosity liquids (< 2 poise) at low temperature (0–5°C), but form self-supporting cubic liquid crystalline gels at room temperature [53]. Therefore, Pluronics solutions at the proper concentration can be used as one-shot, phase change valves. These one shot valves are initially closed and become permanently opened once activated by a lowering of the valve temperature below the Pluronics gel transition temperature. We have found that the presence of Pluronics molecules does not inhibit PCR reactions, and a 9 mm $\times$ 0.25 mm $\times$ 0.25 mm valve made of 30%

Pluronics in a polycarbonate channel can hold 20 psi pressure, well above the 6.8 psi generated during the PCR reaction. The advantages of Pluronics temperature transition valves are their simplicity of implementation and operation. Although in solid gel form, Pluronics gels are not cross–linked and can be easily injected into microfluidic structure to form one-shot valves using a preloaded syringe at room temperature.

7.4.3 Assay in an Integrated Reactor

Two genetic targets were asymmetrically amplified in the assays: *E. coli* K–12 MG1655 gene (221 bp) and *E. faecalis* DNAE gene (195 bp). The A–PCR reaction mixture contained 10 mM Tris–HCl (pH 8.3), 50 mM KCl, 1.5 mM $MgCl_2$, 0.001% gelatin, 250 µg/mL bovine serum albumin, 125 µM each deoxynucleotide triphosphate, 1.2 µM reverse primer, 12 nM forward primer, 25 units/mL AmpliTaq, DNA polymerase (Perkin–Elmer), and *E. faecalis* or *E. coli* genomic DNA (50 pg/mL). The primer set used to amplify a 221 bp segment of *E.coli* gene target was 5′AAC GGC CAT CAA CAT CGA ATA CAT3′ (forward) and 5′[cy3] GGC GTT ATC CCC AGT TTT TAG TGA3′ (reverse). The probe used for hybridization was AAG CGA CAG TTC GGC TTC GTG NH_2 3′. The primer set used to amplify *E. faecalis* gene was 5′GCC AGA TTT TTC GTT CGC TCA T3′ (forward) and 5′[Cy3] AAA TCG GCA ACT TCT CGC TCA G (reverse). The probe used for hybridization is CGG AAG AAA GCT CTG AGC G NH_2 3′. The probe for negative control was AGC TCA CGT GCC TGC AGA AG NH_2 3′. All the oligo probes and PCR primers were ordered from Operon Technologies Inc. (Alameda, CA).

The integrated device is shown in Fig. 7.11. The device contains a PCR chamber (38 µL), a hybridization channel (7 µL), a syringe coupled to a hybridization wash solution channel (20 µL), a waste channel coupled to a waste syringe, four Pluronics trapping reservoirs, one hydrophobic membrane valve, four Pluronics valves, seven reagent introduction holes, and three external syringe pump interface reservoirs. The dimensions of the device are 5.4 mm × 8.6 mm × 0.75 mm, and resemble that of a miniature credit card. The PCR chamber volume (38 µL) is large relative to the current hybridization channel volume (7 µL). But this volume can be utilized in a longer hybridization channel with a higher density array. The hybridization channel was designed to accommodate efficient dispensing of probes, using a SpotBot™ Personal Microarrayer. With this spacing, all 4 pins of the microarrayer were utilized, with no need for device position adjustment. It took the SpotBot™ about ten minutes to dispense 120 probes into the four channels. Pluronics valves were installed before any reagent solution was introduced into the device. The two Pluronics valves adjacent to the PCR chamber enclose the PCR solution during the reaction. The first Pluronics valve (V1) isolates the PCR chamber from the external pump, and the second Pluronics valve (V2) is located between the PCR chamber and the hybridization channel. The third

Pluronics valve (V3) is placed between the hybridization channel and the wash solution channel. The fourth Pluronics valve (V4) isolates the hybridization channel from the waste chamber. PCR mixture and hybridization wash solution were introduced into their corresponding chambers on the device prior to permanent sealing of all reagent access holes by application of 1 layer of adhesive tape and 1 layer of parafilm.

During PCR thermal cycling, only the PCR chamber portion of the device was sandwiched between Peltier thermal electric heating units. After PCR thermal cycling, the two Pluronics valves adjacent to the PCR chamber were cooled to 5°C with a Peltier thermal electrical device, and the syringe

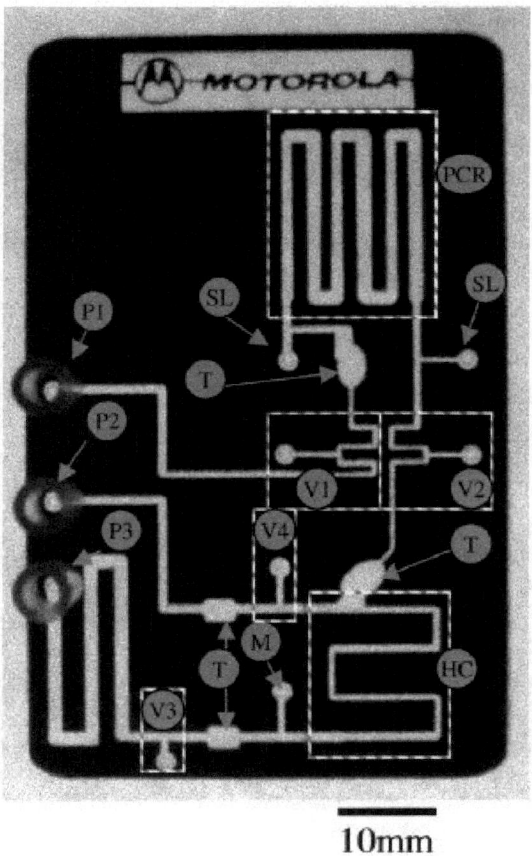

Fig. 7.11. Monolithic, integrated DNA assay device. Legend: Serpentine PCR channel (PCR), hybridization channel (HC), Pluronics valves (V1–4), Pluronic traps (T), Hydrophobic air permeable membrane (M), PCR reagent loading holes (SL), Sample driving syringe pump P1, waste withdrawing syringe pump (P2), wash syringe pump (P3)

pump (P1) was then used to push the Pluronics valve solution and PCR amplification solution toward the DNA hybridization channel. When Pluronics solution entered the Pluronics traps, located outside of the Peltier cooler zone, the Pluronics solution resolidified into a solid gel state and did not travel any further. This prevented the Pluronics gel from blocking the connecting channel to the hybridization chamber. The amplified PCR sample solution was then continuously pushed into the hybridization channel. The air permeable hydrophobic membrane vent at the end of the hybridization channel allowed air from the channel to flow out of the device, while sealing target solution that flowed into the channel. Because of the small dimension of the fluidic channel, target DNA molecules were confined in close proximity to the capture probes. Assuming a target diffusion coefficient of 1.7×10^{-7} cm^2/s, it was estimated that it would take only about 30 minutes for a 200 base target to reach capture probes from the top of the channel by diffusion. We determined experimentally that one hour reaction time is sufficient for detection of hybridization event. Further improvement of hybridization efficiency could be realized when in-channel target solution oscillation is implemented in the future design [20]. The Peltier device, underneath the hybridization chamber, was maintained at 50°C during the one-hour hybridization reaction. After hybridization, valve three (V3) and valve four (V4) were opened by activation of the syringe pumps, P2 and P3. Since the pressure–holding requirement for V3 and V4 is not as high as for V1 and V2, V3 and V4 were designed to hold less pressure and allow activation by syringe pumps alone. The first 10 μL of the wash solution was pushed into the hybridization channel, while the waste syringe withdrew the target solution. The next 10 μL of the wash solution was left in the hybridization channel, to incubate for 20 minutes. The wash solution was manually removed by the waste syringe before scanning. We attribute the successful integration of multiple functions on a monolithic device to the implementation of the Pluronics valves. Plastic devices containing only fluidic channel structures are very inexpensive when produced in large quantities by injection molding. However, the cost of the device will increase if an additional complicated fabrication process is needed for addition of mechanical valves. The implementation of Pluronics phase change valves does not require additional fabrication steps and thus is desirable for low cost, disposable chip solutions. Since the device is preloaded with all of the necessary reagents needed for the assay, potential contamination from human interference is eliminated, and automation is made possible.

Three different types of probes (*E. coli*, *E. faecalis*, and control) were dispensed in four horizontal sections of the serpentine hybridization channel. The probe layout was identical in each of the four horizontal sections. Two sets of a 1 × 5 array of each type of probe was located in each horizontal channel section, with a total of eight sets in the entire serpentine hybridization channel. Assymetrical PCR protocol was used to produce single stranded DNA targets. Depending on which target DNA template molecules were present in the PCR chamber, the corresponding probe sites were detectable by fluorescence after successful hybridization. Figure 7.12a is the fluorescent image of the hybridization chamber, using the *E. coli* 221 bp gene as amplification target. Hybridization reactions occurred at the sites of *E. coli* probes. Two sets of hybridization sites were enlarged for better view. The fluorescent signals of corresponding probes in the same array were very uniform. Interestingly, the fluorescent background inside the fluidic channel is lower than that from the surrounding ridges. One possible explanation is that thermal bonding causes increased roughness at the bonding interface and therefore causes an increase in scattered light during scanning. These integrated devices were also tested

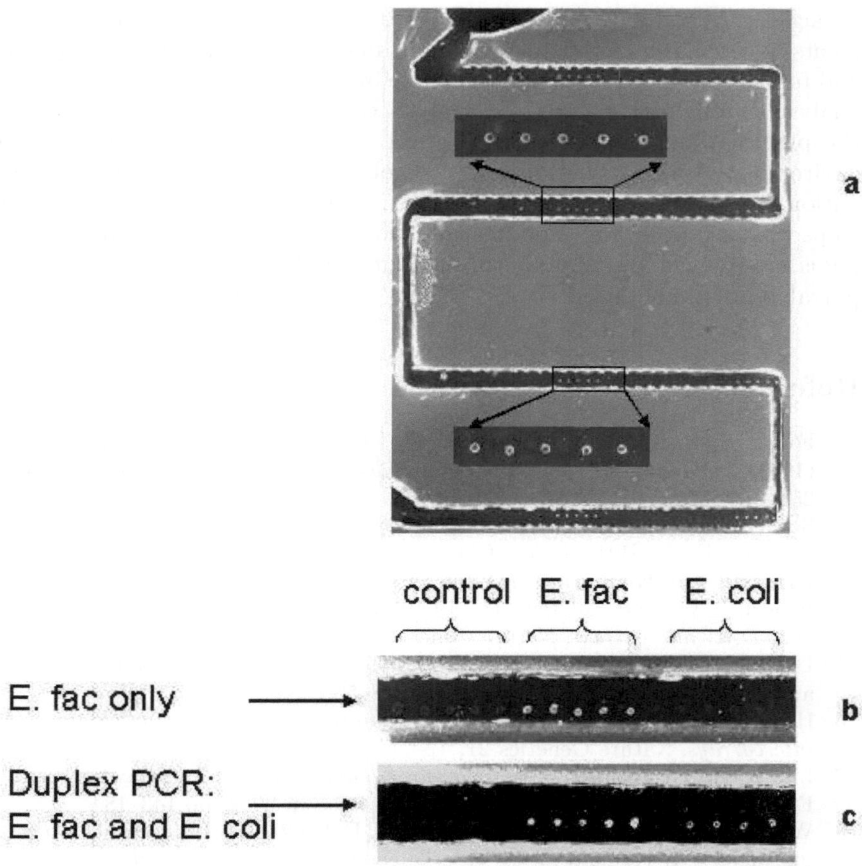

Fig. 7.12. PCR hybridization results from monolithic integrated device, **(a)** *E. coli* 221 bp hybridization after amplification. Portions of the biochannel were enlarged for better viewing, **(b)** Fluorescent image of portion of biochannel after *E. fae* amplification and hybridization, **(c)** Fluorescent image of portion of biochannel after multiplex (E. fae and *E. Coli*) amplification and hybridization

for *E. faecalis* gene (195 bp) amplification and detection, and for multiplex PCR (*E. coli* and *E. faecalis*) amplification and detection. All amplification and hybridization reactions were successful, as shown in Fig. 7.12(b,c).

7.5 Summary and Conclusions

Microarray hybridization technologies have become indispensable tools in the studies of gene mapping, gene expression, and single nucleotide polymorphisms. The microarray field has enjoyed tremendous progress in the last decade, resulting in successful commercialization of several chip approaches [1,2,12,13,17,24]. However there is still significant room for improvements, particularly in the areas of assay kinetics, on-chip sample preparation and further functional integration. These improvements will not only increase analysis throughput and reduce analysis cost, but will enable broadening of the practical applications to such areas as doctor's office diagnostics, field environmental monitoring, and rapid biothreat recognition. A clever combination of existing microarray techniques and newly developing microfluidic chips promise powerful analytical solutions where high parallelism of sensing is complimented with high throughput, rapid assay kinetics and compact, portable instrumentation.

References

1. Fodor, S. P., Read, J. L., Pirrung, M. C., Stryer, L., Lu, A. T., Solas, D. (1991) *Light-directed, spatially addressable parallel chemical synthesis.* Science 251, 767–773
2. Schena, M. *Microarray Biochip Technology.* Eaton Publishing: Natick, MA, 2000
3. Duggan, D. J., Bittner, M., Chen, Y., Meltzer, P., and Trent, J. M. (1999) *Expression profiling using cDNA microarrays.* Nature Genetics 21, 10–14
4. Brown, P. O., and Botstein, D. (1999) *Exploring the new world of the genome with DNA microarrays.* Nature Genetics 21, 33–37
5. Debouck, C., and Goodfellow, P. N. (1999) DNA microarrays in drug discovery and development. Nature Genetics 21, 48–50
6. Hacia, J. G. (1999) *Resequencing and mutational analysis using oligonucleotide microarrays.* Nature Genetics 21, 42–47
7. Liu, R. H., Grodzinski, P. In *Microfluidic and BioMEMS Applications*; Tay, F. E. H., Ed.; Kluwer Academic Publishers: Boston, 2002, pp 143–184
8. Wittwer, C. T., Fillmore, G.C., Garling, D.J. (1990) *Minimizing the time required for DNA amplification by efficient heat transfer to small samples.* Anal Biochem 186, 328–331
9. Belgrader, P., Benett, W., Hadley, D., Richards, J., Stratton, P., Mariella Jr. , R., Milanovich, F., (1999) *PCR detection of bacteria in seven minutes.* Science, 284, 449–450
10. Hühmer, A. F. R. (2000) *Noncontact infrared–mediated thermocycling for effective polymerase chain reaction amplification of DNA in nanoliter volumes.* Analytical Chemistry 72, 5507–5512

11. Southern, E., Mir, K., and Shchepinov, M. (1999) *Molecular interactions on microarrays.* [Review] [37 refs]. Nature Genetics 21, 5–9
12. Edman, C. F., Raymond, D. E., Wu, D. J., Tu, E., Sosnowski, R. G., Butler, W. F., Nerenberg, M., and Heller, M. J. (1997) *Electric field directed nucleic acid hybridization on microchips.* Nucleic Acids Res 25, 4907–14
13. Radtkey, R., Feng, L., Muralhidar, M., Duhon, M., Canter, D., DiPierro, D., Fallon, S., Tu, E., McElfresh, K., Nerenberg, M., and Sosnowski, R. (2000) *Rapid, high fidelity analysis of simple sequence repeats on an electronically active DNA microchip.* Nucleic Acids Res 28, E17
14. Fan ZH, M. S., Granzow R, Heaney P, Ho W, Dong Q, Kumar R. (1999) *Dynamic DNA hybridization on a chip using paramagnetic beads.* Anal Chem 71, 4851–4859
15. Fan, Z. H., Kumar, R. (2001) In *Biochip Technology*, Cheng, J., Kricka, L., Ed., Harwood Academic Publishers, Philadelphia, PA, pp. 291 – 307
16. Chee, M., Yang, R., Hubbell, E., Berno, A., Huang, X. C., Stern, D., Winkler, J., Lockhart, D. J., Morris, M. S., and Fodor, S. P. (1996) *Accessing genetic information with high–density DNA arrays.* Science 274, 610–4
17. Cheek, B.J., Steel, A.B., Torres, M.P., Yu, Y-Y., Yang, H. (2001) *Chemiluminescence Detection for Hybridization Assays on the Flow–Thru Chip, a Three–Dimensional Microchannel Biochip.* Anal Chem 73, 5777–5783
18. Liu, R. H., Chen, H., Luehrsen, K. R., Ganser, D., Weston, D., Blackwell, J., Grodzinski, P., In *Technical Digest of International MEMS Conf.*, Interlaken, Switzerland, Jan. 21–25 2001; 439–442
19. Liu, R., Lenigk, R., Luehrsen, K. R., Yu, H., Chen, H., Ganser, D., Bonanno, J., Grodzinski, P., In *Micro Total Analysis Systems 2001*; Ramsey, J. M., Van den Berg, A., Eds.; Kluwer Academic Publishers: Monterey, CA, 2001, 465–467
20. Lenigk, R., Liu, R., Athavale, M., Chen, Z., Ganser, D., Yang, J., Rauch, C., Liu, Y., Chan, B., Yu, H., Ray, M., Marrero, R., Grodzinski, P. (2002) *Plastic biochannel hybridization devices: A new concept for microfluidic DNA arrays.* Analytical Biochemistry 311, 40–49
21. Jackman, R. J., Brittain, S. T., Adams, A., Prentiss, M. G., Whitesides, G. (1998) *Design and Fabrication of Topologically Complex, Three–Dimensional Microstructures.* Science 280, 2089–2091
22. Jo, B. H., Lerberghe, L. M. V., Motsegood, K. M., Beebe, D. J. (2000) Three–dimensional micro–channel fabrication in *Polydimethylsiloxane(PDMS) elastomer.* Journal of Micro–electro–mechanical Systems 9, 76–81
23. Umek RM, L. S., Vielmetter J, Terbrueggen RH, Irvine B, Yu CJ, Kayyem JF, Yowanto H, Blackburn GF, Farkas DH, Chen YP. (2001) *Electronic detection of nucleic acids: a versatile platform for molecular diagnostics* J Mol Diagn 3, 74–84
24. Steel, A., Torres, M., Hartwell, J., Yu, Y.-Y., Ting, N., Hoke, G., Yang, H. In *Microarray Biochip Technology*; Schena, M., Ed.; Eaton Publishing, 2000, pp 87–117
25. Corporation, C. R. (2001) CFD–ACE+ User's Manual, CFD Research Corporation, Huntsville, AL
26. Elder, S. A. (1959) *Cavitation Microstreaming.* J. Acoust. Soc. Am. 31, 54–62
27. Liu, R.H., Lenigk, R., Ganser, D., Bonanno, J., Sanchez, B., Singal, P., Grodzinski, P., *Improvement of DNA Microarray Biochips using Microfluidic Mixing Technique*, presented at Solid–State Sensor and Actuator Workshop, Hilton Head Island, SC, June, 2002

28. Liu, R. H., Lenigk, R., Yang, J., Druyor-Sanchez, R., Singal, P., Grodzinski, P. (2003) *Hybridization Enhancement by Bubble–induced Acoustic Microstreaming.* Analytical Chemistry, 75 1911-1917
29. Nyborg, W. L. (1958) *Acoustic streaming near a boundary.* J. Acoust. Soc. Am. 30, 329–339
30. Kolb, J., Nyborg, W. L. (1956) *Small–Scale Acoustic Streaming in Liquids.* J. Acoust. Soc. Am. 28, 1237–1242
31. Bard, A. J., Faulkner, L. R. *Electrochemical Methods*; Wiley: New York, 1980
32. White, F. M. *Viscous Fluid Flow*; McGraw-Hill: New York, 1974
33. Jonsson, U. L., al., e. (1991) *Real–time biospecific interaction analysis using surface plasmon resonance and a sensor chip technology.* BioTechniques 11, 620–627
34. Garland, P. B. (1996) *Optical evanescent wave methods for the study of biomolecular interactions.* Q. Rev. Biophys. 29, 91–117
35. Anderson, R. C., Su, X., Bogdan, G. J., Fenton, J. (2000) *A miniature integrated device for automated multistep genetic assays.* Nucleic Acids Res 28, E60
36. Harrison, D. J., Fluri, K., Seiler, K., Fan, Z., Effenhauser, C. S., Manz, A. (1993) *Micromachining a Miniaturized Capillary Electrophoresis–based Chemical Analysis System on a Chip.* Science 261, 895–897
37. Wang, J. (2000) *From DNA biosensors to gene chips.* Nucleic Acids Res 28, 3011–3016
38. Ramsey, J. M., Jacobson, S. C., Knapp, M. R. (1995) *Microfabricated Chemical Measurement Systems.* Nature Medicine 1, 1093–1096
39. Jacobson, S. C., Hergenroder, R., Moore Jr., A. W., Ramsey, J. M. (1994) *Precolumn Reactions with Electrophoretic Analysis Integrated on a Microchip.* Anal. Chem. 66, 4127–4132
40. Woolley, A. T., Hadley, D., Landre, P., deMello, A. J., Mathies, R. A., Northrup, M. A. (1996) *Functional Integration of PCR Amplification and Capillary Electrophoresis in a Microfabricated DNA Analysis Device.* Anal. Chem. 68, 4081–4086
41. Waters, L. C., Jacobson, S. C., Kroutchinina, N., Khandurina, J., Foote, R. S., Ramsey, J. M. (1998) *Microchip Device for Cell Lysis, Multiplex PCR Amplification, and Electrophoretic Sizing.* Anal. Chem. 70, 158–162
42. Burns, M. A., Johnson, B. N., Brahmasandra, S. N., Handique, K., Webster, J. R., Krishnan, M., Sammarco, T. S., Man, P. M., Jones, D., Heldsinger, D., Mastrangelo, C. H., Burke, D. T. (1998) *An integrated nanoliter DNA analysis device.* Science 282, 484–487
43. Sosnowski, R. G., Tu, E., Butler, W. F., O'Connell, J. P., Heller, M. J. (1997) *Rapid determination of single base mismatch mutations in DNA hybrids by direct electric field control.* Proc Natl Acad Sci U S A 94, 1119–1123
44. Taylor, M. T., Belgrader, P., Joshi, R., Kintz, G. A., Northrup, M. A. In *Micro Total Analysis Systems 2001*; Ramsey, J. M., Van den Berg, A., Eds.; Kluwer Academic Publishers: Monterey, CA, 2001, pp 670–672
45. Yuen, P., Kricka, L., Fortina, P., Panaro, N., Sakazume, T., Wilding, P. (2001) *Microchip Module for Blood Sample Preparation and Nucleic Acid Amplification Reactions.* Genome Research 11, 405–412
46. Smith, L., Hok, B., In *Transducers '91*, San Francisco, CA 1991; 1049–1051
47. Barth, P. W. In *Transducers'95*: Stockholm, Sweden, 1995, pp 276–277

48. Jacobson, S. C., McKnight, T. E., Ramsey , J. M. (1999) *Microfluidic Devices for Electrokinetically Driven Parallel and Serial Mixing.* Analytical Chemistry 71, 4455–4459
49. Duffy, D. C., Gillis, H. L., Lin, J., Sheppard, N. F., Kellogg, G. L. (1999) *Microfabricated Centrifugal Microfluidic Systems: Characterization and Multiple Enzymatic Assays.* Analytical Chemistry 71, 4669–4678
50. Unger, M. A., Chou, H. P., Thorsen, T., Scherer, A., Quake, S. R. (2000) *Monolithic microfabricated valves and pumps by multilayer soft lithography.* Science 288, 113–116
51. Peters, E. C., Svec, F., Frechet, J. M. J. (1999) *Rigid Macroporous Polymer Monoliths.* Advanced Materials 11, 1169–1181
52. Beebe, D. J., Moore, J. S., Bauer, J. M., Yu, Q., Liu, R. H., Devadoss, C., Jo, B. H. (2000) *Functional hydrogel structures for autonomous flow control inside microfluidic channels.* Nature 404, 588–590
53. Liu, Y., Rauch, C., Stevens, R., Lenigk, R., Yang, J., Rhine, D., Grodzinski, P. (2002) *DNA Amplification and Hybridization Assays in Integrated Plastic Monolithic Devices.* Analytical Chemistry 74, 3063–3070
54. Chiou, J., Matsudaira, P., Sonin, A., Ehrlich, D. (2001) *A closed-cycle capillary polymerase chain reaction machine.* Anal Chem 73, 2018–2021

8

Labels and Detection Methods

James J. Storhoff, Sudhakar S. Marla, Viswanadham Garimella, and Chad A. Mirkin

8.1 Introduction

The sequencing of the human genome [1, 2] along with other organisms is fuelling the development of new tools for the highly parallel analysis of genomic information. Microarray technology has emerged as a robust methodology for quantitatively analyzing a large number of nucleic acid sequences in parallel, as shown in Fig. 8.1 [3, 4]. High density oligonucleotide [5] or cDNA microarrays have been utilized for measuring the abundance of mRNA transcripts, which is typically referred to as gene expression analysis [6, 7]. Differential gene expression analysis is used to determine which genes are up-regulated or down-regulated during specific cellular processes or in response to environmental stimuli [8]. Cellular responses triggered during specific disease states, or by exposure to drugs, toxins, or other molecules of interest have been studied [9, 10]. Such arrays are currently being developed for diagnosis of specific diseases such as cancer [11] as well as for identifying novel mechanisms of drug action [10]. In addition, microarrays have found applications in identifying single nucleotide polymorphisms (SNPs) or other genetic variations [7, 12–15]. The detection of SNPs associated with genetic disorders has led to the development of diagnostic microarrays for diseases such as cystic fibrosis [16]. For pharmacogenomic applications, SNP arrays are used to identify key mutations in genes that encode for enzymes responsible for drug metabolism [17].

The major factors that have limited the utility of microarrays in the research and diagnostic applications described above are the amount of target needed, detection specificity, as well as cost and reliability of detection equipment and assays. A critical determinant of these parameters is the labelling and detection methodology. While the current gold standard is fluorescence technology, the emphasis on higher sensitivity, specificity, and cost-effective detection instrumentation has spurred the development of a number of new labelling and detection methodologies. Recent reports have demonstrated that fluorophore-labelled dendrimers, up-converting phosphor reporters, electrochemical detection techniques and semiconductor or metal nanoparticle labels

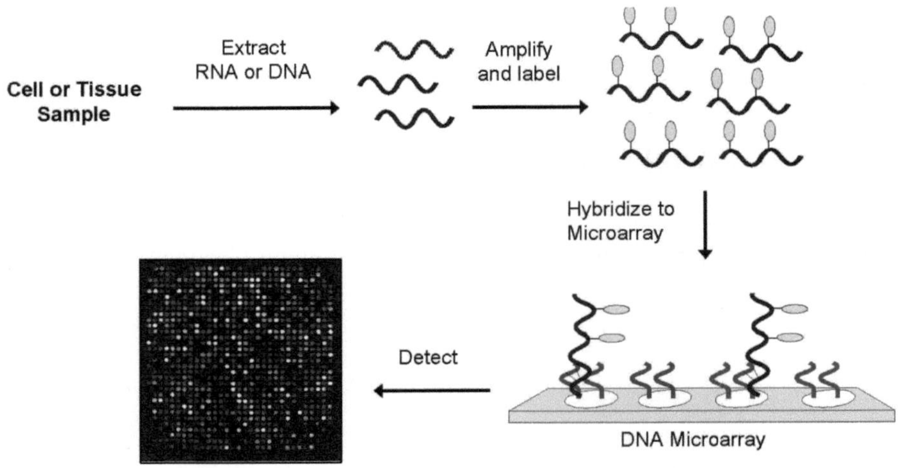

Fig. 8.1. Oligonucleotide microarrays are generated on glass slides via robot deposition ($< 10{,}000$ oligonucleotide/cm^2) or in situ photolithographic synthesis ($>$ 250,000 oligonucleotide/cm^2 possible). Labelling and detection of nucleic acid targets on arrays typically is achieved by using the following procedure. First, target RNA or DNA is extracted from the sample and amplified to generate more copies. During the amplification process, a reporter group (e.g. fluorophore) is incorporated into the target for detection. The labelled targets are subsequently hybridized to a microarray containing the specific gene sequences of interest. The amount of target bound to each location on the microarray is quantified by detecting the attached label with the appropriate instrument

can positively impact sensitivity, specificity, cost and complexity of detection instrumentation. The development of such technologies will not only improve current microarray applications, but also point to new opportunities for microarrays. These emerging labelling and detection methodologies will be the focus of this chapter with particular emphasis on metal nanoparticle probes. The important considerations that will be used to evaluate each microarray label and detection methodology are sensitivity, specificity, dynamic range, cost, and number of distinguishable labels.

8.2 Fluorophore Labelling and Detection Methods

Fluorophore labelling has become the technology of choice for detection on microarrays in both gene expression and SNP analysis [7, 18, 19]. In a fluorescence experiment, photons absorbed by a dye molecule illuminated at a specific wavelength are re-emitted (in part) as radiation at a lower frequency that is measured with a photodetector. A multitude of fluorophore dyes with spectrally unique signatures have been developed for high sensitivity biological labelling studies. The fluorophore labels most commonly used for microarray analysis are Cyanine 3 (Cy3) and Cyanine 5 (Cy5). These fluorophores exhibit efficient quantum yields, moderate photostability, unique excitation and emission spectra which enable multiplexing, and can be efficiently incorporated enzymatically into biomolecules via reverse transcription, ligation, or PCR. The sensitivity and dynamic range of fluorophore labelling is dependent on the detection system. Individual fluorophores have been detected on surfaces using highly sophisticated optical detection equipment [33]. However, the utilization of such detection instrumentation is not practical for microarray analysis. Instead, fluorescence scanners that utilize red and green lasers for Cy3/Cy5 excitation and a photo multiplier tube (PMT) for quantitation of specific fluorophore signals were developed for this purpose [34]. These scanners reliably detect < 1 attomole of fluorophore on a 100 μm diameter spot which translates to a detection limit of < 75 fluorophores/μm^2. The reported dynamic range of quantifiable fluorescence signal was over 3 orders of magnitude using this detection methodology. Today, other types of fluorophore microarray scanners that utilize different methods of illumination and detection have been developed and are commercially available [35]. Recent fluorescence detection data collected with commercially available instrumentation suggest that Cy3 probe densities of ~ 5 Cy3 molecules/μm^2 are detectable (Table 8.1) [25]. Instrument manufacturer specifications for a variety of scanners indicate detection limits of < 1 fluorophore/μm^2 [20]. These values represent the lower limit of a detection system based upon fluorescence. The actual detection limit of an assay is typically limited by the background resulting from any autofluorescence associated with the solid support and non-specific binding of the fluorophore-labelled biomolecule targets.

A detailed description of fluorescence-based labelling for gene expression applications is provided in Chap. 11, and the application to SNP genotyping and genomic analysis is provided in Chaps. 10 and 12, respectively. For all of these applications, a detection label that provides higher sensitivity than fluorescence would enable analysis of smaller target quantities, and for some

Table 8.1. Detection and assay sensitivity for selected probe technologies

Technology	**Probe detection sensitivity (probes /μm^2)**	**Assay sensitivity**	
		Assay description	Detection limits (amount/copies)
Fluorescence/ Phosphorescence		T = Target D = Detection	
Fluorescently-labelled dyes [20]	5	T: Spiked transcripts D: Laser scanning	2.8×10^7 copies (580 fM)

Continued on next page

Technology	Probe detection sensitivity (probes /μm^2)	Assay sensitivity Assay description	Detection limits (amount/copies)
Fluorescently–labelled DNA dendrimers [21]	–	T: Total RNA D: Laser scanning	2.5 μg
Up-converting phosphors [22]	Single particle	T: Labelled DNA fragment D: Modified fluorescence microscope	1 ng/μL ($\sim 1 \times 10^9$ copies/μL)*
Electrochemical			
Electrochemical sensor Motorola [23]	–	T: 74 base DNA strand D: Electron transfer	50nM (3×10^{10} copies/μL)*
Nanoparticles			
Quantum dots [24]	–	T: Single-stranded DNA D: Fluor. microscope	10 nM (6×10^9 copies/μL)*
DNA-modified gold probes with silver amplification (Nanosphere) [25]	Single particle (0.0025)	T: Total genomic DNA/PCR products D: Evanescent wave-based scatter measurements	6×10^6 copies (200 fM) (gen. DNA) 3000 copies (100 aM) (PCR)
Resonant Light Scattering (Genicon) [26,27]	Single particle (0.005)	T: Spiked transcripts D: CCD-based system	8×10^6 copies (170 fM)
Streptavidin-coated gold nanoparticles with silver amplification (AAT) [28]	5	T: PCR products D: Laser illuminated scatter	6×10^7 copies (~1 pM)
High resolution surface plasmon resonance [29]	–	T: Single-stranded DNA D: SPR spectrometer	3.2×10^6 copies (54 fM)
Gold nanoparticle–enhanced SPR [30]	0.5–20	T: Single-stranded DNA D: Scanning angle SPR	10 pM (6×10^6 copies/μL)*
SERS probes [31]	–	T: Single-stranded DNA D: Raman spectroscopy	20 fM (1.2×10^4 copies/μL)*
Gold nanoparticle-based electrical detection [32]	–	T: Single-stranded DNA D: Conductivity	500 fM (3×10^5 copies/μL)*

*NOTE: copies/mL are reported for assays where reaction volumes were unavailable.

applications, eliminate the need for target amplification steps such as PCR. This is a major driver for the development of new labels since it has the potential to lower the cost and complexity of such assays, while increasing data reliability. In addition, the development of a labelling methodology that provides a larger number of distinguishable 'colors' for analysis is highly desirable since it would increase multiplexing capabilities for such applications. Two color Cy3/Cy5 labelling methodology is commonly used for gene expression analysis [6,7], and up to four uniquely colored fluorophore dyes have been used in multiplex SNP genotyping applications [36]. It is important to note that the complexity of the microarray scanner increases with the number of fluorophore dyes since each dye requires a different excitation wavelength. Ultimately, a multicolor, high sensitivity labelling methodology that utilizes low cost and complexity instrumentation is desired for microarray-based applications.

8.3 Enhanced Fluorescence-Based Assays

8.3.1 DNA Dendrimer Technology

One pathway for achieving higher detection sensitivity is to increase the number of labels associated with each cDNA or target nucleic acid bound to a microarray. A number of research groups have explored using branched- or dendrimer-based nucleic acid structures to increase the label density per nucleic acid target [37–40]. For microarray labelling, Genisphere has developed fluorophore labelled nucleic acid dendrimers which are referred to as 3DNA probes [21]. The 3DNA probes are prepared by hybridizing and crosslinking complementary oligonucleotide building blocks to form a 'core' dendritic structure. Cy3 or Cy5 labelled oligonucleotides are subsequently hybridized

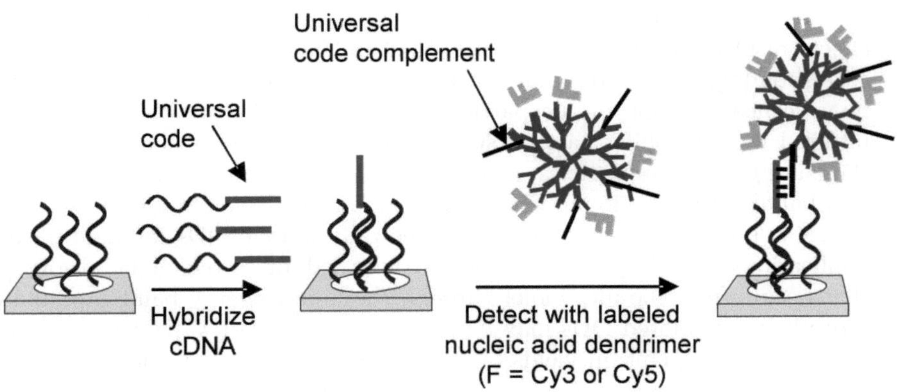

Fig. 8.2. Schematic illustration of cDNA detection on microarrays using fluorophore labelled dendrimers. Note that dendrimers typically contain ∼ 250 fluorophores

and crosslinked to the core dendrimers so that each 3DNA probe contains on average ~ 250 fluorophore dyes. For gene expression monitoring, cDNAs are coded with a single universal sequence tag during transcription which is recognized by a 3DNA probe containing the universal code complement in a second labelling step, Fig. 8.2. The 3DNA probe yields 250 fluorophores/cDNA target compared to ~ 12 fluorophores/cDNA target through direct incorporation of fluorophore labelled dNTPs [6]. This corresponds to a ~ 20 fold increase in the number of fluorophores/cDNA probe over direct enzymatic incorporation. For comparison of experimental detection limits, 1–50 µg of total RNA was transcribed and labelled with both methods. After hybridization to cDNA arrays, the average specific signal for each gene was measured [21]. The specific signal obtained with the 3DNA dendrimer probe using 2.5 µg of total RNA was equivalent to the specific signal obtained with direct Cy3 incorporation using 40 µg of total RNA. This amounts to a ~ 16–fold increase in detection sensitivity which correlates well with the number of fluorophores bound per probe. In addition, it was noted that the 3DNA probe signal was stable to repetitive scanning whereas the Cy3 labelled cDNA signal significantly degraded over time, demonstrating that this labelling technology is more robust. In summary, the use of significantly less RNA starting material, the higher stability of the label compared with conventional fluorophores, and the ready integration of the labels with existing fluorescence microarray scanners make the 3DNA dendrimer labelling technology attractive for gene expression analysis.

8.3.2 Semiconductor Quantum Dots

Semiconductor quantum dots (QDs) have emerged as a new class of fluorophore labels [41–43]. These new labels comprise nanometer sized particles of group II–VI or III–V atoms from the periodic table of elements such as CdSe or InAs that are smaller than the exciton Bohr radius (typically 1–10 nm in size) [44–46]. As a result, the QDs exhibit quantum confinement effects resulting in optical properties that are significantly different than the corresponding bulk material or the atoms that comprise the particle [47]. The fluorescence emission of the particles may be tuned from blue to the near infrared by controlling particle size and chemical composition, which alters the band gap of the particles, as shown in Fig. 8.3. Narrow fluorescence emission bandwidths (25–30 nm FWHM for CdSe QDs) have been observed which makes it possible to generate many spectrally unique QDs for multiplexing applications [45]. Importantly, the particles exhibit broad light absorption that occurs from the ultraviolet through the lowest energy band gap, and as a result, multicolored QDs may be excited by a single UV light source. QDs have exhibited quantum yields of 40–50%, which are slightly lower than the quantum yields for commercial organic fluorophores. This is compensated by the high molar extinction coefficients of such particles at 10^5–10^6 M^{-1} cm^{-1}, which is 10–100 times larger than that for typical organic fluorophores. In

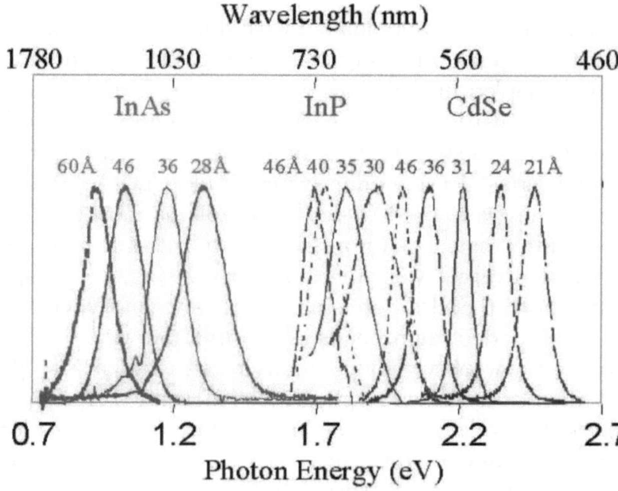

Fig. 8.3. Size- and material-dependant fluorescence emission spectra of several surfactant-coated semiconductor nanocrystals. (Reprinted with permission from Professor Paul Alivisatos at UC Berkeley)

one study, it was estimated that single ZnS-capped CdSe QDs are ~ 20 times brighter than rhodamine 6G organic dye molecules [43]. The QDs are also highly stable against photo–bleaching.

These fluorescence properties are suitable for many biological applications such as cellular imaging [48] and fluorescent in situ hybridization (FISH) [49] where stable fluorophores are desired, and microarray labelling where multiple colored probes are possible using a single excitation source [24]. However, a limitation of QD technology to date has been the difficulty in functionalizing the particles with biomolecules for robust labelling in such applications [24]. As a result, only a few examples of direct QD labelling for biological applications have been reported. An innovative approach designed to utilize the unique fluorescent properties of QDs for optical coding of biomolecules was recently reported by Nie and coworkers [50]. In this approach, 1.2 µm polymeric microbeads are optically encoded by embedding different colored QDs at defined ratios for use in 'liquid arrays', where a biomolecule attached to the surface of the microbead is barcoded by the unique signal from a single QD or a group of different QDs in a well-defined ratio. Beads encoded with organic fluorophores were previously reported for liquid array applications [51]. However, major benefits of QDs for barcoding include a single UV excitation source for bead detection, and the potential for many more unique colors. The number of codes in this approach is defined by the number of unique colors and intensities, such that n intensities and m colors generate $n^m - 1$ codes. The authors suggest that six spectrally unique QDs at 6 intensity levels is feasible, generating approximately 10,000 recognizable QD codes [50]. Experimentally, polymeric beads were loaded with differing amounts of a single color QD, and the fluorescence intensity was quantified from individual beads using wavelength resolved fluorescence spectroscopy. Using a single color bead, the fluorescence intensity scales linearly with the number of QDs/bead from loadings of 640 to $\sim$ 50,000 QDs/bead, with 10 distinct intensity levels at 3 standard deviations. In a model system, three DNA labelled beads with unique QD barcodes were used to detect complementary DNA sequences in a multiplex detection assay. The fluorescence intensities from each bead and the fluorophore labelled target were measured using single bead spectroscopy. In a more recent report, 5 SNPs were simultaneously genotyped from a single solution using QD encoded microbeads in conjunction with flow cytometry [52].

Alivisatos and coworkers have demonstrated that direct QD labelling for conventional microarray based applications is feasible [24]. Four separate gold substrates were derivatized with four unique DNA sequences for study, along with four spectrally unique QDs, each derivatized with the complement of one of the surface bound targets. In these studies, an argon ion laser was used for excitation, and the fluorescence emission was captured with a CCD camera through a 60× objective. Sequence specific hybridization of each color QD was demonstrated by exposing each substrate to a mixture of the four QDs, which predominantly resulted in hybridization of only the perfectly complementary QD. The surface density required for detectable signal was not reported, but 10–100 nM concentrations of the QDs are needed to produce detectable signal. The low sensitivity probably stems from poor functionalization or unoptimized assays given that the high quantum yields of QDs should yield signal intensities that are at least comparable to organic dye labels. Therefore, further work in labelling will likely significantly improve detection capabilities of QDs for microarray applications. An additional focus area is the preparation of QDs made of more environmentally benign materials than CdSe which is toxic. The benefits of more robust and reliable multicolor detection with simplified instrumentation are attractive if this can be achieved. Quantum Dot Corporation and others are currently marketing semiconductor quantum dot probes for a variety of biological labelling applications.

8.4 Phosphor Reporters

Autofluorescence background on microarray substrates negatively impacts the sensitivity of fluorophore labelling. A novel approach devised to eliminate autofluorescence utilizes up-converting phosphor labels that absorb two photons of lower frequency light in the infrared region and emit a single photon at a higher frequency in the visible region [22,53–55]. Up-converting phosphor (UP) labels typically comprise submicron-sized yttriumYoxysulfide particles (0.2–0.4 µm diameter) that are doped with lanthanide ions such as Ytterbium and Erbium for excitation and emission [55]. Phosphorescence from the lanthanide ions persists for $> 10^{-8}$ seconds after excitation while organic fluorophores emit light for $< 10^{-9}$ seconds after illumination; thus, phosphorescence is distinguishable from autofluorescence using time resolved fluorescence spectroscopy [56,57]. Up-converting phosphor labels with different emission colors are generated by using the same absorber ion with different lanthanide ions for emission. For example, the Ytterbium/Erbium excitation/emission pair emits green light while the Ytterbium/Thulium pair emits blue light. UP materials also are characterized by narrow emission bandwidths (25–50 nm), which has enabled the development of over six spectrally unique emission colors [22]. Additionally, infrared excitation is advantageous for microarrays since other assay components (e.g. substrates) do not absorb infrared light resulting in lower overall background. A detection limit of $\sim$ 12 UP particles in a 30–40 mm^2 well was achieved using infrared laser excitation and detection with a photomultiplier tube [22]. Therefore, the theoretical detection limit of a few UP labels per microarray spot is orders of magnitude better than molecular Cy3 fluorophores with detection limits of $\sim$ 1–5 fluors/μm^2 (Table 8.1).

In an actual microarray labelling experiment, the specific and non-specific binding properties of the UP particles to the array surface and kinetics of binding play a role in determining assay sensitivity. In a recent study conducted by Tanke and coworkers, arrays containing a serial dilution of human elongation factor (HEF) probes ($\sim$ 1000 base–pair) were hybridized to a biotin labelled (HEF) target, followed by staining with Cy5-labelled avidin, and subsequent labelling with an UP particle for comparison [22] (Fig. 8.4). The limit of detection (LOD) for Cy5 was $\sim$ 4 ng/μL using a laser-based microarray scanner, while the LOD of the UP labels was $\sim$ 1 ng/μL recorded with a fluorescence microscope modified for infrared excitation. Therefore, the assay sensitivity was increased 4–fold when compared to conventional fluorescence detection. The phosphor luminescence was found to correlate linearly with probe and target concentration over a concentration range of $\sim$ 3 orders of magnitude, which was comparable to Cy5 labelling.

Phosphor technology offers a greater number of 'colors' for labelling (currently 6) than fluorescence with single source infrared excitation [22]. For microarrays, this translates to a greater potential for multiplexing with simplified and lower cost detection instrumentation. Despite the sensitive UP probe detection capabilities, the assay sensitivity is currently in the same range as molecular fluorophore probes. This is likely due to the large particle size, which results in poor diffusion, steric hindrance, and large van der Waals forces between surface and substrate. The use of smaller particles and improved conjugation methods offers a potential route to higher sensitivity detection. More recently, glass microbarcodes with lanthanide ion emitters were reported for multiplexed DNA detection assays [58]. The combination of single source excitation and multiple colors is also extremely attractive for barcoding applications.

8.5 Electrochemical Detection

Electrochemical and electronic detection have received significant interest as a viable means for microarray labelling since inexpensive and robust instrumentation may be used for detection [23,59,60]. In addition, such sensors can function in complex sample environments such as blood where optical sensing is difficult. Electrochemical sensors are used in point–of–care diagnostic applications such as glucose testing, where inexpensive electronic circuitry rapidly quantifies glucose levels in blood samples [61]. The biggest drawback of electrochemical detection is low sensitivity. For molecules such as glucose that are present at micromolar concentrations in vivo, sensitivity is not an issue [61]. However, genetic targets such as nucleic acids are typically present at significantly lower concentrations presenting a significant challenge for this technology. Electrochemical detection platforms that utilized redox active probes,

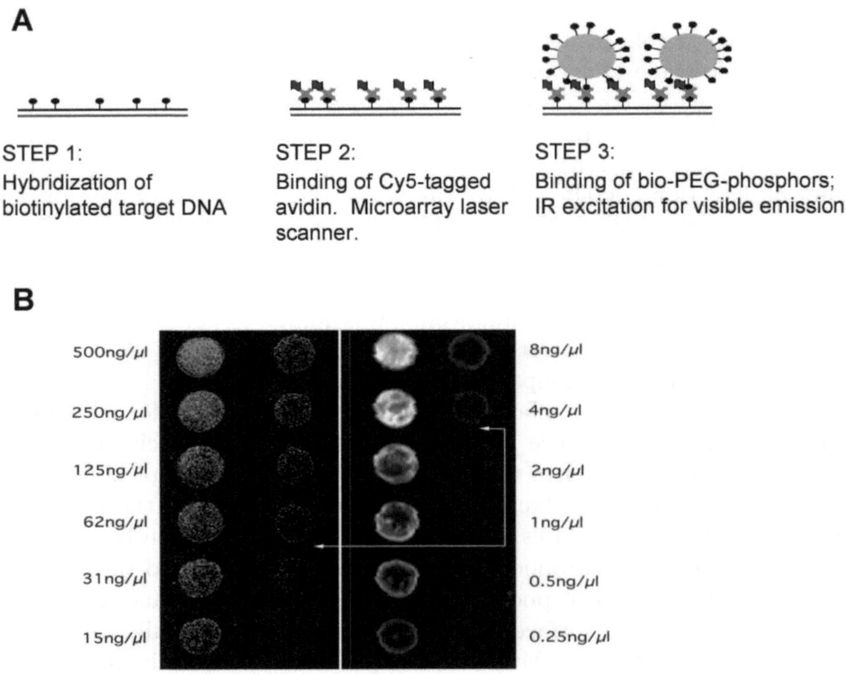

Fig. 8.4. **(a)** Schematic outline of the experiment. **(b)** Model low complexity microarray hybridization with biotin HEF–DNA detected with avidin–Cy5 and laser scanning (right panel) and subsequent detection with Bio–PEG UPT (left panel) (Reprinted with permissions from [22]. Copyright 2001 Macmillian Magazines Limited)

redox active intercalators, or the inherent redox active properties of DNA have been developed for nucleic acid analysis on microarrays.

The first platform, which has been under development at Motorola's Clinical Microsensors division, utilizes ferrocene labelled nucleic acid probes in a low density array format that is geared towards clinical diagnostic applications [23, 62]. Electrochemical detection of ferrocene labelled probes, which contain Fe^{II}/Fe^{III} redox centers, is achieved in a sandwich hybridization assay format on gold electrodes (Fig. 8.5). Disposable low density arrays of gold electrodes ($\sim$ 250–500 μm diameter) are fabricated via conventional printed circuit board technology, and individual electrodes are derivatized with a monolayer that contains specific thiol modified oligonucleotide sequences for target capture. When target is bound to capture strands on the electrode surface, a reporter nucleic acid probe containing multiple ferrocene moieties hybridizes to the target/capture complex. When a given potential is applied to the electrode, electron transfer occurs between the ferrocene labels and the gold electrode. The current generated by the ferrocene labels is used to quantify the amount of nucleic acid present. It should be noted that only ferrocene labels hybridized to the surface generate signal so that hybridization and detection may be performed in a single solution without the removal of excess probes.

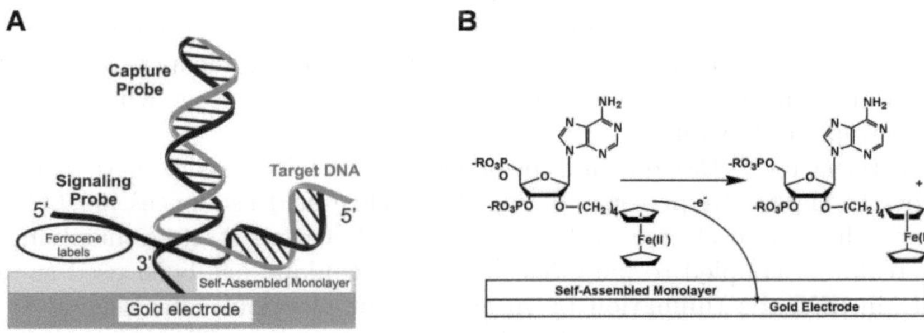

Fig. 8.5. **(a)** Schematic illustration of electrochemical detection of nucleic acids on gold substrates. **(b)** Scheme depicting electrochemical oxidation of ferrocene groups at an electrode surface

Using this approach, 50 nM solutions of single stranded nucleic acid target are detectable in a sandwich hybridization assay, but double stranded targets of similar concentration yield almost no signal [23]. As a result, asymmetric PCR is used to generate high concentrations of single stranded nucleic acid for detection. Genotyping of the C282Y mutation of the *Hfe* gene was achieved using asymmetric PCR by comparing signal intensities from wild type and mutant capture probes. Asymmetric RT–PCR also demonstrated that gene expression monitoring of a small number of genes (5 in the reported example) is feasible with this approach, although the dynamic range was not reported.

The most important attributes of this detection approach are the simple and in-expensive electronic detection system and electrode chips, and the integration of hybridization and detection into a single step. Additionally, multiple reporter groups have been developed by tuning the redox potential of the ferrocene moiety. A disadvantage of this approach is the relatively low sensitivity compared to fluorescence detection and the inability to detect double stranded DNA targets (Table 8.1).

The second electrochemical detection platform utilizes the electron transfer properties of nucleic acids for detection, eliminating the need for nucleic acid probes labelled with redox active groups [59, 60]. Thorp and coworkers developed a label free electrochemical detection strategy that utilizes mediated electron transfer from guanine in target nucleic acids bound to an electrode [60]. The amount of peak current correlates with the number of guanine residues in the target. Detection limits of ~ 26 molecules/μm^2 were reported for a 1497 bp PCR amplicon deposited directly onto an Indium Tin Oxide (ITO) electrode [63]. In a proof–of–concept study for gene expression, RT–PCR amplified RAK gene products from six breast tissue samples were quantified via fluorescence and then measured via electrochemical detection [64]. Overexpression in the breast cancer samples was correctly identified by measuring the peak current associated with each PCR amplicon on individual ITO electrodes. ITO microelectrode arrays with gene specific capture sequences have been applied to low density gene expression applications [65]. The benefits are label-free detection in addition to the inexpensive detection hardware, but low detection sensitivity is still a disadvantage when compared to fluorophore labelling.

Alternatively, Barton and coworkers have utilized redox active intercalators to signal the presence of specific nucleic acid sequences [59, 66]. The approach uses an electrocatalytic signal amplification strategy involving the intercalators coupled to $[Fe(CN)_6]^{3-}$. This technique has been used successfully in SNP discrimination by measuring the electrochemical signal at the electrodes containing matched and mismatched probes. The electrochemical response from the mismatched hybrids is diminished owing to the disrupted electron transfer between the electrode and the intercalator, allowing the identification of the perfectly matched hybrid. In addition, electrocatalytic signal amplification strategies offer the potential to improve detection sensitivity by generating more electrochemical signal per target. GeneOhm Sciences, Inc. is currently developing this technology for SNP analysis.

8.6 Metal Nanoparticle Labels and Metal Thin Films for Microarrays

8.6.1 Introduction to Metal Nanoparticle Based Detection Methodologies

Gold nanoparticles have been utilized as labels for cellular imaging [67] as well as detection of proteins [68, 69] and nucleic acids [70], but it was not until recently that nanoparticle labelling was applied to biomolecule detection on microarrays [71]. Recent interest in metal nanoparticles as labels has been fuelled by the development of reliable preparation methods [72] and robust functionalization techniques with nucleic acids or proteins [69, 73]. Metal nanoparticles exhibit unique optical, catalytic, and electronic properties owing to their size, and therefore, can be used in a variety of detection schemes based on different modes of signal transduction. An explosion of research in this area has led to a number of different approaches that may be utilized for detecting such particles in both optical and electrical detection formats. Reported detection formats include colorimetric changes [73, 74], silver enhanced imaging [28, 71], surface plasmon resonance imaging [30, 75], light scatter [76, 77], surface enhanced Raman spectroscopy [31], photo–thermal imaging [78], electrical detection [32], and scanning electrochemical microscopy [79].

The preparation and properties of colloidal gold particles were studied in the early 1800s by Faraday [80]. Reproducible methods have now been developed for preparing highly monodisperse 1–100 nm diameter gold particles which are available through commercial sources. Although gold is the easiest metal to prepare in nanoparticle form, the synthesis of other metal particles such as silver have now been realized. The method used for functionalizing the nanoparticles with biomolecules for detection is critical, as it dictates the binding properties of the resulting label, as well as the application of the labels for detection. Both direct and indirect nanoparticle labelling strategies have been developed for nucleic acid detection on microarrays (Fig. 8.6). For indirect nucleic acid labelling, metal nanoparticles are functionalized with antibodies such as antibiotin or streptavidin which passively adsorb to the surface (Fig. 8.6a). For detection, haptens are incorporated into the nucleic acid target and bound to the microarray, followed by labelling with the complementary antibody-labelled gold nanoparticle in a separate step. The advantage of this method is that a single particle may be used for detection of all nucleic acid sequences. Disadvantages include compromised sensitivity due to passive adsorption, and the requirement of incorporating a label into the target sequence of interest. Direct nanoparticle probe labelling was pioneered by Mirkin and coworkers [73]. In this approach, oligonucleotides are covalently anchored to the nanoparticle surface using thiol linkers (Fig. 8.6b). For detection, the DNA-modified gold nanoparticle probes are hybridized to nucleic acid targets in a sandwich assay format. The probes exhibit high stability toward thermal fluctuations as well as elevated concentrations of salt [73] and are typically used directly in the assay for detection. Multiple oligonucleotides attached to each nanoparticle confer unique properties to the probes when compared to molecular fluorophores (Fig. 8.7) [81]. These include an elevated melting temperature (Tm) and an unusually sharp melting transition, which provides for enhanced sequence discrimination and enables higher stringency hybridizations. In addition, this approach does not require the incorporation of labels into the target, which simplifies direct detection of nucleic acid sequences. A potential disadvantage is that multiple probes may be required for analysis of multiple sequences, although universal nucleic acid labelling strategies are well established and feasible. Experimental data for each of these labelling approaches will be described in the ensuing sections on detection methodologies.

8.6.2 Scatter-Based Detection of Metal Nanoparticle Probes on Microarrays

Gold and silver nanoparticles are characterized by a plasmon resonance absorption band that gives rise to intensely colored solutions. The absorption band is due to electrons confined at the particle surface that collectively oscillate at a specific frequency, which is commonly referred to as the surface plasmon resonance frequency. According to Mie theory, the plasmon frequency is defined by particle composition, size, shape, and the dielectric medium, which determines the maximal absorption wavelengths, and therefore, the resulting color of the particle solutions. For example, the plasmon band of a 20 nm Ag

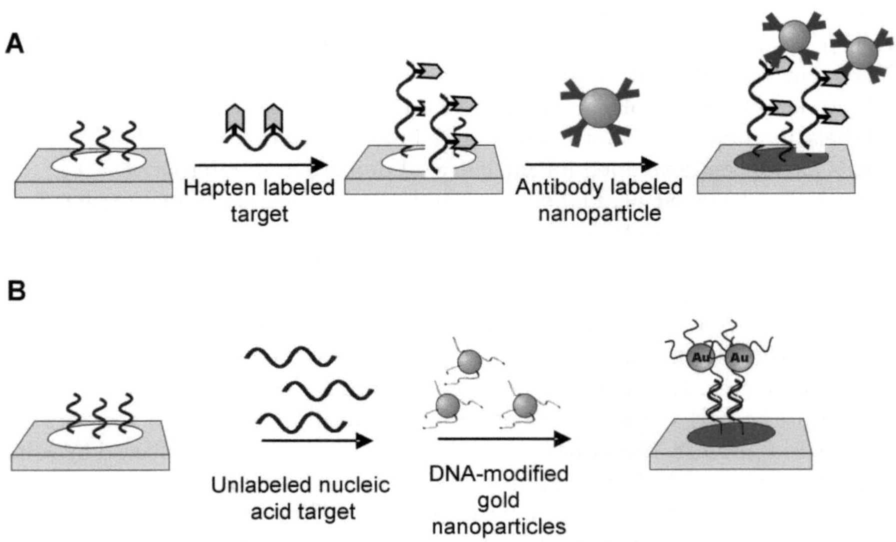

Fig. 8.6. Nucleic acid detection on microarrays using metal nanoparticles. **(a)** Antibody modified gold nanoparticle labels. **(b)** DNA-modified gold nanoparticle probes

particle is centered at 395 nm resulting in a yellow solution, while a 20 nm Au particle absorbs at 520 nm resulting in a red solution [82]. The plasmon bands are typically broad (50–100 nm bandwidths for 50 nm diameter particles), and absorption extends from the plasmon band to higher energy into the UV. The extinction coefficient of the plasmon bands of gold and silver nanoparticles scales with particle volume and is extremely large at 10^8–10^{11} M^{-1} cm^{-1} for 15–100 nm diameter particles [83]. As a result, the particles can be visualized by absorbance in solution at nanomolar to picomolar concentrations. Silver amplification techniques have been developed to enhance visualization of gold particles at lower concentrations for immunochemistry applications [84].

A more sensitive method for detecting larger metal nanoparticles (> 30 nm diameter) is to monitor scattering. When illuminated with white light, metal nanoparticles in the 20–120 nm diameter size range scatter light of a specific color at the surface plasmon resonance frequency (Fig. 8.8) [26,83,85]. This has been referred to as resonance light scattering (RLS) or plasmon resonance scattering by various groups. The light scattering scales with particle volume as observed for absorbance, but the scattered light is detectable at much lower concentrations than the absorbed light. For example, light scattered by a solution of 80 nm diameter gold particles is detectable down to

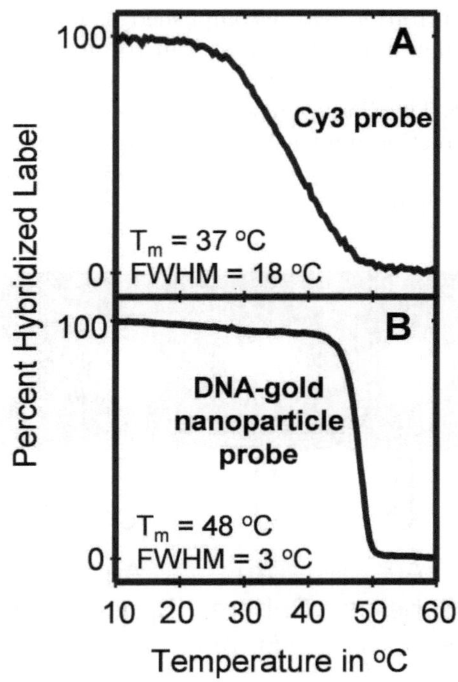

Fig. 8.7. (a) Melting analysis of Cy3 labelled probes in a sandwich hybridization assay. **(b)** Melting analysis of DNA-modified gold nanoparticle probes in same assay

5 fM concentration [86] which is roughly 1000–fold lower concentration than detectable by absorbance. In a direct comparison with fluorescence, a single 60 nm diameter gold particle emitted roughly the same amount of light as 5×10^5 fluorescein molecules [86]. The enhanced detection sensitivity of this approach is attractive for microarray labelling applications. Additionally, the metal particles produce a stable signal and do not photobleach or quench upon prolonged illumination as observed for fluorophore labels.

DNA-Modified Gold Nanoparticle Probes

Mirkin and co-workers were the first to report the use of nanoparticle labels for microarrays [71]. The initial method employed 15 nm diameter gold particles labeled with oligonucleotides in a sandwich assay format (see Fig. 8.6b). These probes are visible at high surface coverages on glass and provide sufficient sensitivity to allow detection of targets in the nanomolar concentration range [87]. A simple and elegant method was devised to improve their optical detection by using these gold nanoparticle probes to promote the reduction of Ag (I) to silver metal (Fig. 8.9) [71]. Briefly, after the sandwich hybridization assay, catalytic reduction of silver onto the gold nanoparticle surface was promoted by the reducing agent hydroquinone, which intensified the visual signal. Signal quantitation was accomplished using grayscale intensity from an ordinary flatbed scanner. A greater than 10^5–fold sensitivity improvement was achieved by silver enhancement with a reported detection limit of 50 fM nucleic acid target. This limit of detection was ~ 2 orders of magnitude better than the 5 pM detection limit achieved under the same conditions with Cy3 labelling and fluorescence-based confocal scanning. In addition, the dynamic range of the assay spanned ~ 2 orders of magnitude with a single silver de-

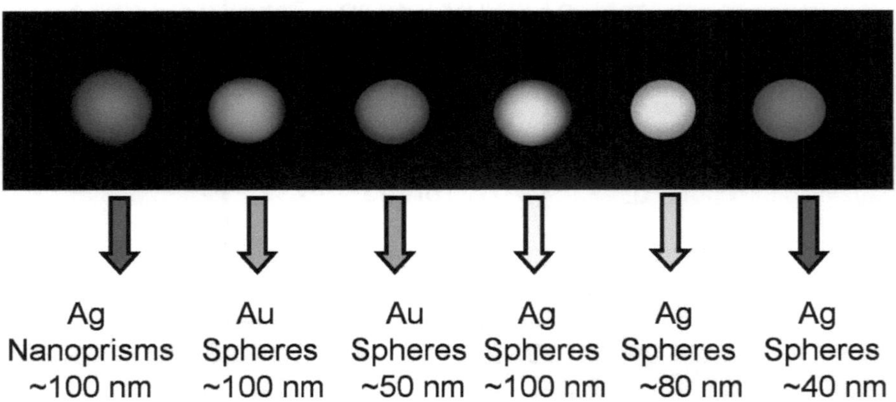

Fig. 8.8. Scatter of metal nanoparticle, based on particle size, shape and composition of matter. All particles were aqueous suspensions. (Reprinted with permission from [85]. Copyright 2001 American Association for the Advancement of Science)

velopment, but could be extended to 6 orders of magnitude by using three consecutive silver development steps.

The selectivity of the oligonucleotide functionalized nanoparticle probes was compared initially to a corresponding fluorescence-based system in a sandwich assay [71]. Detection of nanoparticle probes by flatbed scanner was approximately four fold better than detection of Cy3 fluorocescence by confocal scanning with regard to discriminating the A:T match from the difficult to resolve G:T wobble pair, with signal ratios of 10:1 and 2.6:1 respectively. In addition, the nanoparticle probe system also demonstrated a much sharper melting transition and higher melting temperature (see Fig. 8.7), which may be attributed to the multiple equivalent sites made available to the target by the nanoparticle probe [81]. More recently, it has been demonstrated that the selectivity factor in the case of DNA-modified gold nanoparticle probes in conjunction with electrical-based detection could be increased to 500,000:1 with a salt-based stringency wash at room temperature.

Nanosphere Inc. is currently developing the silver-amplified gold probe technology originally reported by Mirkin and coworkers [71] for diagnostic applications. A major distinction of Nanosphere's work is the development of a scattering-based detection system for the silver amplified gold nanoparticles [25]. The detection system illuminates the glass slide with a planar waveguide and captures the scatter of the silver amplified gold particles with a CMOS detector. A single image of the entire slide is recorded by this optical configuration, which eliminates the need for moving parts and image stitching. The detection limit of the silver amplified nanoparticle probes was determined by spotting a serial dilution of the respective probes on standard glass slides. After silver amplification of the nanoparticle probes a scatter signal from < 0.0025 probes/μm^2 could be detected with 95% confidence above background using the Nanosphere detection system. This is roughly 2–3 orders of magnitude better than detection of Cy3 molecules, Table 8.1. In addition, the dynamic range recorded with the new detection system covers greater than 3 orders of magnitude, which is an order of magnitude improvement over the flatbed scanner discussed previously.

An assay sensitivity of 100 aM (3000 total copies) was demonstrated on a Factor V Leiden gene SNP array (1691 G $\rightarrow$ A), Fig. 8.10. This assay sensitivity is greater than 3 orders of magnitude better than other nanoparticle-based

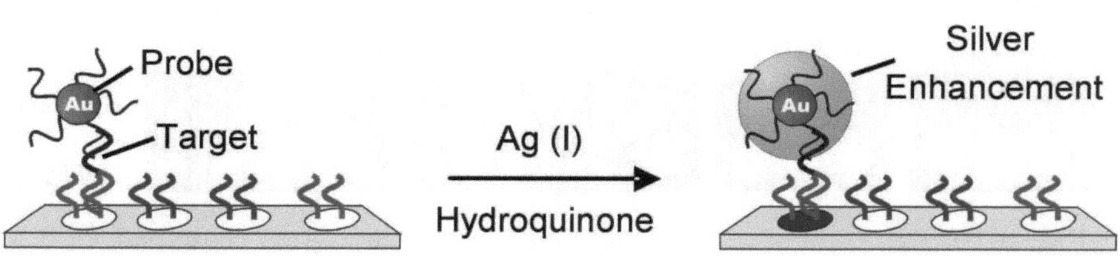

Fig. 8.9. Illustration of silver amplification of gold nanoparticle probes which leads to signal enhancement

detection systems that utilize larger metal particles labelled with antibodies (Table 8.1). In addition, the high melting temperature and sharp melting transitions of the DNA-modified nanoparticle probes have enabled single base mismatch discrimination, even at 100 aM detection levels. Direct detection in human genomic DNA also requires high specificity since the complexity is extremely high at 1×10^9 bp [88]. DNA-modified nanoparticle probes permit operation at high stringencies owing to the sharp melting transitions. For initial testing, the MTHFR gene sequence was targeted in a sandwich hybridization format using a single step hybridization reaction (Fig. 8.11). Remarkably, the MTHFR gene was directly detected from a 20 μg sample of human genomic DNA (6×10^6 copies) using the DNA-modified gold probes in conjunction with silver amplification [25]. Assays for discriminating SNPs in unamplified genomic DNA are currently under development. In addition, mecA gene detection in *Staphylococcus* genomic DNA samples has been used to determine methicillin resistance status with DNA-modified gold nanoparticle probes [25].

This work demonstrated the potential of DNA-modified gold probes in conjunction with silver amplification for microarray-based applications. In addition to a detection sensitivity of 10^3 copies, the DNA-modified gold probes provide enhanced specificity, which has enabled improved SNP discrimination and sequence identification in complex genomic DNA samples. The combination of increased specificity and sensitivity is unique to the Nanosphere labelling and detection methodology and is leading to the development of nucleic acid diagnostic assays for infectious diseases and SNPs that do not require target amplification or complexity reduction. Additionally, the low complexity and cost of the detection instrumentation is well suited for diagnostic applications and also separates Nanosphere's detection platform from other nanoparticle labelling strategies that utilize more complex instrumentation. The main limitation of this system is that the silver amplification methodology yields a single color format when scatter based detection is utilized. For applications requiring more than a single color, larger DNA-modified gold nanoparticle probes (50–100 nm diameter) have been developed in conjunction with scatter-based detection for two color labelling [76]. In this detection

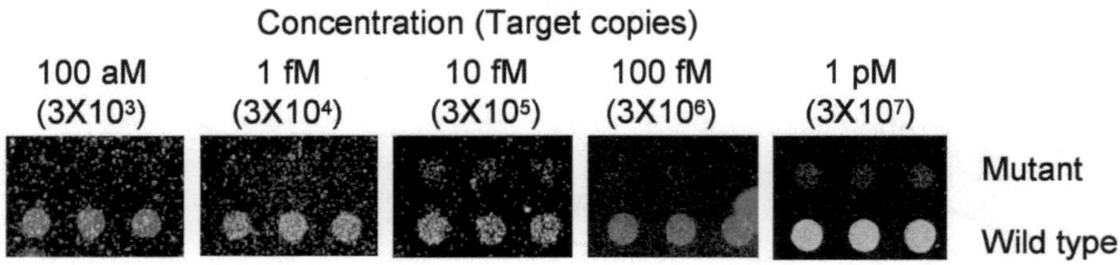

Fig. 8.10. Detection of a single nucleotide mismatch in a 250 base pair PCR amplicon of the factor VLeiden gene. Panel shows a serial dilution of amplicon in an overnight hybridization reaction. The assay detection limit is 100 aM

methodology, two different probe colors are achieved by controlling particle size, shape, and chemical composition, which determines the color of scattered light in the absence of silver amplification [76, 85].

Antibody-Functionalized Metal Nanoparticles

Yguerabide and coworkers and Genicon Sciences Corporation first reported the use of antibody labelled metal nanoparticles with resonant light scattering (RLS) detection for microarray applications [86]. Light scattered from 60 nm diameter gold particles deposited onto glass microarray surfaces was detectable at 0.005 particles/μm^2 using white light illumination and CCD based imaging (Table 8.1) [86]. This detection sensitivity is 2–3 orders of magnitude better than the corresponding 1–5 Cy 3 molecules/μm^2 using a standard fluorescence microarray scanner.

Bao et al. have reported the use of 80 nm diameter metal nanoparticles conjugated with anitibiotin (RLS labels) for gene expression [77]. A human gene cDNA array consisting of $\sim$ 2000 genes was employed to test the sensitivity and specificity of the RLS labels in comparison with Cy3. cDNA probes prepared from human poly(A) RNA were co-labelled with biotin and Cy3 and hybridized to the human gene array. The Cy3 fluorescence signal for each expressed gene was quantified using a confocal fluorescence scanner, followed by incubation with the RLS labels and detection using a CCD-based imaging system. Both labelling technologies detected nearly 100% of the genes when the cDNA arrays were challenged with 500 ng of target, but the RLS labels outperformed the Cy3 at lower target dilutions, allowing detection of 10–300 times as many genes when challenged with 1–5 ng of target. By comparison, approximately 20 times the amount of target was required for Cy3 labelling to detect an equivalent number of genes. Comparable reproducibility was observed when 100 ng of the co-labelled target was hybridized to two separate slides using the procedure described above, and the net hybridization signals for each label were evaluated. The data from this experiment also indicated comparable dynamic range for the two labels at $>$ 2 orders of magnitude. A strong correlation in differential gene expression levels was observed for leukemia samples using single color fluorescence or RLS labelling verifying

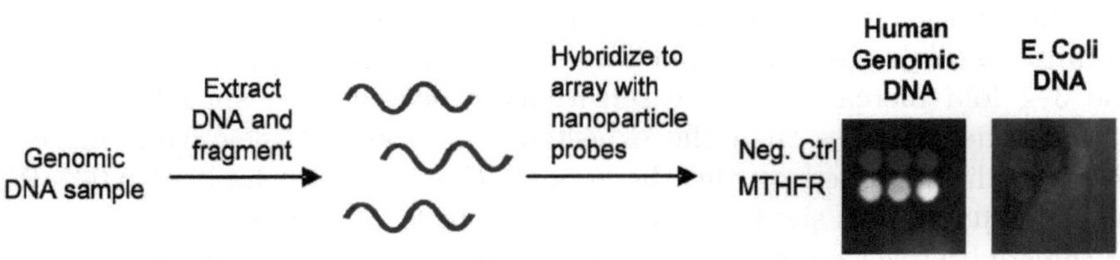

Fig. 8.11. Detection of a specific gene from a human genomic DNA sample using an oligonuceotide array and DNA-modified gold nanoparticle probes with silver amplification

the signal specificity. A more recent report by Genicon Sciences assessed the dynamic range and limit of detection in gene expression studies using known amounts of specific cDNA transcripts that were spiked into complex cDNA samples [107]. The reported lower limit of detection (LLOD) was 8.2×10^6 copies ($\sim$ 170 fM, 80 uL) with a 3.3 log dynamic range. By comparison, a Cy3 label had an LLOD of 2.8×10^7 copies ($\sim$ 580 fM, 80 uL) with a 3.2 log dynamic range. It should be noted that although a single color was reported in this study, two color nanoparticle labelling is now available using silver and gold particles [27].

In an approach similar to Genicon, Schultz and coworkers have utilized large silver nanoparticles referred to as plasmon resonant particles (PRPs) and resonant scatter based detection for microarray labelling [89]. The 40–100 nm diameter silver particles are prepared by solution-based reduction of silver onto small gold particle seeds ($\sim$ 5 nm diameter). The PRPs scatter light based on the position of the surface plasmon band as observed for gold particles [90]. In this study, 55 ± 17 nm diameter particles which exhibit maximum scatter at $\sim$ 430 nm were utilized. The particles were derivatized with mouse anti-biotin antibodies for detection of biotin labelled targets, Fig. 8.6a. For detection, the slide is illuminated in dark field using a halogen lamp, and a high resolution image of each microarray spot is captured using a CCD camera through a 10× or 100× dark-field/bright-field objective lens on an optical microscope [89]. Individual plasmon resonant particles provide a scattering signal that is distinguishable from other sources of scatter, thereby enabling particle counting to be used for measuring the amount of total signal from each microarray spot. This unique detection methodology was applied on a small model array containing positive and negative control capture sequences. A biotin labelled 30–mer target was hybridized to the array overnight followed by overnight incubation with the antibiotin labelled silver particles. A detection limit of 1×10^6 target copies (830 fM, 2 uL) was achieved, which was $\sim$ 10–fold better than obtained by measuring average scatter intensity (1×10^7 copies) using this illuminaton/detection technique. The improvement in sensitivity is attributed to the elimination of background pixels that decrease the average scattering signal on microarray spots that are not completely coated with particles. This labelling technology is under development at Seashell Corporation for microarray applications [89].

The above cited literature clearly demonstrates that RLS labels hold promise as high sensitivity labelling systems for gene expression. However, the 3–4 fold increase in assay sensitivity observed with spiked transcripts was significantly less than the expected 2–3 orders of magnitude improvement predicted based on the theoretical RLS detection limit of 0.005 RLS particles/μm^2. This significant disparity is likely attributed to the large size of the gold particles required, which presents steric and kinetic limits to the number of particles bound to each cDNA probe. In addition, the passive adsorption of the antibiotin antibodies to the nanoparticle surface may be prone to desorption [77]. Particle counting may be used to increase sensitivity in such detection systems, but this strategy requires the use of slower and more complex instrumentation [89]. While the use of smaller metal nanoparticles for labelling can enhance hybridization kinetics and relieve steric issues, this approach will result in lower scattering intensity. Alexandre et al. in collaboration with Advanced Array Technology (AAT) have employed streptavidin coated $\sim$ 10 nm gold nanoparticles in conjunction with silver amplification for nucleic acid analysis on microarrays [28,91]. In a direct comparison, this approach yielded detection limits (0.1 fmol, $\sim 6 \times 10^7$ copies) equivalent to a Cy3 labelled target. These detection limits in comparison to the DNA-modified nanoparticle probes (Table 8.1) indicate that in addition to the smaller size of the nanoparticles, the functionalization strategy, antibody or DNA, must play an important role in determining assay sensitivity.

In summary, scatter-based nanoparticle detection enables single particle detection capabilities. Therefore, the major determinant of assay sensitivity in these detection strategies is background, target binding affinity, particle binding kinetics and sterics. A detection limit of $\sim$ 100 aM (3000 target copies) has been achieved using Nanosphere's oligonucleotide-modified gold nanoparticle (15 nm diameter)–silver amplification technology in conjunction with simple optical detection instrumentation for nucleic acid detection. Antibody-labelled gold or silver particles ($>$ 60 nm diameter) without silver amplification have achieved fM to pM detection limits ($\sim 10^6$–10^7 target copies) in nucleic acid detection assays. The higher sensitivity achieved with the Nanosphere strategy is likely a combination of the small particle size which increases binding kinetics and limits sterics, the use of covalent DNA particle modification which enhances target binding affinity, and the use of silver amplification which results in a higher signal per nanoparticle probe due to increased particle size. More importantly, the assay sensitivity achieved with the Nanosphere technology is roughly 3 orders magnitude more sensitive than a comparable assay with fluorescently-labelled dyes (Table 8.1) which has enabled direct detection of genomic DNA samples.

8.6.3 Surface Plasmon Resonance Detection

Surface plasmon resonance (SPR) spectroscopy is a detection methodology that enables measurement of changes in thickness and/or index of refraction of organic or biomolecular thin films at noble metal surfaces (Au, Ag, or Cu) [75]. This technology has been reviewed extensively [75,92,93], therefore discussion will be limited to recent advancements in using SPR with microarrays. Surface plasmons are generated by conduction electrons at the metal surface that collectively oscillate at a specific frequency. The surface plasmon resonance frequency is sensitive to the metal/dielectric medium interface such that the adsorption of biomolecules at the surface interface results in changes in the SPR which can be measured by scanning angle SPR, SPR wavelength shift, or SPR imaging [75]. The scanning angle SPR technique is the most commonly employed method, and instruments are commercially available through Biacore and others. This method utilizes a single wavelength such as a HeNe laser for excitation, and measures the percent reflectance change at the surface of a gold thin film (~ 50 nm thick) as a function of incident angle. Theoretical Fresnel calculations are used to model changes in reflectivity at gold surfaces [75]. Figure 8.12 shows theoretical SPR changes for the adsorption of a 5 nm film of refractive index 1.45 onto a gold thin film. Recent reports have demonstrated that SPR imaging is a powerful technique for monitoring biomolecule interactions on microarrays [94, 95]. SPR has been used for in situ, label-free, optical detection of antibody–antigen binding, DNA hybridization, and protein DNA interactions [96, 97]. Sensitivity limits achieved by using this technology are in the nanomolar range for DNA detection. Recent improvements in instrumentation and signal amplification strategies have significantly improved the limits of detection. Zhou and coworkers have developed higher resolution SPR spectrometers to enhance detection sensitivity [29]. With commercially available instrumentation, SPR angle shifts are measurable to ~ 0.001 degrees. The high resolution SPR spectrometer measures angle shifts down to 10^{-4}–10^{-5} degrees. In a model DNA assay, a 30–mer oligonucleotide capture probe was immobilized on a gold thin film, and the hybridization of a 47 base single stranded target oligonucleotide was monitored in real time using the high resolution SPR spectrometer. A detection limit of 54 fM was achieved in ~ 5 minutes utilizing this detection methodology.

Nanoparticle amplified surface plasmon resonance (SPR) utilizes gold nanoparticle labels to enhance detection sensitivity, with a > 1000 fold improvement in nucleic acid detection [30]. The sensitivity enhancement is due to an enhanced shift in SPR reflectivity as a combined result of greatly increased surface mass, high dielectric constant of the gold particles, and electromagnetic coupling between the gold nanoparticles and the gold film. To measure detection sensitivity as a function of particle size, particles were spotted onto a surface as a dilution series, and the corresponding SPR signal was measured [98]. Using 12 nm gold particles, surface densities of 20 particles/μm^2 were detectable with a signal to noise ratio of 10, which could be improved 40–fold to 0.5 particles/μm^2 by using larger 45 nm gold particles. This detection limit is roughly equivalent to Cy3 detection (Table 8.1). In a model DNA array, a gold thin film (48 nm thick) and DNA modified gold probes (12 nm diameter) were utilized to detect a 24 base oligonuceotide target in a sandwich hybridization assay with a reported detection limit of ~ 10 pM (≤ 8 oligonucleotides/μm^2) [30]. Although currently not as sensitive as Cy3 labelling (Table 8.1), the sensitivity of nanoparticle-amplified SPR should improve significantly through the use of larger probes which offer greater detectability, or in combination with the aforementioned high resolution SPR spectrometer.

This work provides a sound basis for future SPR-based microarray labelling applications. The potential for rapid, label-free biomolecule detection is intriguing. The high lateral spatial resolution (~ 10 μm) is conducive to arrays and miniaturization. In addition, recent advancements in sensitivity

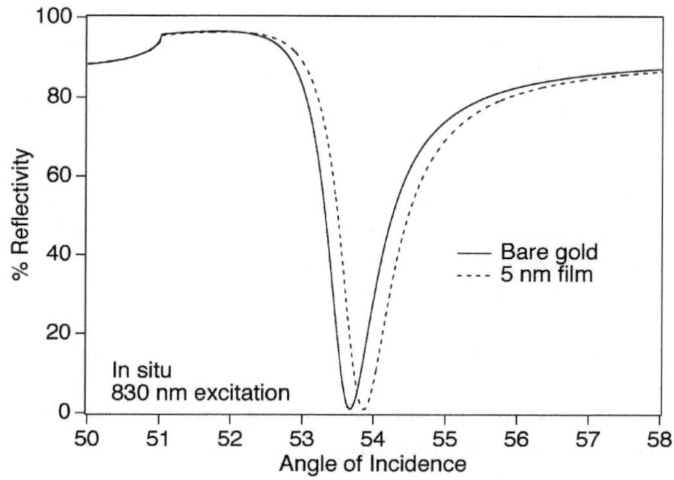

Fig. 8.12. Calculated Surface plasmon resonance (SPR) curves at 830-nm excitation for a (*solid line*) three layer system composed of an SF–10 glass prism (n= 1.711), a 45.0 nm-thick Au film (n= 0.165 + 5.205i), and an infinite layer of water (n= 1.327) and (*dashed line*) a four layer system composed of an SF–10 glass prism (n= 1.711), a 45.0 nm-thick Au film (n= 0.165 + 5.205i), a 5.0 nm-biopolymer film (n= 1.45), and an infinite layer of water (n= 1.327). (With permission from [75] and the Annual Review of Physical Chemistry, Volume 51, Copyright 2000, by annual reviews, www.annualreviews.org)

using nanoparticle amplification or higher resolution SPR spectrometers significantly enhance the capabilities of this detection methodology for DNA microarrays. Furthermore, recent reports have demonstrated that real time SPR measurements can be performed on ensembles or even single metal nanoparticles for biomolecule detection applications offering prospects for even further miniaturization and increased sensitivity [95, 99].

8.6.4 Surface-Enhanced Raman Scattering

Numerous reports have demonstrated that metal surfaces with nanometer scale roughness may be used to amplify Raman scattering signals of adsorbed molecules [100–102]. This technique is commonly referred to as surface enhanced Raman spectroscopy (SERS). Raman scattering enhancement factors of up to 10^8 have been reported for molecules adsorbed onto roughened metal surfaces [103]. Additionally, Raman scattering signals from single rhodamine 6G dye molecules adsorbed onto silver nanoparticles have been detected demonstrating that Raman enhancement factors on the order of 10^{14}–10^{15}, and thus detection of single molecules, is achievable [104]. Graham and coworkers first reported a SERS platform for nucleic acid sequence detection that utilized silver nanoparticles tagged with Raman active dye labelled nucleic acids [105]. Using a modification of the silver amplification methodology for microarrays outlined in Fig. 8.9, Mirkin and co-workers have recently developed a SERS-based detection system for microarray analysis. Raman dye labelled oligonucleotide probes attached to 15 nm diameter Au particles are designed to label specific nucleic acid sequences in a sandwich hybridization assay format (Fig. 8.13). At nucleic acid target concentrations less than 1 nM, the gold probes hybridized to the glass surface are spectroscopically silent since isolated spherical gold probes are not adequate SERS promoters. Catalytic reduction of silver onto the gold probe surface enhances the Raman scattering signal of the attached dye labels. For SERS detection on microarrays, spots on the glass slide are illuminated with 633 nm laser excitation, and the Raman scattering signal from each spot is measured. Using a Cy3 labelled oligonucleotide as a Raman tag on the nanoparticle probe, a detection limit of 20 fM was achieved for the hybridization of a 30 base oligonucleotide target on an arrayed glass slide. One advantage of this approach over previously reported scatter-based detection approaches is the reduction of background signal since silver particles and slide defects do not significantly contribute to the Raman scattering signal.

The multiple vibrational signatures for each dye create a spectroscopic finger–print for the DNA sequence present. The vibrational signatures are characterized by narrow emission bandwidths of 15–30 cm^{-1} which has enabled the development of multiple dyes with different spectroscopic signatures for barcoding or multicolor detection applications. For applications such as expression profiling, a specific vibrational mode for each dye may be chosen for multicolor detection. In initial studies performed by Mirkin, DNA modified gold probes were designed to identify six different pathogens in a sandwich hybridization format. Each gold probe was encoded with a unique Raman dye for detection. All of the Raman tagged probes specifically hybridized at the appropriate array locations and were correctly identified by their unique Raman spectra. By monitoring a specific vibrational mode of two spectrally unique dyes, two color signal ratioing on a single microarray spot was demonstrated by spiking in known ratios of single base mismatched targets. The Raman scattering signal ratios of the two dyes correlated well with the input target ratio providing a proof–of–concept demonstration of two color detection.

For microarrays, this detection methodology offers the high sensitivity and high selectivity of the silver amplified DNA modified gold nanoparticles with the added benefits of multicolor detection and signal ratioing capabilities. In addition, a single excitation source may be used for a variety of Raman dyes in this SERS approach, simplifying detection instrumentation and accelerating analysis. It is also important to note that background may be minimized in this detection system since only Raman active components produce signal, eliminating background scattering signal due to surface defects or silver in scatter-based detection systems.

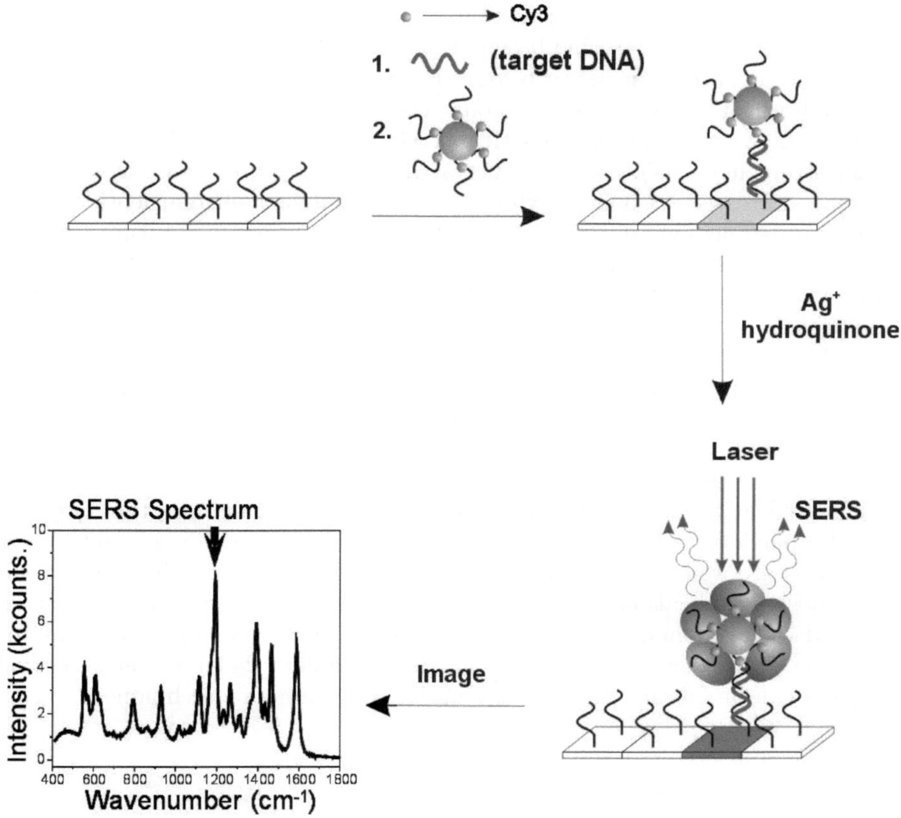

Fig. 8.13. Scheme showing Raman spectroscopic detection using nanoparticle probes with silver amplification. (Reprinted with permission from [31]. Copyright 2002 American Association for the Advancement of Science)

8.6.5 Electrical-Based Detection of Metal Nanoparticles

Conductivity measurements of metal nanoparticle aggregates [106] and silver amplified gold nanoparticles [107] have demonstrated that electrical properties of metal nanoparticles offer a viable route to biomolecule detection. Recently, Mirkin and coworkers reported the development of an oligonucleotide array-based electrical detection format that utilizes DNA-modified gold nanoparticle probes for nucleic acid detection, Fig. 8.14 [32]. Oligonucleotide probe sequences were deposited in a 20 micron gap between pairs of gold microelectrodes on glass supports, and used to capture nucleic acid targets in a sandwich hybridization with DNA-modified gold probes in the electrode gap. Silver amplification of the gold particles created a conductivity bridge between the electrodes, which results in a measurable change in conductivity.

Initial testing was performed on a model SNP array. Capture sequences containing the four possible base permutations (A, C, G, and T) at the SNP site were deposited in between four electrode pairs. An assay was performed by hybridization of a 10 nM oligonucleotide target solution to the electrode array, followed by gold nanoparticle labelling and silver development. After a salt stringency wash, the resistance at the perfectly matched oligonucleotide probe decreased to 500 Ω, while the 3 mismatched oligonucleotide probes show resistances greater than 200 MΩ. Therefore, the match:mismatch signal ratio in this detection format translates to greater than 500,000:1. An unoptimized lower limit of detection of 500 fM target was achieved via this detection approach.

This electrical detection format combines the benefits of robust and inexpensive electronic detection hardware with the high sensitivity and specificity of gold nanoparticle probes. In addition, the use of salt based stringency offers a method for performing hybridization assays without the need for temperature control. In principle, the sensitivity of this system can be substantially increased by reducing electrode gap size, which will minimize the number of probe particles required to obtain a measurable signal. These combined attributes are well suited for clinical diagnostics and potentially point–of–care diagnostic applications. In order to achieve this, the system will need to be tested with genomic DNA or RNA samples in more complex sample environments. This detection format is also highly scalable since larger microelectrode arrays can be fabricated using conventional lithographic techniques.

8.7 Conclusions

The various microarray labelling and detection methodologies discussed offer specific advantages in sensitivity, specificity, dynamic range, cost, or number of distinguishable labels when compared to traditional organic fluorophore labelling and detection. Therefore, the ideal labelling and detection strategy is highly dependent on the specific needs of the microarray application. For high sensitivity gene expression applications, Genisphere's 3DNA dendrimer technology and RLS nanoparticle labels exhibit superior sensitivity to conventional direct Cy3 labelling in a two color labelling format. With both tech-

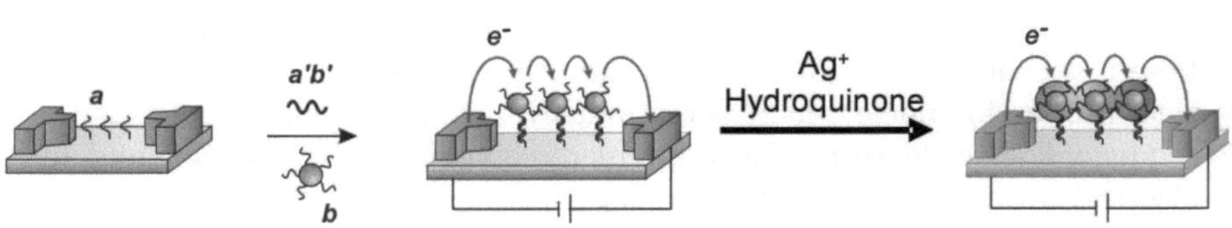

Fig. 8.14. Scheme showing electrical detection of nucleic acids using silver amplification of gold nanoparticle probes. (Reprinted with permission from [31]. Copyright 2002 American Association for the Advancement of Science)

nologies, it has been demonstrated that up to 10–fold less RNA is required for detection [21,77]. Up-converting phosphor labels demonstrate marginally better sensitivity than Cy3 labelling in gene expression applications to date [22]. Even greater detection sensitivity is attainable through further optimization of these nascent labelling and detection strategies.

Certain nanoparticle and phosphor-based labelling methodologies offer a larger number of distinguishable colors than conventional organic dyes, combined with simplified and lower cost instrumentation (e.g. single source excitation). The potential for enhanced multiplexing capability is especially important for liquid-based array and barcoding applications. Quantum dots (i.e. semiconductor nanoparticles) offer at least 6 distinguishable colors with a single excitation source, using particles of different size and composition [42], providing the potential for thousands of unique codes through combination of various colors and intensities, all with higher photostability [50]. One drawback of this labelling methodology is the toxicity of CdSe particles, which requires careful handling and disposal. Phosphor technology also offers the potential for more colors (six spectrally unique colors reported) with the added benefits of single source infrared excitation and longer decay times, which minimizes background fluorescence [55]. Finally, nanoparticle probe-based SERS labels offer the greatest potential for multiplexing combined with high sensitivity [31].

The electrochemical detection platform [23, 59, 60] offers one of the most robust and lowest cost detection strategies, yet, sensitivity limitations in current assays necessitates the use of target amplification, thereby increasing assay complexity and cost. For applications in clinical diagnostics such as SNP detection and infectious disease identification, the elimination of target amplification represents a holy grail, since it would increase assay reliability, significantly reduce cost and assay complexity, and save time. Assuming that a drop of blood is a reasonable target source, sensitivities of $< 10^6$ target copies are required for detection of single copy targets in total human genomic DNA in microarray type applications without target amplification. Detection limits of ~ 100 aM (3000 target copies) have been achieved using Nanosphere's DNA-modified gold nanoparticle (15 nm diameter) technology in conjunction with simple optical detection instrumentation for nucleic acid detection. However, for hybridization based detection of SNPs or mutational sequence changes of just a few bp, specificity is even more critical than sensitivity in the absence of complexity reduction [88]. Here too the higher specificity of DNA-modified gold probes conferred by the sharp melting transitions has enabled detection of gene sequences within unamplified human genomic DNA samples using oligonucleotide microarrays [25]. It is envisioned that strategies such as this will result in broad-based genetic disease diagnostics, with equal potential for infectious disease identification. However, in the latter case a single life cell can be detected by conventional microbiological procedures ('gold standard'), making the necessity of a short culture period likely for some bacterial diagnostic applications where $< 10^3$ copies of the organism are present. Finally, point–of–care diagnostic applications will require not only high sensitivity and specificity, but also simple, rapid, and robust detection assays. Gold nanoparticle probe-based electrical detection systems that lend themselves to assay miniaturization and planar device integration have demonstrated that these goals may be achievable in the not too distant future.

References

1. Venter JC, Adams MD, Myers EW et al., 2001. *The sequence of the human genome.* Science 291, 1304–
2. Lander ES, Linton LM, Birren B et al., 2001. *Initial sequencing and analysis of the human genome.* Nature 409, 860–921
3. Chee M, Yang R, Hubbell E et al., 1996. *Accessing genetic information with high–density DNA arrays.* Science 274, 610–614
4. Lockhart DJ, Winzeler EA, 2000. *Genomics, gene expression, and DNA arrays.* Nature 405, 827–836
5. Fodor SPA, Read JL, Pirrung MC et al., 1991. *Light–Directed, Spatially Addressable Parallel Chemical Synthesis.* Science 251, 767–773
6. Duggan DJ, Bittner M, Chen Y et al., 1999. *Expression profiling using cDNA microarrays.* Nature genetics supplement 21, 10–14
7. Lipshutz RJ, Fodor SPA, Gingeras TR et al., 1999. *High density synthetic oligonucleotide arrays.* Nature genetics supplement 21, 20–24
8. Spellman PT, Sherlock G, Zhang MQ et al., 1998. *Comprehensive identification of cell cycle–regulated genes of the yeast Saccharomyces cerevisiae by microarray hybridization.* Mol. Biol. Cell 9, 3273–3297
9. Clarke PA, Poele RT, Wooster R et al., 2001. *Gene expression microarray analysis in cancer biology, pharmacology, and drug development: progress and potential.* Biochem. Pharmacol. 62, 1311–1336
10. Debouck C, Goodfellow PN, 1999. *DNA microarrays in drug discovery and development.* Nature Genetics 21, 48–50
11. Golub TR, Slonim DK, Tamayo P et al., 1999. *Molecular classification of cancer: Class discovery and class prediction by gene expression monitoring.* Science 286, 531–537
12. Chakravarti A, 1999. *Population genetics–making sense out of sequence.* Nature genetics supplement 21, 56–60
13. Wang DG, Fan JB, Siao CJ et al., 1998. *Large–scale identification, mapping, and genotyping of single– nucleotide polymorphisms in the human genome.* Science 280, 1077–1082
14. Hacia JG, 1999. *Resequencing and mutational analysis using oligonucleotide microarrays.* Nature Genetics 21, 42–47
15. Hacia JG, Fan JB, Ryder O et al., 1999. *Determination of ancestral alleles for human single–nucleotide polymorphisms using high–density oligonucleotide arrays.* Nature Genetics 22, 164–167
16. Cronin MT, Fucini RV, Kim SM et al., 1996. *Cystic fibrosis mutation detection by hybridization to light– generated DNA probe arrays.* Hum. Mutat. 7, 244–255
17. Longenbach-Huber S, Safgren S, Raich T et al., 2001. *Cytochrome P450 (CYP450) genotyping on the CodeLink (TM) bioarray.* Clin. Chem. 47, 37

18. Guo Z, Guilfoyle RA, Thiel AJ et al., 1994. *Direct Fluorescence Analysis of Genetic Polymorphisms by Hybridization with Oligonucleotide Arrays on Glass Supports.* Nucleic Acids Res. 22, 5456–5465
19. Hacia JG, Brody LC, Chee MS, et al. (1996) *Detection of heterozygous mutations in BRCA1 using high density oligonucleotide arrays and two–colour fluorescence analysis.* Nature Genetics 14:441–447
20. Bowtell DD, Sambrook JF (Eds.) (2002) *DNA Microarrays: A Molecular Cloning Manual.* Cold Spring Harbor Laboratory Press, New York
21. Stears RL, Getts RC, Gullans SR (2000) *A novel, sensitive detection system for high–density microarrays using dendrimer technology.* Physiological Genomics 3:93–99
22. van de Rijke F, Zijlmans H, Li S, et al. (2001) *Up–converting phosphor reporters for nucleic acid microarrays.* Nature Biotechnology 19:273–276
23. Umek RM, Lin SW, Vielmetter J, et al. (2001) *Electronic detection of nucleic acids – A versatile platform for molecular diagnostics.* Journal of Molecular Diagnostics 3:74–84
24. Gerion D, Parak WJ, Williams SC, et al. (2002) *Sorting fluorescent nano-crystals with DNA.* Journal of the American Chemical Society 124:7070–7074
25. Storhoff JJ, Marla SM, Bao P, et al. (2004) *Gold nanoparticle–based detection of genomic DNA targets on microarrays using a novel optical detection system.* Biosensors and Bioelectronics 19:875–883
26. Yguerabide J, Yguerabide EE (1998) *Light–scattering submicroscopic particles as highly fluorescent analogs and their use as tracer labels in clinical and biological applications – II. Experimental characterization.* Analytical Biochemistry 262:157–176
27. See white papers at www.geniconsciences.com
28. Alexandre I, Hamels S, Dufour S, et al. (2001) *Colorimetric silver detection of DNA microarrays.* Analytical Biochemistry 295:1–8
29. Song F, Zhou F, Wang J, et al. (2002) *Detection of oligonucleotide hybridization at femtomolar level and sequence–specific gene analysis of the arabidopsis thaliana leaf extract with an ultrasensitive surface plasmon resonance spectrometer.* Nucleic Acids Research 30:e72
30. He L, Musick MD, Nicewarner SR, et al. (2000) *Colloidal Au–enhanced surface plasmon resonance for ultrasensitive detection of DNA hybridization.* Journal of the American Chemical Society 122:9071–9077
31. Cao YWC, Jin RC, Mirkin CA (2002) *Nanoparticles with Raman spectroscopic fingerprints for DNA and RNA detection.* Science 297:1536–1540
32. Park SJ, Taton TA, Mirkin CA (2002) *Array–Based Electrical Detection of DNA with Nanoparticle Probes.* Science 295:1503–1506
33. Nie SM, Zare RN (1997) *Optical detection of single molecules.* Annual Review of Biophysics and Biomolecular Structure 26:567–596
34. Cheung VG, Morley M, Aguilar F, et al. (1999) *Making and reading microarrays.* Nature genetics supplement 21:15–19
35. Brown CS, Goodwin PC, Sorger PK (2001) *Image metrics in the statistical analysis of DNA microarray data.* Proceedings of the National Academy of Sciences of the United States of America 98:8944–8949
36. Lindroos K, Sigurdsson S, Johansson K, et al. (2002) *Multiplex SNP genotyping in pooled DNA samples by a four–color microarray system.* Nucleic Acids Research 30:e70

37. Collins ML, Irvine B, Tyner D, et al. (1997) *A branched DNA signal amplification assay for quantification of nucleic acid targets below 100 molecules/ml.* Nucleic Acids Research 25:2979–2984
38. Shchepinov MS, Udalova IA, Bridgman AJ, et al. (1997) *Oligonucleotide dendrimers: synthesis and use as polylabelled DNA probes.* Nucleic Acids Research 25:4447–4454
39. Nilsen TW, Grayzel J, Prensky W (1997) *Dendritic Nucleic Acid Structures.* Journal of Theoretical Biology 187:273–284
40. Wang J, Jiang M, Nilsen TW, et al. (1998) *Dendritic Nucleic Acid Probes for DNA Biosensors.* Journal of the American Chemical Society 120:8281–8282
41. Alivisatos AP (2001) *Less is more in medicine – Sophisticated forms of nanotechnology will find some of their first real–world applications in biomedical research, disease diagnosis and, possibly, therapy.* Scientific American 285:66–73
42. Bruchez M, Moronne M, Gin P, et al. (1998) *Semiconductor nanocrystals as fluorescent biological labels.* Science 281:2013–2016
43. Chan WCW, Nie SM (1998) *Quantum dot bioconjugates for ultrasensitive nonisotopic detection.* Science 281:2016–2018
44. Weller H (1993) *Colloidal Semiconductor Q–Particles – Chemistry in the Transition Region between Solid–State and Molecules.* Angewandte Chemie–International Edition in English 32:41–53
45. Chan WCW, Maxwell DJ, Gao XH, et al. (2002) *Luminescent quantum dots for multiplexed biological detection and imaging.* Current Opinion in Biotechnology 13:40–46
46. Murray CB, Norris DJ, Bawendi MG (1993) *Synthesis and Characterization of Nearly Monodisperse Cde (E = S, Se, Te) Semiconductor Nanocrystallites.* Journal of the American Chemical Society 115:8706–8715
47. Alivisatos AP (1996) *Semiconductor clusters, nanocrystals, and quantum dots.* Science 271:933–937
48. Wu XY, Liu HJ, Liu JQ, et al. (2003) *Immunofluorescent labeling of cancer marker Her2 and other cellular targets with semiconductor quantum dots.* Nature Biotechnology 21:41–46
49. Pathak S, Choi SK, Arnheim N, et al. (2001) *Hydroxylated quantum dots as luminescent probes for in situ hybridization.* Journal of the American Chemical Society 123:4103–4104
50. Han MY, Gao XH, Su JZ, et al. (2001) *Quantum–dot–tagged microbeads for multiplexed optical coding of biomolecules.* Nature Biotechnology 19:631–635
51. Ferguson JA, Steemers FJ, Walt DR (2000) *High–density fiber–optic DNA random microsphere array.* Analytical Chemistry 72:5618–5624
52. Xu HX, Sha MY, Wong EY, et al. (2003) *Multiplexed SNP genotyping using the Qbead (TM) system: a quantum dot–encoded microsphere–based assay.* Nucleic Acids Research 31:e43
53. Ostermay.Fw (1971) *Preparation and Properties of Infrared–to–Visible Conversion Phosphors.* Metallurgical Transactions 2:747
54. Hampl J, Hall M, Mufti NA, et al. (2001) *Upconverting phosphor reporters in immunochromatographic assays.* Analytical Biochemistry 288:176–187
55. Zijlmans H, Bonnet J, Burton J, et al. (1999) *Detection of cell and tissue surface antigens using up– converting phosphors: A new reporter technology.* Analytical Biochemistry 267:30–36

56. Verwoerd NP, Hennink EJ, Bonnet J, et al. (1994) *Use of Ferroelectric Liquid–Crystal Shutters for Time–Resolved Fluorescence Microscopy.* Cytometry 16:113–117
57. Beverloo HB, Vanschadewijk A, Zijlmans H, et al. (1992) *Immunochemical Detection of Proteins and Nucleic–Acids on Filters Using Small Luminescent Inorganic Crystals as Markers.* Analytical Biochemistry 203:326–334
58. Dejneka MJ, Streltsov A, Pal S, et al. (2003) *Rare earth–doped glass microbarcodes.* Proceedings of the National Academy of Sciences of the United States of America 100:389–393
59. Boon EM, Ceres DM, Drummond TG, et al. (2000) *Mutation detection by electrocatalysis at DNA–modified electrodes.* Nature Biotechnology 18:1096–1100
60. Napier ME, Loomis CR, Sistare MF, et al. (1997) *Probing biomolecule recognition with electron transfer: Electrochemical sensors for DNA hybridization.* Bioconjugate Chemistry 8:906–913
61. Wang J (2001) *Glucose biosensors: 40 years of advances and challenges.* Electroanalysis 13:983–988
62. Yu CJ, Wan YJ, Yowanto H, et al. (2001) *Electronic detection of single–base mismatches in DNA with ferrocene–modified probes.* Journal of the American Chemical Society 123:11155–11161
63. Armistead PM, Thorp HH (2000) *Modification of indium tin oxide electrodes with nucleic acids: Detection of attomole quantities of immobilized DNA by electrocatalysis.* Analytical Chemistry 72:3764–3770
64. Armistead PM, Thorp HH (2002) *Electrochemical detection of gene expression in tumor samples: Overexpression of Rak nuclear tyrosine kinase.* Bioconjugate Chemistry 13:172–176
65. Popovich ND (2001) *Mediated electrochemical detection of nucleic acids for drug discovery and clinical diagnostics.* IVD Technology April
66. Boon EM, Kisko JL, Barton JK (2002) In Redox Cell Biology and Genetics, Pt B, Vol. 353, pp. 506–522
67. Handley DA (1991) In Colloidal Gold; Principles, Methods, and Applications, Vol. 1 (Ed, Hayat, M. A.) Academic Press, San Diego, pp. 1–12
68. Rohringer R (1991) In Colloidal Gold: Principles, Methods, and Applications, Vol. 2 (Ed, Hayat, M. A.) Academic Press, San Diego, pp. 398–411
69. Niemeyer CM (2001) *Nanoparticles, Proteins, and Nucleic Acids: Biotechnology Meets Materials Science.* Angewandte Chemie–International Edition 40:4128–4158
70. Storhoff JJ, Mirkin CA (1999) *Programmed Materials Synthesis with DNA.* Chemical Reviews 99:1849–1862
71. Taton TA, Mirkin CA, Letsinger RL (2000) *Scanometric DNA array detection with nanoparticle probes.* Science 289:1757–1760
72. Handley DA (1991) In Colloidal Gold; Principles, Methods, and Applications, Vol. 1 (Ed, Hayat, M. A.) Academic Press, San Diego, pp. 13–32
73. Mirkin CA, Letsinger RL, Mucic RC, et al. (1996) *A DNA–based method for rationally assembling nanoparticles into macroscopic materials.* Nature 382:607–609
74. Elghanian R, Storhoff JJ, Mucic RC, et al. (1997) *Selective colorimetric detection of polynucleotides based on the distance–dependent optical properties of gold nanoparticles.* Science 277:1078–1081

75. Brockmann JM, Nelson BP, Corn RM (2000) *Surface Plasmon Resonance Imaging Measurements of Ultrathin Organic Films.* Annual Reviews of Physical Chemistry 51:41–63
76. Taton TA, Lu G, Mirkin CA (2001) *Two-color labeling of oligonucleotide arrays via size-selective scattering of nanoparticle probes.* Journal of the American Chemical Society 123:5164–5165
77. Bao P, Frutos AG, Greef C, et al. (2002) *High-Sensitivity Detection of DNA Hybridization on Microarrays Using Resonance Light Scattering.* Analytical Chemistry 74:1792–1797
78. Boyer D, Tamarat P, Maali A, et al. (2002) *Photothermal imaging of nanometer-sized metal particles among scatterers.* Science 297:1160–1163
79. Wang J, Song FY, Zhou FM (2002) *Silver-enhanced imaging of DNA hybridization at DNA microarrays with scanning electrochemical microscopy.* Langmuir 18:6653–6658
80. Faraday M (1857) *Experimental relations of gold (and other metals) to light.* Philos Trans R Soc London 147:145
81. Jin RC, Wu GS, Li Z, et al. (2003) *What controls the melting properties of DNA-linked gold nanoparticle assemblies?* Journal of the American Chemical Society 125:1643–1654
82. Link S, Wang ZL, El-Sayed MA (1999) *Alloy formation of gold-silver nanoparticles and the dependence of the plasmon absorption on their composition.* Journal of Physical Chemistry B 103:3529–3533
83. Yguerabide J, Yguerabide EE (1998) *Light-scattering submicroscopic particles as highly fluorescent analogs and their use as tracer labels in clinical and biological applications – I. Theory.* Analytical Biochemistry 262:137–156
84. Scopsi L (1991) In Colloidal Gold: Principles, Methods, and Applications, Vol. 1 (Ed, Hayat, M. A.) Academic Press, San Diego, pp. 251–295
85. Jin RC, Cao YW, Mirkin CA, et al. (2001) *Photoinduced conversion of silver nanospheres to nanoprisms.* Science 294:1901–1903
86. Yguerabide J, Yguerabide EE (2001) *Resonance Light Scattering Particles as Ultrasensitive Labels for Detection of Analytes in a Wide Range of Applications.* Journal of Cellular Biochemistry Supplement 37:71–81
87. Taton TA, Mucic RC, Mirkin CA, et al. (2000) *The DNA-mediated formation of supramolecular mono- and multilayered nanoparticle structures.* Journal of the American Chemical Society 122:6305–6306
88. Lander ES (1999) *Array of hope.* Nature genetics supplement 21:3–4
89. Oldenburg SJ, Genick CC, Clark KA, et al. (2002) *Base pair mismatch recognition using plasmon resonant particle labels.* Analytical Biochemistry 309:109–116
90. Schultz S, Smith DR, Mock JJ, et al. (2000) *Single-target molecule detection with nonbleaching multicolor optical immunolabels.* Proceedings of the National Academy of Sciences of the United States of America 97:996–1001
91. Alexandre I, Houbion Y, Collet J, et al. (2002) *Compact disc with both numeric and genomic information as DNA microarray platform.* Biotechniques 33:435
92. Liedberg B, Nylander C, Lundstrom I (1995) *Biosensing with surface plasmon resonance – how it all started.* Biosensors and Bioelectronics 10:i–ix
93. Homola J, Yee SS, Gauglitz G (1999) *Surface plasmon resonance sensors: review.* Sensors and Actuators B 54:3–15

94. Peterlinz KA, Georgiadis RM (1997) *Observation of hybridization and dehybridization of thiol– tethered DNA using two–color surface plasmon resonance spectroscopy.* Journal of the American Chemical Society 119:3401–3402
95. Haes AJ, Van Duyne RP (2002) *A nanoscale optical biosensor: Sensitivity and selectivity of an approach based on the localized surface plasmon resonance spectroscopy of triangular silver nanoparticles.* Journal of the American Chemical Society 124:10596–10604
96. Jordan CE, Frutos AG, Thiel AJ, et al. (1997) *Surface plasmon resonance imaging measurements of DNA hybridization adsorption and streptavidin/DNA multilayer formation at chemically modified gold surfaces.* Analytical Chemistry 69:4939–4947
97. Thiel AJ, Frutos AG, Jordan CE, et al. (1997) *In situ surface plasmon resonance imaging detection of DNA hybridization to oligonucleotide arrays on gold surfaces.* Analytical Chemistry 69:4948–4956
98. Goodrich GP, Nicewarner SR, He L, et al. (2001) *Nanoparticle–amplified surface plasmon resonance for detection of DNA hybridization.* Proceedings of SPIE 2:80–85
99. Raschke G, Kowarik S, Franzl T, et al. (2003) *Biomolecular Recognition Based on Single Gold Nanoparticle Light Scattering.* Nano Letters 3:935–938
100. Freeman RG, Grabar KC, Allison KJ, et al. (1995) *Self–Assembled Metal Colloid Monolayers – an Approach to Sers Substrates.* Science 267:1629–1632
101. Mulvaney SP, Keating CD (2000) *Raman spectroscopy.* Analytical Chemistry 72:145R–157R
102. Isola NR, Stokes DL, Vo-Dinh T (1998) *Surface–Enhanced Raman Gene Probe for HIV Detection.* Analytical Chemistry 70:1352–1356
103. Haynes CL, Van Duyne RP (2003) *Plasmon–Sampled Surface–Enhanced Raman Excitation Spectroscopy.* Journal of Physical Chemistry B 107:7426–7433
104. Nie SM, Emery SR (1997) *Probing single molecules and single nanoparticles by surface– enhanced Raman scattering.* Science 275:1102–1106
105. Graham D, Mallinder BJ, Whitcombe D, et al. (2002) *Simple multiplex genotyping by surface–enhanced resonance Raman scattering.* Analytical Chemistry 74:1069
106. Brust M, Bethell D, Schiffrin DJ, et al. (1995) *Novel Gold–Dithiol Nano–Networks with Nonmetallic Electronic– Properties.* Advanced Materials 7:795–
107. Velev OD, Kaler EW (1999) In *situ assembly of colloidal particles into miniaturized biosensors.* Langmuir 15:3693–3698

9

Marker-free Detection on Microarrays

Matthias Vaupel, Andreas Eing, Karl-Otto Greulich, Jan Roegener, Peter Schellenberg, Hans Martin. Striebel, and Heinrich F. Arlinghaus

9.1 Introduction

The binding of oligomers or DNA are usually detected by fluorescence. To this end at least one binding partner is labelled with a fluorescence marker. This detection method cannot be used for protein reactions since biological and chemical properties of proteins are often changed by a bound marker. Marker-free observation of a protein reaction is favorable. We discuss imaging ellipsometry, as well as imaging surface plasmon resonance (SPR) and compare the results of both methods with scanning probe microscopy (SPM) and detection using fluorescence markers.

Intrinsic ultraviolet (UV) fluorescence is presented as an alternative method to classical two dimensional gel electrophoresis.

Time–of–flight secondary ion mass spectrometry (TOF–SIMS) is investigated in comparison to ultraviolet matrix-assisted laser desorption / ionization mass spectrometry (MALDI–MS) for read-out of peptide nucleic acid (PNA) microarray chips. All presented marker-free detection methods are intended for the development of a marker-free microarray reader for cancer detecting protein biochips [1].

9.2 Imaging Ellipsometry and Imaging Surface Plasmon Resonance on Biochips

9.2.1 Imaging Null Ellipsometry

Ellipsometry is a non-destructive, label-free optical method for determining thickness and optical properties of thin films [2]. It measures the change in polarization of the light reflected by the surface of the film. Fast ellipsometry methods, single or multi-wavelength, have been adopted for monitoring film growth in situ, allowing the precise control of film deposition processes [3]. Commercial imaging ellipsometers, e.g. I–Elli2000 and EP^3 from Nanofilm

Technologie operate on the principle of null ellipsometry (Fig. 9.1b). The laser beam is elliptically polarized after passing through a linear polarizer (P) and a quarter-wave plate (Compensator, C). The elliptically polarized light is then reflected off the sample (S) through a second polarizer (analyzer, A) and imaged by a CCD camera through a long working distance objective. In this configuration (PCSA), the angles of P and C with respect to the plane of incidence are chosen in such a way that the elliptically polarized light is completely linearly polarized after it is reflected off the sample. The null or minimum of intensity is detected when A is perpendicular with respect to the polarization axis of the reflected light. The angles of P, C and A at the null of intensity determine the ellipsometric parameters Delta and Psi. The tangent of the angle Psi is the ratio of the reflection coefficients of both polarization components (p and s, perpendicular and parallel to the plane of incidence), while Delta is the relative phase shift of these polarization components upon

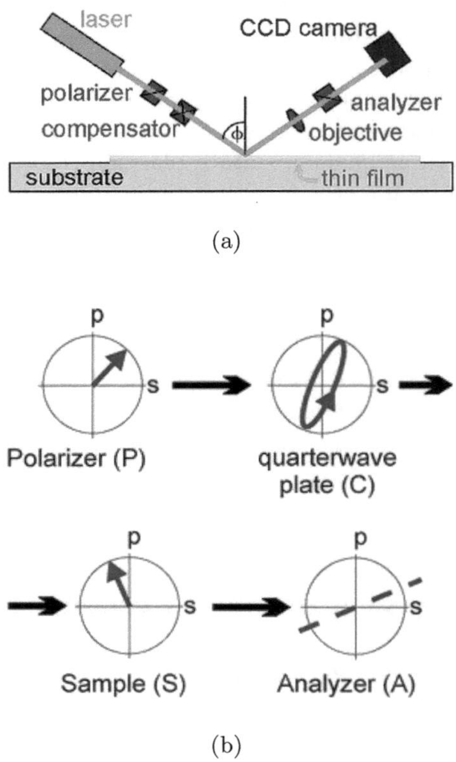

Fig. 9.1. Set-up of the imaging ellipsometers I–Elli2000 and EP^3 from Nanofilm Technologie **(a)** and the propagation of the polarization through a null ellipsometer **(b)**

reflection. Reduction of the measured Delta and Psi with computerized optical modelling leads to a deduction of the optical properties of the sample (complex refractive indices) and the film thickness.

Imaging ellipsometry (Fig. 9.1) combines ellipsometry with microscopy. Spots on the sample, which have different optical properties, e.g. film thickness, have different reflection coefficients and thus different angles of P, C, and A of null intensity. The ellipsometric image of the sample shows null intensity only in spots with the same optical properties. Other spots appear brighter. The contrast in an image is typically such that a 10 pm high step on the sample is observable. The lateral resolution of an image is typically 1 µm, which is given by the numerical aperture of the objective.

9.2.2 Imaging Surface Plasmon Resonance

Conventional surface plasmon resonance (SPR) technology is a very sensitive method to measure the adsorption kinetics of ligands on immobilized substances. It can be used to detect the binding of antibodies to their antigens or the binding of proteins to their reaction partners. In an SPR–cell, e.g. from Nanofilm Technologie, a polarized beam propagates in glass and is reflected from a thin gold film (Fig. 9.2) whose reflection coefficient is highly sensitive on the optical properties of a thin reaction layer on the gold. The reflection coefficient of p–polarization has a minimum at the resonance angle of the SPR. The resonance angle is shifted proportional to the mass of a substance adsorbing on the surface [4].

An ellipsometer measures the ellipsometric parameters Psi and Delta instead of just the reflection coefficient of p–polarization as it is done in classical SPR–devices, e.g. from Biacore. The tangent of Psi is proportional to the re-

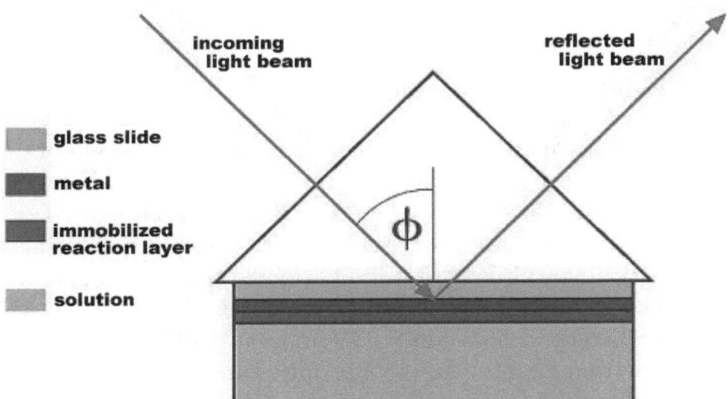

Fig. 9.2. Sketch of an SPR–cell with the incoming and outgoing light beam and the angle of incidence ϕ

flection coefficient of p–polarization. Thus the parameter Psi is analogous to the reflected intensity in classical SPR whereas the phase shift Delta provides additional information exceeding classical SPR.

Sensitivity regarding thickness or mass, respectively, is proportional to the derivative (slope) of the measured parameter. At the resonance angle of SPR the slope of Psi (δ Psi / $\delta\phi$) is limited, where the slope of Delta (δ Delta / $\delta\phi$) is unlimited (Fig. 9.3). Thus a measurement of Delta can be much more sensitive than a measurement of Psi or classical SPR. The sensitivity of the classical SPR approach (δ Psi / $\delta\phi$) on the other hand is solely determined by the physico–chemical properties of the layer system and cannot be increased.

9.2.3 Quality Control on Micro Arrays

All spots of immobilized biological macromolecules on a biochip should have a homogeneous shape and the same size and a defined mass. If these requirements are fulfilled, the amount of material that can hybridize is quantified correctly and the results are reliable. Ideally, one displays the quality of the spots before a hybridization process to avoid the loss of expensive probes on less than optimal biochips. Many techniques for quality control either need very time consuming staining processes or destroy the biochips. With imaging ellipsometry one can check the shape and the size of all spots without staining or before the hybridization takes place, and evaluate the results afterwards. As an example, a non-hybridized oligonucleotide spot is displayed in Fig. 9.4.

In another example we have observed non-specific binding of DNA with the imaging ellipsometer. Ellipsometric thickness maps and scanning probe microscopic (e.g. AFM, SFM, STM) maps have usually comparable thickness resolution. Ellipsometric thickness maps have two advantages: much larger field of view (up to some cm) and much faster recording time (30 sec). But only scanning microscopes offer lateral resolution below the wavelength of light.

Thickness and Mass Quantification

A monolayer of bovine serum albumin with a molecular weight of 67 kDa typically has a surface capacity of ~ 3 ng/mm^2 and a thickness between 2 to 3 nm depending on the surface density (18000–27000 molecules/μm^2) [3]. Thus approximately 1 nm thickness is measured with an ellipsometer at 1 ng/mm^2 surface density. Typical detection limits representing the smallest detectable relative thickness change are 40 pm (Organic on Glass), 10 pm (Organic on Gold or on Silicon), < 0.03 pm (!)(Organic on Gold–SPR–sensor measured with ellipsometry, Fig. 9.5). An electro–optic tunable Gold–SPR–sensor array is under development [1] in order to further decrease the detection limit towards the range of 1 fm or 1 fg/mm^2 which is the sensitivity of fluorescence readers. This sensitivity enables single molecule detection.

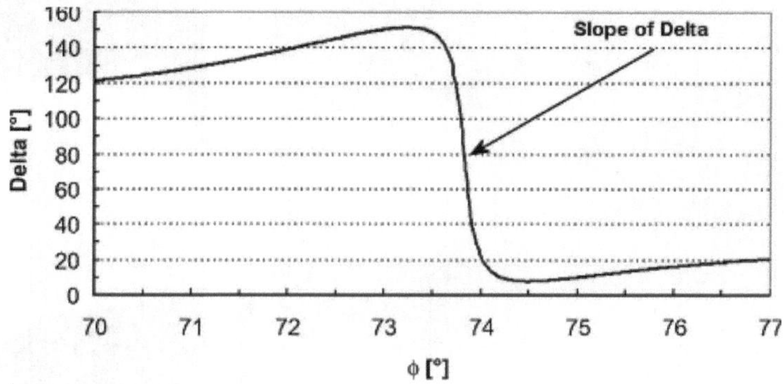

(a)

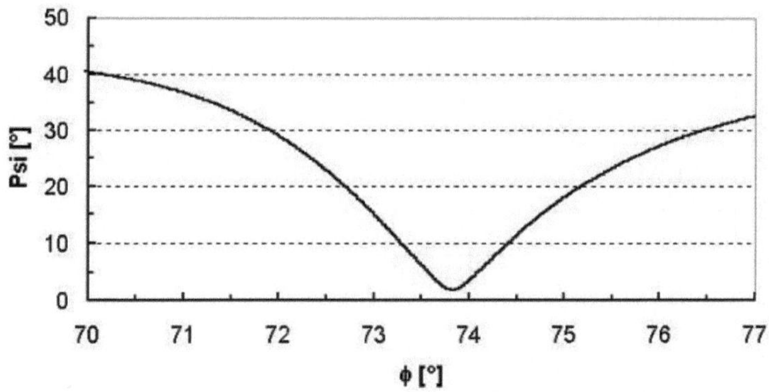

(b)

Fig. 9.3. The slope of Psi corresponds to the slope of the reflection coefficient of p–polarization in classical SPR. The slope of Delta becomes infinite if the minimum of Psi approaches zero. This is the case if the wavelength and the gold–layer thickness are suitable

The surface capacity of immobilized oligonucleotides (fragments of single stranded DNA) is in the range of 70 000 molecules/μm^2 [5–7]. Thus, depending on the strand length, the thickness of such a layer is 0.8 nm (20–mer oligo), 1.9 nm (50–mer oligo) and 5.8 nm (150–mer oligo). Assuming a hybridization yield of 33% [6] the medium thickness increase due to the binding of the complementary oligonucleotides is 0.3 nm (20–mer oligo), 0.6 nm (50–mer oligo) and 1.9 nm (150–mer oligo).

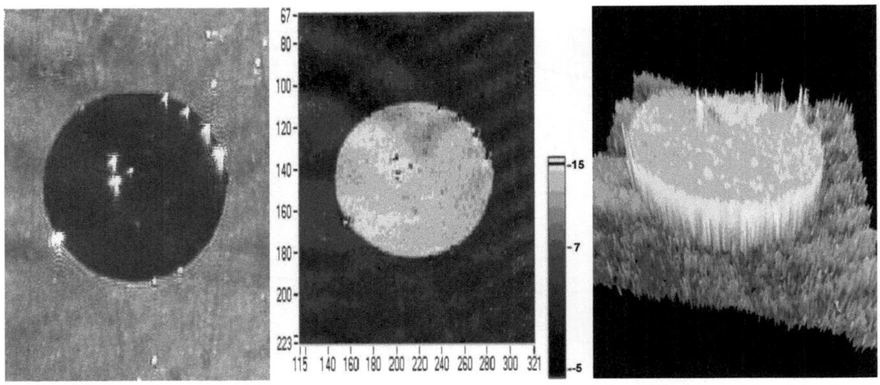

Fig. 9.4. Images of a spot of immobilized non-hybridized 50–mer oligonucleotides produced by Advalytix AG (Brunnthal, Germany). The spot diameter is 150 μm. **(a)** Ellipsometric contrast, **(b)** thickness–map [z in nm and x/y in pixel] and **(c)** the corresponding 3D–profile

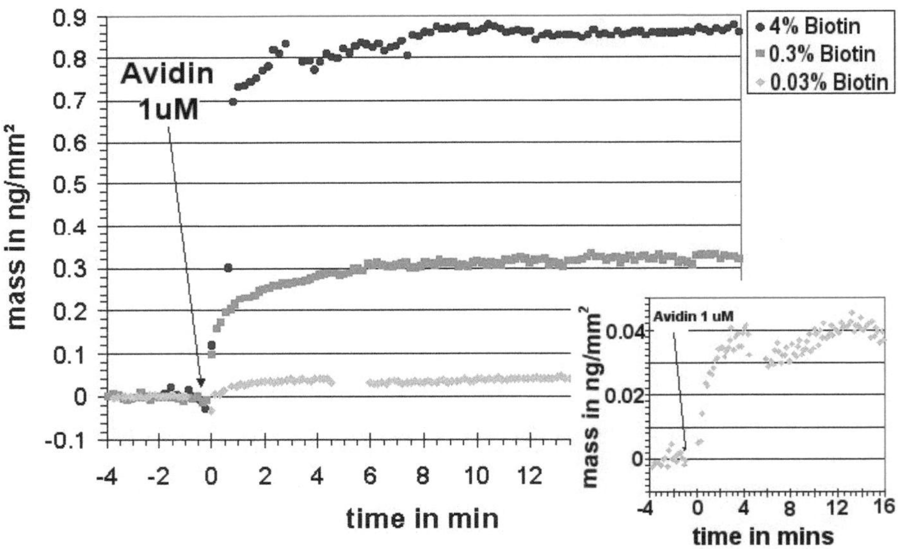

Fig. 9.5. Reaction kinetics of avidin binding on biotin spots with different concentrations, sample kindly provided by Graffinity, Heidelberg, Germany

Protein Spots and the Influence of the Spotting Procedure on the Spot–Shape

Figure 9.6 represents an example of protein spots with the typical 'donut shape' which results from the applied spotting or printing–technique. This kind of non-perfect spotting is still a problem. The influence of additives in the spotting solution on the shape or the homogeneity of the resulting protein spot can be evaluated and give versatile information about the optimal spotting conditions right after the spotting process (Fig. 9.6). Time consuming and expensive steps to visualize the spots, e.g. binding of fluorescent substances, are not required.

The imaging ellipsometer can characterize the homogeneity of the surface–layer, e.g. a streptavidin–layer (Fig. 9.7), which is rather non-homogeneous in comparison with the dextran layer in Fig. 9.6. Homogeneity is a quality

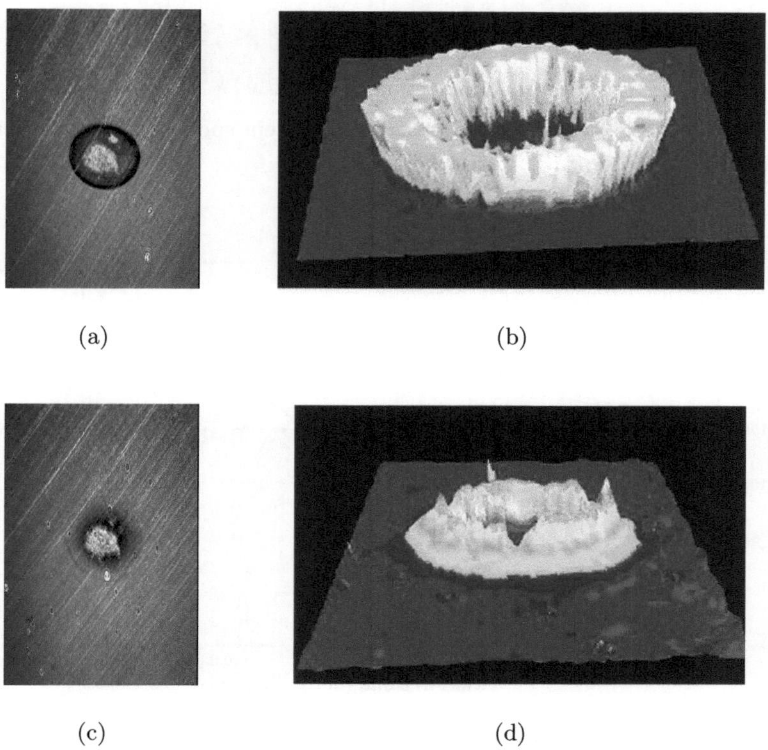

Fig. 9.6. Images of protein–spots on gold (diameter 200 μm) without **(a, b)** and with an additive in the spotting solution **(c, d)**. (Ellipsometric contrast images (a and c) and the corresponding 3D–profile of the thickness (b and d))

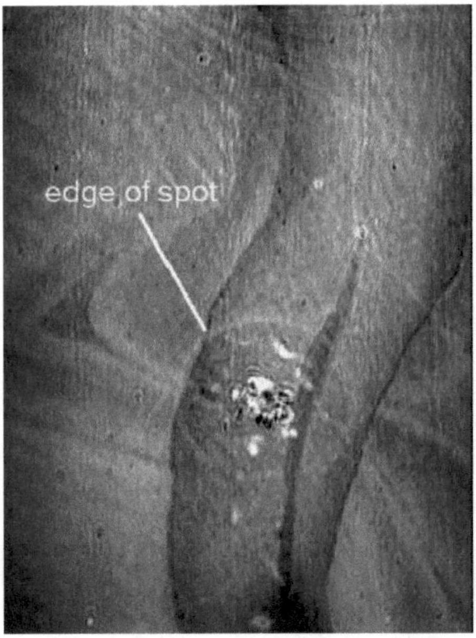

Fig. 9.7. Ellipsometric contrast image of a protein spot (diameter 200 μm) on a streptavidin–surface

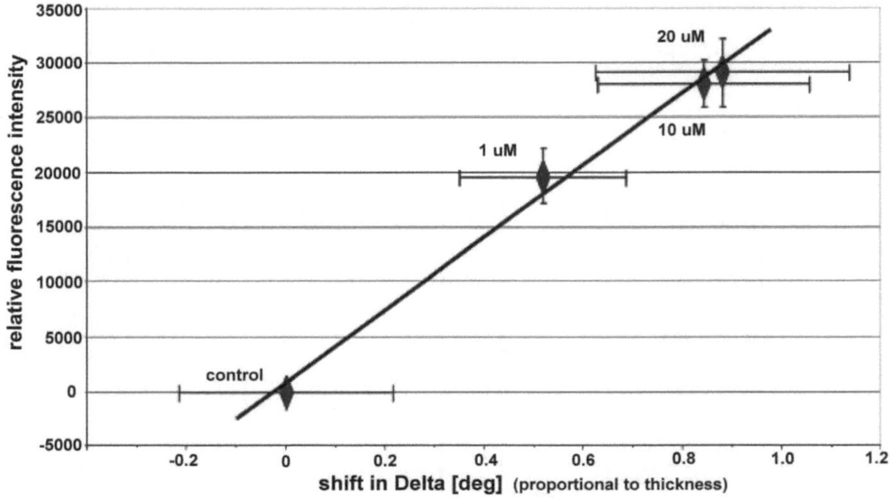

Fig. 9.8. Linear regression of the ellipsometric parameter Delta with the relative fluorescence intensity of hybridized DNA spots. The DNA for the hybridization has been labelled with Cy 5. The measurement of the control spot is defined as the reference in both techniques. Error bars represent standard deviation of a minimum of 15 spots. The standard deviation of the linear regression is much smaller. (Samples and data kindly supplied by PicoRapid Technologie GmbH, Bremen, Germany)

criterion of the surface because it determines the amount of the substance that can bind to the surface.

Comparison of Hybridized DNA Spots Visualized with Fluorescence and Imaging Ellipsometry

In the conventional evaluation of microarrays the fluorescence signal of a control spot is compared to test spots where hybridization takes place. At the control, no hybridization occurs because the oligonucleotides are not complementary. For simplicity, we assume that the fluorescence intensity is proportional to the amount that binds to the spot.

To determine whether ellipsometry is comparable to fluorescence, identical DNA spots have been evaluated with both methods. With the ellipsometer, the parameter Delta yields the signal. The difference in Delta between the control spot and the diverse hybridized test–spots is displayed versus the relative fluorescence intensity of the identical spots (Fig. 9.8). It is observed that the shift in Delta is proportional to the relative fluorescence intensity.

The proportionality between the fluorescence signal and the ellipsometric parameter demonstrates that both methods yield equivalent results, but the fluorescence signal cannot be transferred into the amount of bound material directly. In contrast, the layer thickness can be calculated from the ellipsometric parameter Delta. The layer thickness is related to the mass of adsorbed material, which can be transferred into molecules per area.

In Situ Reaction Kinetics

Imaging ellipsometry can display simultaneously all reaction channels fitting in the field of view. An array with 2500 spots (100 μm diameter) on a 1 cm^2 field of view could be observed with the large area EP^3EP^3 from Nanofilm Technologie. With this imaging ellipsometer, 8 spots with different biotin concentrations (Fig. 9.9) on a gold–SPR–sensor before and after binding of avidin were recorded. To this end the beam at 594 nm from the ellipsometer was coupled through a prism into the glass slide (refractive index $n = 1.7$) (Fig. 9.9), which was coated with a 35 nm thick gold film and spotted with biotin. Wavelength and refractive index of glass were chosen in order to minimize Psi at the resonance angle of the SPR and to optimize the sensitivity. The recording of the phase shift Delta of spots with different biotin concentrations as a function of time is shown in Fig. 9.5. While 4% biotin concentration yields 900 pg/mm^2 (almost half of a monolayer), the smallest concentration of 0.03% yields (40 ± 3) pg/mm^2. The 3 pg/mm^2 noise is caused by chemical fluctuations on the sensor surface and refractive index fluctuations in the solution, where the repeatability (relative error bar) of such an imaging ellipsometer is up to 100 times more precise.

9.3 Intrinsic UV Fluorescence for Chip Analysis of Rare Proteins

9.3.1 Introduction

Disease phenotypes are governed mainly by proteins, but less directly by DNA. Therefore protein chip analysis promises to be more efficient. Usually, protein chips carry commercially available antibodies, enzymes or regulatory proteins. At most a few thousand human proteins are readily available, but 30,000–40,000 different proteins can be expected from the human genome sequence. The majority of human proteins have still to be produced, for example by gene technology. They will often be available only in small quantity. In order to use such proteins on protein chips, methods for low-amount– but high-yield–preparation are required. Chemical modification such as fluorescence labelling is, in that sense, counterproductive and should thus be avoided. Imaging ellipsometry, surface plasmon resonance and mass spectrometry are suitable label free methods. This trio of techniques is complemented by the use of intrinsic protein UV fluorescence originating from the aromatic amino acids tryptophan and tyrosine. Fluorescence detection is one of the most sensitive techniques to probe molecules, with detection limits often down to the single molecule level. It is therefore straightforward to use intrinsic fluorescence methods to test ligand binding to protein chips. Of 1,026,890 proteins with molecular masses larger than 10 kD found in the data base NCBlnr 9.23, more than 99% contain at least one tryptophan or tyrosine and hence are detectable by UV fluorescence.

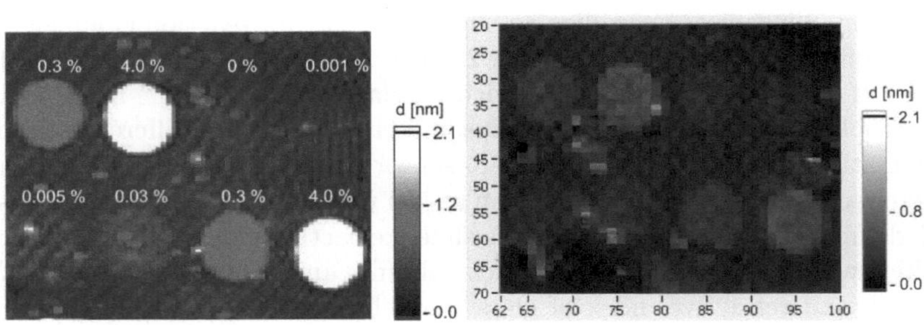

Fig. 9.9. Biotin spots (0.6 mm diameter) with different biotin concentrations on gold–SPR–sensor before and after binding of avidin (1 μM solution in HEPES–buffer pH 7.4), Thickness maps recorded with large area I–Elli2000 from Nanofilm Technologie, sample kindly provided by Graffinity, Heidelberg, Germany

9.3.2 Materials and Methods

UV Protein Fluorescence to Detect Proteins and Protein – Ligand Binding

Detecting UV fluorescence on protein microarrays is a new approach. In contrast to DNA, proteins excited in the UV at 280–290 nm reveal considerable intrinsic fluorescence. Particularly tyrosine (molar absorption coefficient $\epsilon = 1200$ Mol^{-1} cm^{-1}, fluorescence yield $\Phi = 0.065$) and tryptophan ($\epsilon = 5600$ Mol^{-1} cm^{-1}, $\Phi = 0.16$ - 0.21) contribute to the total intrinsic fluorescence. Although extrinsic fluorescence dyes with 50 000–100 000 Mol^{-1} cm^{-1} and yields up to 0.8 are better suited in this respect, intrinsic fluorescence is sufficiently strong for analysis [8]. A first step in utilizing intrinsic protein fluorescence in chip technology is the mere detection of protein spots by steady state illumination. More informative will be the detection of protein–protein binding, since this allows the finding of potential partners of a protein in a signalling cascade, which may be upset in a disease process. In some favorable circumstances, protein–protein binding may be detected by spectral shifts. For example, when tryptophan, originally exposed to solvent, becomes buried in the interior of a newly formed protein pair, its fluorescence maximum shifts from 355 nm to 330 nm [9,10].

More generally applicable are fluorescence lifetimes, which are sensitive to interactions between the probe molecules on the chip substrate and target proteins. In a trivial case, distinct lifetimes of the two proteins may just be averaged upon binding. More sophisticated is the Foerster mechanism that alters lifetimes by energy transfer to neighboring amino acids or to other chromophores. The energy transfer rate is reduced due to the proximity of acceptors upon binding to other proteins, or is adjusted due to changes of the protein folding structure. Furthermore, changes in the dynamics of the protein solvent cage as a result of folding can also lead to an alteration in the internal conversion rate, which modifies fluorescence lifetime. Note that these mechanisms also modulate fluorescence quantum yield and therefore fluorescence intensity. However, one has to work with very well defined quantities to detect these changes, which is difficult to achieve. On the other hand, fluorescence lifetime is a very robust parameter, not influenced by concentration. There are still difficulties to overcome when utilizing fluorescence decay time measurements to probe binding to a chip. Typically, several tyrosines and tryptophans are present in a protein, each with its specific fluorescence lifetime or even an inhomogeneous distribution of lifetimes [11,12]. Similarly, not all amino acids are influenced equally by modifications of the protein environment. Due to this effect the change in lifetime may be small and has to be measured with high accuracy.

In the present work the frequency tripled output from a self mode-locked Ti:Sa laser was used for excitation. The time resolved fluorescence was detected by time correlated single photon counting (TCSPC) with a time resolution of 50 ps. Alternatively, a streak camera may be used, thereby improving time resolution to about 2 ps and reducing measuring times. The background fluorescence from the substrate has to be low and/or with very different decay time constants compared to the spot. With either set–up, one spot location on the chip is probed at a time, and the chip has to be moved after each measurement to a new spot position, which is rather time consuming. Alternatively, the whole chip may be probed at once with a set–up including a gated UV-sensitive CCD–detector with gating windows of about 200 ps. Although this is the method of choice for automated processes, it has worse time resolution and requires a rather large change of the fluorescence lifetime of the system.

Attaching Proteins to a Surface: Finding the Right Turn

Unlike DNA molecules with their comparatively uniform structures and outlined sets of established methods for their successful surface immobilization, proteins require much more custom tailored surface immobilization techniques, simply because of their highly distinct properties.

In order to retain native shape and functionality of immobilized proteins, surface chemistry has become an important aspect of protein array development. In this regard, glass plays a central role as a basic support, and as a starting point for subsequent chemical derivatization. In order to combine glass supports and proven protein immobilization chemistries, coating these supports with suitable materials is a practical option [13]. Depending on the intended detection physics, different coating materials may be applied [14]. Artificial polymers have the advantage of being made up from a chemical matrix that may be modified to some extent in order to adapt to special protein immobilization needs. The artificial polymer used most often as a coating material is polyacrylamide. Other artificial polymer coating materials are based on derivatives of polymethacrylate.

Natural polymers, like agarose or cellulose, combine a number of properties that provide an advantageous environment for immobilized proteins in their native states. On the one hand, both materials may be dissolved in appropriate solvents (agarose in hot water, cellulose in dimethylsulfoxide), and spread over glass supports to yield thin, non-fluorescing layers once the solvent has evaporated. These layers are able to retain water in considerable amounts, which makes them ideally suited to enclose proteins in an environment preserving native protein structures. On the other hand, agarose and cellulose are chemically versatile materials. Particularly agarose can be converted into matrices containing high densities of aldehyde functions by careful oxidation with sodium periodate [15]. Matrices carrying high densities of aldehyde functions may then be used to immobilize polymer layers on aminated glass slides, as well as to immobilize proteins via the amino groups on the surface of the protein. Therefore, sandwich-like structures may be generated, featuring glass slides as a basic support, which is covered by a biopolymer layer that in turn has proteins immobilized on its surface.

Native Protein Binding Techniques

Depending on their surface characteristics, binding proteins on surfaces may occur in three ways: covalently, electrostatically, and by affinity. These mechanisms have been discussed in detail in Chaps. 2 and 3. Binding by affinity interactions requires ligands with high specificity towards the protein to be immobilized. This may be accomplished best with antibodies and their respective antigens, or special protein–ligand pairs like biotin and avidin [16]. In many cases, the anchoring ligands are proteins themselves, transferring the need of protein immobilization to just another protein species.

Dependent on the chemical structures of protein surfaces, there are a few basic methods for covalent protein binding [17]. As the majority of proteins are water soluble, they feature patterns of acidic or basic amino acid side chains on their surfaces which provide points of attack for immobilization reagents. Acid side chains, usually provided by amino acids like glutamic or aspartic acid, may be coupled to primary amino functions via EDC [1–ethyl–3–(–3–dimetylaminopropyl)carbodiimide]. Amine side chains provided by amino acids like arginine, asparagine, or glutamine may be coupled to aldehyde–function bearing substrates directly by amine–aldehyde chemistry.

Proteins exposing free thiol functions, generated for example by reducing antibodies with DTT (dithiothreitol), may be immobilized either on gold surfaces, or other thiol binding functions.

9.3.3 Results

In order to provide substrates suitable for the study of protein–protein interaction on their surfaces by UV-based detection methods, supports of protein arrays need to reveal low fluorescence background. Plastic supports are therefore not recommendable since even UV transparent materials still reveal some fluorescence when excited at 280 nm. Glass or, even better, fused silica supports are suited best. Surfaces should provide an environment for protein immobilization that is optimally suited to binding proteins in their native states. This requirement may be accomplished by coating glass supports with layers of natural polymers.

For study of protein–protein interactions, two binding processes have to be considered: first, immobilization of a probe protein on the support, and second, subsequent docking of a suspected ligand protein to the probe protein without non-specific binding to the areas not covered by immobilized probe proteins. Blocking of areas not covered by protein spots may be a solution, performed by saturation of active binding sites with neutral proteins like BSA, but this very likely interferes with the UV–detection process for protein–protein interaction. Non-fluorescing blocking agents are desirable for this purpose.

Another possibility may be the design of immobilization chemistries that selectively bind probe proteins but not sample proteins.

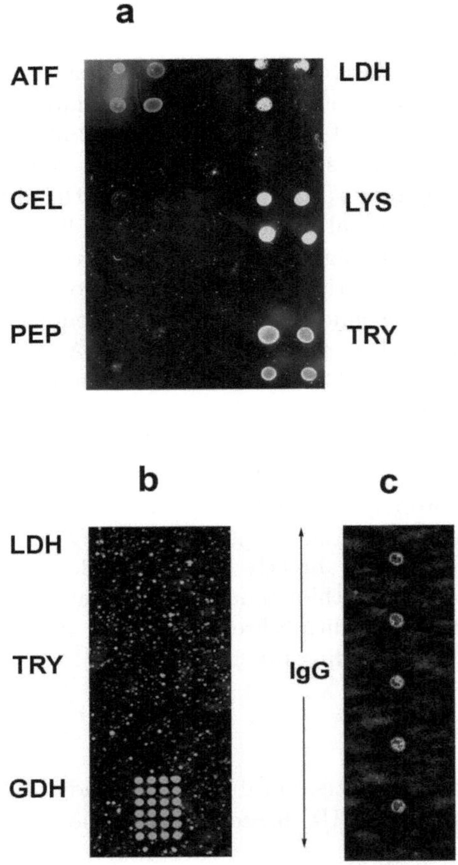

Fig. 9.10. ATF = aminotransferase; CEL = cellulase; GDH = glutamatedehydrogenase; IgG = Immunoglobuline G; LDH = lactatedehydrogenase; LYS = lysozyme; PEP = pepsin; TRY = trypsin. **(a)** Surface made up from oxidized agarose; proteins are bound to aldehyde functions via aldehyde–amine chemistry; surplus aldehyde functions at areas not covered by immobilized probe proteins were then saturated by small amine reactants, in this case tris buffer. Spot diameters are 1000 µm, spot detection occurred at 493 nm (excitation), and 517 nm (emission) after staining with FLUOS. **(b)** Cation exchange surface, generated by reacting a coating of oxidized agarose with glycine. Proteins are bound by electrostatic interaction. **(c)** Array of immobilized IgG; proteins are bound to amino functions on a commercially available support (SCIENION AG, Berlin). Spot diameters of (b) and (c) are 300 µm; detection at (b) and (c) occurred at 280 nm (excitation), and 300–375 nm (emission)

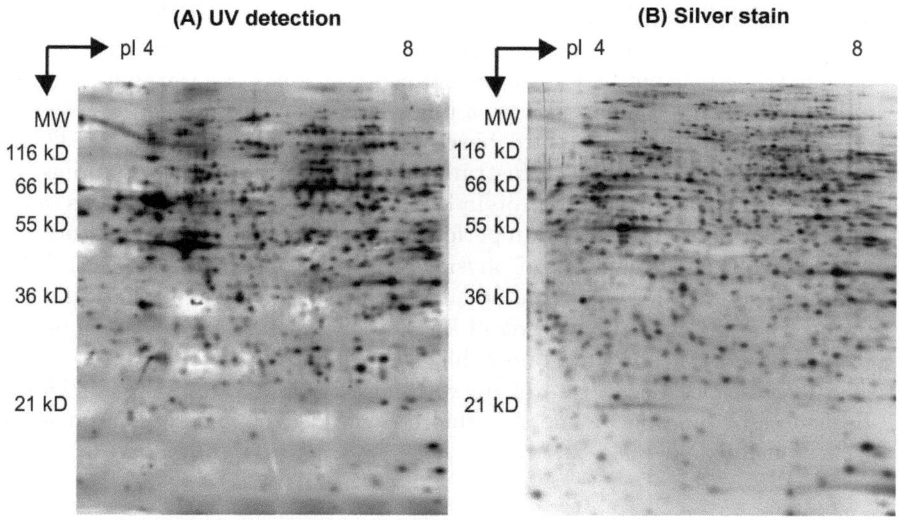

Fig. 9.11. Two–dimensional polyacrylamide gel electrophoresis (2D–PAGE). Spot pattern of EA.hy 926 epithelial cells whole cell lysate. Size 6 × 7 cm^2. **(A)** UV detection, inverse contrast representation. Exposure was 35 mJ/cm^2. **(B)** Silver stain on identically prepared gel (Reprinted with permission from [18]. Copyright 2003 Academic Press Inc Elsevier Science)

In order to detect a change in fluorescence lifetime by the mechanisms discussed above, there may be several possibilities for the realization of a working chip for protein analysis. Figure 9.10 shows three differently designed supports for protein immobilization, intended to selectively bind certain protein species, whereas others are not bound. Comparison shows that LDH and trypsin are bound solely on surface 9.10a, but not on 9.10b.

An interesting side aspect is that UV fluorescence can also be used in 2D gels, an alternative proteomic technique. Fig. 9.11 shows that detecting intrinsic protein fluorescence in 2D gels has a sensitivity comparable to silver staining or staining with fluorescent dyes in the visible range.

This was unexpected, since the brightness of intrinsic fluorescence is only a fraction compared to that of fluorescent dyes. This possibly provides an alternative to previous staining methods, where selective staining of proteins without producing background signals is a problem.

An additional strategy that may be applied to separate fluorescence signals of bound proteins from those of their ligands is electrophoresis on spots prior to fluorescence detection. This may be performed by covering the chip with a polyacrylamide gel, breaking antigen–antibody binding by low pH or increased temperature, and then moving the proteins by electrophoresis to the side. Previously bound antigens are now in some distance to the spot and can be detected there without fluorescence background from antibodies. This may give additional information about antigen binding and mobility.

Fluorescence lifetime measurements were performed by utilizing the frequency tripled output of a Ti:Sa laser tuned to 280 nm. The system consists of a Spectra Physics Tsunami Laser, a pulse picker to increase the pulse to pulse spacing to 250 ns, and a third harmonic generator from GWU. A TCSPC device (SPC–300, Edinburgh Instruments) was used for signal scanning.

We tested binding of the protein kinesin to microtubules, a process important for cellular motion. Kinesin performs vesicle transport along microtubules and is involved in a number of physiological processes and diseases [19]. Figure 9.12a shows fluorescence decay curves of surface-attached microtubules as probe with and without binding of kinesin as target. Figure 9.12b shows the reverse case with kinesin as immobilized probe and microtubules as target. The fluorescence decay rate of microtubules is slower than that of kinesin. Influences of binding partners to fluorescence decay times on the array are clearly visible.

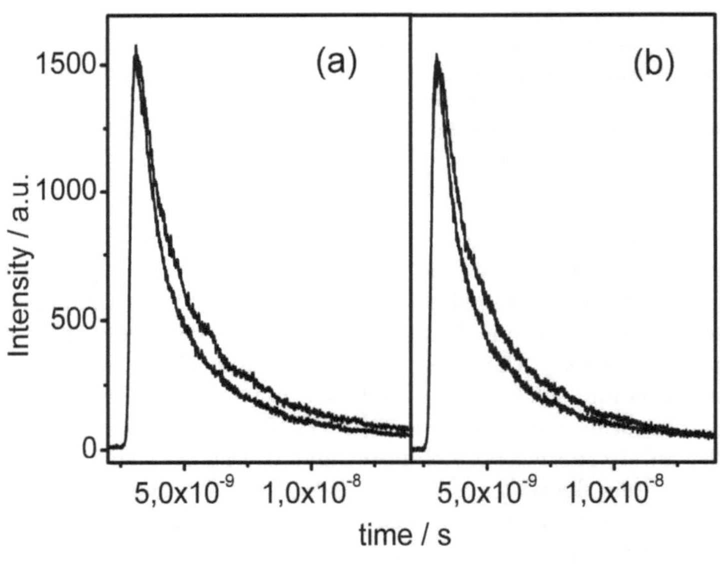

Fig. 9.12. **(a)** Fluorescence decay curve for microtubules immobilized on the chip surface (slow decay), and after binding of kinesin to the microtubules (fast decay). The decay rate is faster upon binding of kinesin, since the isolated kinesin has a shorter fluorescence lifetime, which causes the observed faster overall decay. **(b)** Decay curve for the immobilized kinesin (fast decay), and after binding of microtubules to the kinesin probe (slow decay)

9.3.4 Discussion

The fabrication of protein microarrays is challenging, because due to the generally high variations between proteins and their binding needs, every single protein has to be checked for its own optimal immobilization conditions. This will be a time and resource consuming task, particularly if protein arrays with many proteins in their native states are under consideration.

Additional difficulties have to be overcome if non-modifying protein detection methods are to be applied. A new non-modifying detection method is intrinsic UV–detection.

Apparently this saves material and costs. Furthermore, omitting of staining speeds up the whole procedure, an important aspect with regard to pharmaceutical screening purposes. Also, working with unaltered proteins reduces errors caused by the staining process. This includes malfunctions due to dyes and tags or false quantification due to variations in staining yield. Detection of intrinsic fluorescence is faster and cheaper than mass spectrometry. Sample handling and reproducibility is comparative to standard fluorescence detection procedures. Its great advantages derive from economic material use, short analysis times, and handling of samples in native, non-modified states.

9.4 Genetic Diagnostics with Unlabelled DNA

In recent years, nucleic acid chip technology has been a subject of growing interest for clinical diagnostics as well as for sequencing DNA and cDNAs, for partial sequencing of clones, for single nucleotide polymorphism (SNPs) studies, and for identification of expressed genes. Nucleic acid chips are based on the method of sequencing by hybridization, where unknown DNA fragments are hybridized to complementary nucleic acid sequences, which are immobilized on a solid surface in an array format. The main variables in this process are the attachment of the nucleic acid sequences to a solid surface, the conditions for hybridization, and the detection of the hybridized DNA sequences.

Currently, various techniques are used to detect hybridized DNAs/RNAs, many described in other chapters of this book. Most of them use PCR for amplification, and labelling procedures such as fluorescent, colorimetric or radioactive tags for detection. Also, a number of approaches have been made using stable isotope as tags [20,21]. Indirect methods such as ultraviolet matrix-assisted laser desorption / ionization mass spectrometry (MALDI–MS) [22–26] limit the size of the DNA samples examined to around 50 to 80 bases.

These disadvantages can be avoided by using peptide nucleic acid (PNA) microarray chips [27–36]. With this microarray chip, label-free and PCR-free DNA diagnostics should become possible [3, 37–42]. PNA is a synthesized DNA analog in which both the phosphate and the deoxyribose of the DNA backbone are replaced by polypeptides (see Fig. 9.13). These DNA analogs possess the ability to hybridize with complementary DNA or RNA sequences.

Thus, PNA–chips can be used in the same way as DNA chips. Two major advantages of PNA over DNA are the neutral backbone and the increased strength of the PNA/DNA pairing. The lack of charge repulsion also improves the hybridization properties of DNA/PNA duplexes as compared to DNA/DNA duplexes; the increased binding strength usually leads to a higher sequence discrimination for PNA–DNA hybrids compared to DNA–DNA hybrids, which is particularly important for SNP studies [43–48]. In contrast to the DNA backbone, which contains phosphates, the PNA backbone is free of phosphates; therefore, a technique that identifies the presence of these phosphates in a molecular surface layer would allow the use of even unmodified genomic DNA for hybridization on a microarray chip, rather than using amplified DNA fragments labelled with radioisotopes, stable isotopes, or fluorescent probes.

The detection of unlabelled DNA fragments hybridized to complementary PNAs via the detection of negative phosphate ions (PO_2^- and PO_3^-) or phosphate sugar compound fragments can be achieved in a very efficient way with time–of–flight secondary ion mass spectrometry (TOF–SIMS). In comparison to MALDI–MS, where a laser is used to desorb molecules [49,50], TOF–SIMS analysis utilizes a technique in which the sample is bombarded with a fo-

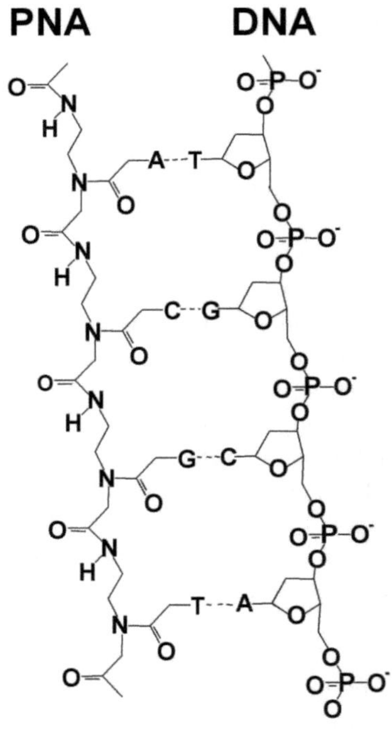

Fig. 9.13. Structure of deoxyribonucleic acid (DNA) and peptide nucleic acid (PNA)

cused, energetic ion beam that sputters atoms, clusters or large molecules (up to 10,000 amu) off the surface [51–54]. Most of these originate from the top monolayer. The ionized sputtered secondary particles can be directly detected with a time–of–flight mass spectrometer (TOF–MS).

Two types of ion source are particularly suited for TOF–SIMS. The first one produces positive noble gas ions (usually argon or xenon) either by electron impact (EI) or in a plasma created by a discharge. The ions are then extracted from the source region, accelerated to the chosen energy and focused in an electrostatic ion optical column. More recently it has been shown that the use of primary polyatomic ions such as SF_5^+, created in EI sources, could enhance the molecular secondary ion yield by several magnitudes [38, 55].

The second type of ion gun produces positive ions from a liquid metal (gallium, indium or gold) [56]. Because the ion production occurs in a very small volume, gallium liquid metal ion sources have a very high brightness. As a result, the ion beam may be focused to a fine spot, resulting in a spot size of 0.2 μm at 8–10 keV or about 20 nm at 30 keV, while being pulsed at frequencies of up to 50 kHz and rastered at the same time.

All ion gun optical columns provide deflection plates for scanning the ion beam over areas adjustable from many square millimeters to a few square micrometers. They have been adapted for pulsing by the introduction of deflection plates, which rapidly sweep the beam across an aperture. Applying an ion beam bunching technique, ion pulses of less than 1 ns width can be produced.

In a TOF mass analyzer (Fig. 9.14), all sputtered ions are accelerated with an extraction voltage of U_0 to a given potential, so that all ions possess the same kinetic energy. The ions are then allowed to drift through a field-free drift path of a given length L before striking the detector. According to the equation $(mL^2)/(2t^2) = qU_0$, light ions travel the fixed distance through the flight tube more rapidly than identically charged heavy ions. Thus, the measurement of the time, t, of ions with mass–to–charge ratio, m/q, provides a simple means of mass analysis with $t^2 = (mL^2)/(2qU_0) \propto m/q$. Because a very well defined start time is required for the flight time measurement, the primary ion gun has to be operated in a pulsed mode in order to be able to deliver discrete primary–ion packages [57]. Electric fields (e.g., ion mirrors [58, 59] or electrical sectors [60, 61]) are used in the drift path in order to compensate for different incident energies and angular distributions of the secondary ions. For good mass resolution, the flight path must be sufficiently long (1–1.5 m), and very sophisticated high frequency pulsing and counting systems must be employed to time the flight of the ion to within a sub–nanosecond. One great advantage of TOF–MS is its ability to provide simultaneous detection of all masses of the same polarity. Charge compensation for insulator analysis is possible using pulsed low-energy electrons, which are introduced during the time interval between ion pulses. With such a TOF–SIMS instrument, the useful mass range is extended beyond 10,000 amu; the mass resolution,

$m/\Delta m$, is $\approx 10,000$ with simultaneous detection of all masses; and within each image, all masses can be detected.

In our development of PNA microarrays, thiols such as alkanethiols or dithiobissuccinimidyl propionate (DTSP) [37, 41, 51] have proven to be the molecules of choice in the formation of self-assembled monolayers (SAMs) [62], which are the basis for PNA immobilization. This has been confirmed during our study of SAMs with TOF–SIMS, as they are simple to handle and can easily be detected on gold- or silver-coated glass slides or Si–wafers.

We investigated different methods in the construction of these PNA microarrays. One method used to immobilize PNA on a gold surface is to build up a thiol–SAM, where the thiol contains a functional end–group. This functional end–group can be a carboxylic acid or an amino group. Next, the PNA is attached to this SAM by using a coupling reagent, which can either link an $-NH_2$ group to a $-COOH$ group [63, 64] or two $-NH_2$ groups together. Examples of such coupling reagents are EDC (1–ethyl–3–(3–dimethylaminopropyl)–carbodiimide hydrochloride) and DSC (disuccinimidyl carbonate). The second method uses PNA synthesized with a thiol linker, which can be readily immobilized or spotted onto a gold surface. In a second step, the surface is covered

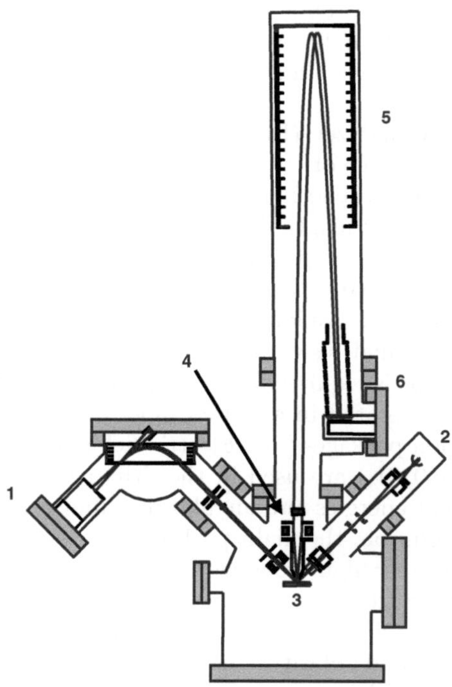

Fig. 9.14. Conceptional diagram of a TOF–SIMS instrument; **(1)** electron impact ion gun (Ar^+ or Xe^+); **(2)** liquid metal ion gun (Ga^+); **(3)** sample holder; **(4)** secondary ion optics; **(5)** reflectron; **(6)** detector

with a layer of other thiol molecules, preferably with a shorter chain length than the linker molecule of the synthesized PNA. These thiols that are used for saturating the surface contain a negatively charged end–group (e.g. a carboxylic acid) in order to prevent DNA, which is also negatively charged, from associating and non-specifically binding to the gold surface.

TOF–SIMS was used to characterize and optimize the various immobilization processes, which depend on a variety of parameters such as immobilization time and concentration. These must be iteratively optimized in order to achieve good hybridization conditions. Preliminary investigations of DNA and PNA fragments immobilized on silanized surfaces have shown that negative mass spectra can be used to identify DNA and PNA fragments [37, 42].

Figure 9.15 depicts parts of negative TOF–SIMS spectra obtained from immobilized DNA and PNA layers. The figure on the left shows the signal obtained from the DNA layer. Besides the deprotonated $(M–H)^-$ signals of the bases cytosine, thymine, adenine and guanine, there are two prominent phosphate peaks visible, PO_2^- and PO_3^-. The figure on the right shows a negative spectrum for immobilized PNA. Again, the deprotonated $(M–H)^-$ signals of the bases cytosine and thymine are visible. Note, however, that the two major DNA-specific phosphate peaks are very small in comparison to the DNA spectrum and are mainly due to contaminants. Some ion peaks caused by contaminants such as bromine are also observed. However, these do not cause any interference because they can be simply separated out by using a mass spectrometer with high mass resolution. A comparison between the PNA and the DNA spectrum demonstrates that the masses corresponding to PO_2^-, PO_3^- provide the best way for detecting the presence of DNA; they can be used to precisely distinguish between DNA and PNA.

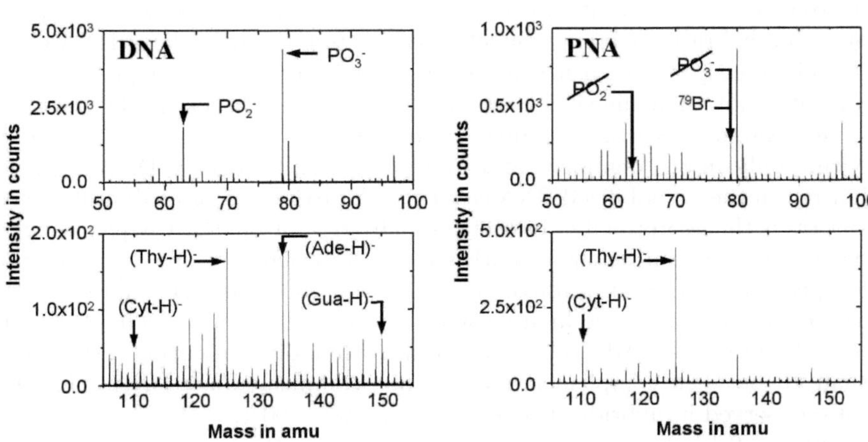

Fig. 9.15. Negative TOF–SIMS spectra (50 to 155 amu) obtained from immobilized PNA and DNA layers. DNA sequence: 5′–ACATGCTGCTAGC–3′; PNA sequence: 5′–TTTTCCCTCTCTC–3′.

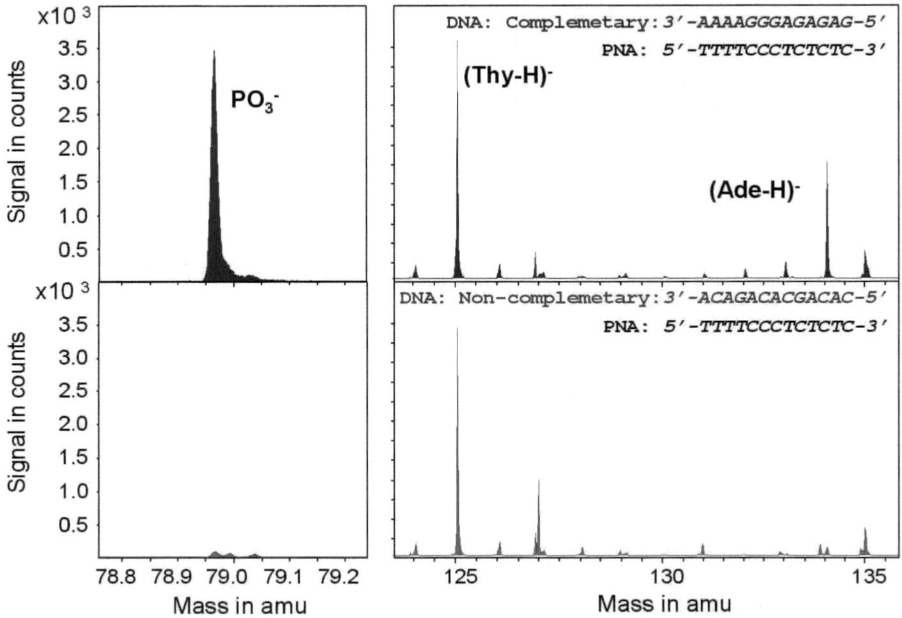

Fig. 9.16. Negative TOF–SIMS spectra (60 to 130 amu) obtained from hybridization experiments in which complementary and non-complementary DNA sequences were hybridized to a PNA sequence

After optimizing the immobilization steps, hybridization experiments were carried out. Partially complementary DNA sequences were removed from the microarray by appropriate washing techniques. Figure 9.16 shows mass spectra of such a hybridization experiment. As expected, the hybridized DNA can be unambiguously distinguished from the PNA by the dominant peaks of PO_3^- and $(Ade–H)^-$. Adenine can be used as an identifier for positive hybridization in this example because the sequences were selected in such a way that adenine occurs only in the complementary and non-complementary DNA sequences but not in the immobilized PNA sequence. A further major peak visible in the spectra is the deprotonated base signal $(Thy\text{-}H)^-$ of the PNA sequence, which has similar concentrations in both spectra. Interferences due to some minor ion peaks occurring at approximately the same mass as PO_3^- and $(Ade–H)^-$ are separated out by the high mass resolution of $m/\Delta m > 7000$. Integration over the PO_3^- and $(Ade–H)^-$ peaks resulted in a discrimination ratio of more than 10 between complementary and non-complementary DNAs. The best ratio observed in hybridization experiments was 200:1.

Figure 9.17 shows an image obtained from a PNA microarray that was produced by Hoheisel et al., DKFZ, Heidelberg. Two different PNA sequences with different concentrations and three different types of spacer were immobilized directly onto a gold surface using robotic spotting techniques [36].

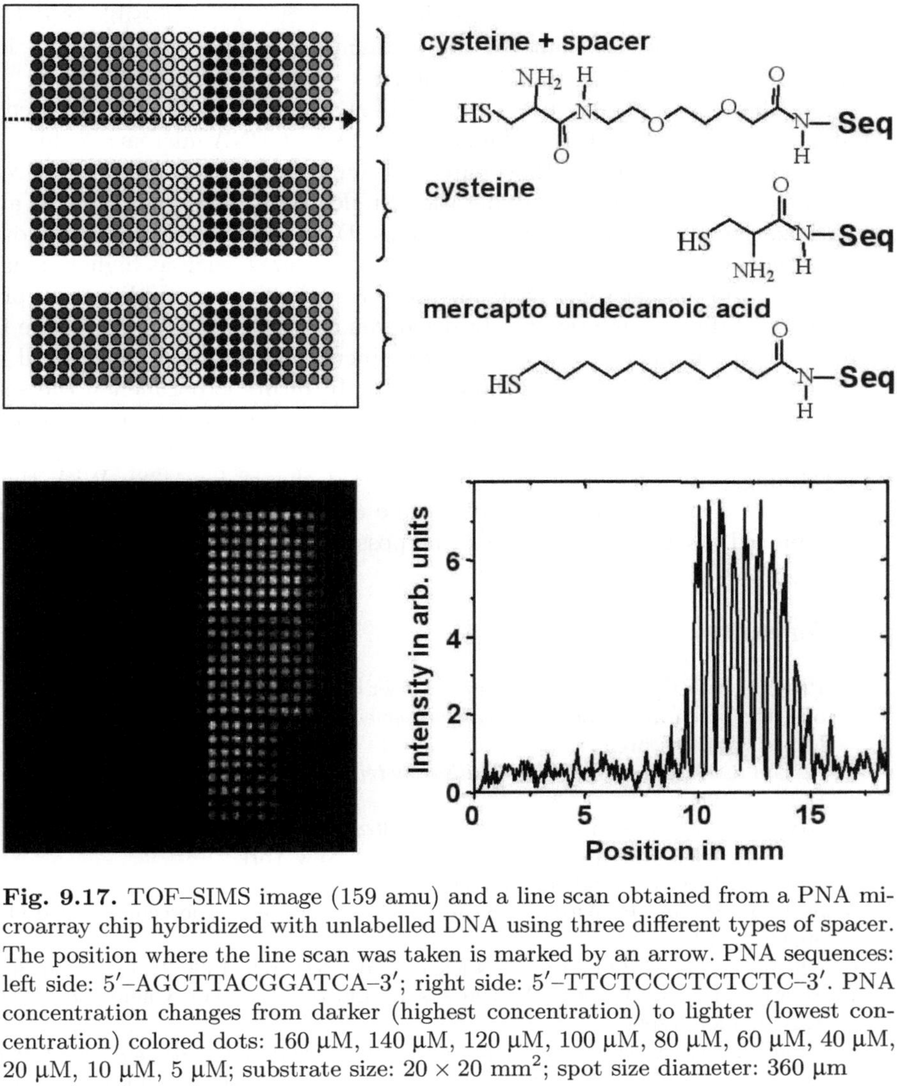

Fig. 9.17. TOF–SIMS image (159 amu) and a line scan obtained from a PNA microarray chip hybridized with unlabelled DNA using three different types of spacer. The position where the line scan was taken is marked by an arrow. PNA sequences: left side: 5′–AGCTTACGGATCA–3′; right side: 5′–TTCTCCCTCTCTC–3′. PNA concentration changes from darker (highest concentration) to lighter (lowest concentration) colored dots: 160 µM, 140 µM, 120 µM, 100 µM, 80 µM, 60 µM, 40 µM, 20 µM, 10 µM, 5 µM; substrate size: 20×20 mm^2; spot size diameter: 360 µm

Unlabelled DNA, which was complementary to one of these sequences, was hybridized to this chip. The TOF–SIMS analysis shows that hybridized unlabelled DNA could be detected with good discrimination at the complementary PNA positions. The highest signal could be obtained at those PNA positions that had the longest spacer length. Also, the line scan shows that at these positions, no significant changes between the eight highest PNA immobilization concentrations were observed. Note that only a small fraction of a monolayer was needed for analysis using an Ar^+ ion beam with a spot size of 30 µm in diameter. Additional experiments showed that it is even possible to detect DNA in an area of less than 100 nm in diameter using a focussed Ga^+ ion beam, corresponding to attomole sensitivity.

The data clearly show that TOF–SIMS is a powerful technique for identifying unambiguously hybridized unlabelled DNA on PNA microarray chips by detecting the phosphate or phosphate-containing compounds present in DNA. It is also very suitable for studying the complexity of the immobilization and hybridization process. Employing unlabelled DNA has several advantages over using fluorescent and radioactive labelling procedures, such as higher signal–to–noise ratio, higher sensitivity, absence of a labelling or amplification procedure, and direct analysis of hybridized genomic DNA. Particularly, the increase in the number of phosphates with increasing sequence length will be advantageous for sequencing genomic DNA. In future experiments, the spot size will be reduced to smaller than $10 \times 10\ \mu m^2$, the repetition rate will be increased up to 200 kHz, and the sensitivity will be further improved by using polyatomic primary ions such as SF_5^+ or gold cluster ions [38]. With these experimental improvements, analysis time of only a few minutes for 10,000 immobilized PNA spots should become possible for genetic diagnostics.

References

1. Project "SCREEN" for foundations of laser based (marker–free) screening methods, for which all authors of this chapter acknowledge funding by the German Ministry of Research
2. Azzam RMA and Bashara NM, *Ellipsometry and Polarized Light.* North Holland Physics, 1987
3. Arwin H in: "Physical Chemistry of Biological Interfaces", Baszkin A., Norde W. (ed.), 2000, 577–607, Marcel Dekker Inc, New York
4. Silin V, Plant A, Trends in Biotech, 1997 15: 353–359
5. Sojka B, Piunno P, Wust C., Krull U., Appl. Biochem. Biotechnol. 2000 Oct, 89(1): 85–103
6. Chrisey L, Lee G, O'Ferrall E, Nucleic Acids Research, 1996 24(15): 3030–3039
7. Strother T, Cai W, Zhao X, Hamers R, Lloyd M, J. Am. Chem. Soc., 2000 122(6): 1205–1209
8. Greulich KO, 1998. *Intrinsic fluorescence techniques for studies on protein–protein and protein–RNA interaction in RNP Particles.* In: Schenkel J (ed), Lab Manual: RNP Particles, splicing and autoimmune diseases. Springer, Heidelberg, 48–71
9. Lakowicz JR, 1983. *Protein Fluorescence, Principles of Fluorescence Spectroscopy.* Plenum Publishing Coorporation, New York, 354–357
10. Schwarzwald R, Greulich KO, 1988. *Tyrosine fluorescence energy transfer as a probe for protein–DNA interactions.* Ber Bunsenges Phys Chem 92: 447
11. Alcala JR, Gratton E, and Prendergast FG, 1987. *Fluorescence lifetime distributions in proteins.* Biophys J 51: 597–604
12. Petrich JW, Chang MC, McDonald DB, and Fleming GR, 1983. *On the origin of nonexponential fluorescence decay in tryptophan and its derivatives.* J Am Chem Soc 105: 3815–3832

13. Raghavachari N, Bao YP, Li G, Xie X, and Müller UR, 2003. *Reduction of autofluorescence on DNA microarrays and slide surfaces by treatment with sodium borohydride.* Anal Biochem 312: 101–105
14. Hermanson GT, 1996. *Bioconjugate Techniques.* Academic Press Inc, San Diego, CA
15. Afanassiev V, Hanemann V, Wölfl S, 2000. *Preparation of DNA and protein micro arrays on glass slides coated with an agarose film.* Nucleic Acids Res 28: e66
16. Peluso P, Wilson DS, Do D, Tran H, Venkatasubbaiah M, Quincy D, Heidecker B, Pointdexter K, Tolani N, Phelan M, Witte K, Jung LS, Wagner P, and Nock S, 2003. *Optimizing antibody immobilization strategies for the construction of protein microarrays.* Anal Biochem 312: 113–124
17. Hermanson GT, Mallia AK, and Smith PK, 1992. *Immobilized Affinity Ligand Techniques.* Academic Press Inc., San Diego, CA
18. Roegener J, Lutter P, Reinhardt R, Bluggel M, Meyer HE, and Anselmetti D, 2003. *Ultrasensitive Detection of Unstained Proteins in Acrylamide Gels by Native UV Fluorescence.* Anal Chem 75: 157–159
19. Böhm KJ, Stracke R, Mühlig P, and Unger E, 2001. *Motor protein–driven unidirectional transport of micrometer–sized cargoes across isopolar microtubule arrays.* Nanotechnology 12: 238–244
20. Arlinghaus HF, Kwoka MP, Guo XQ, Jacobson KB, 1997. *Multiplexed DNA Sequencing and Diagnostics by Hybridization with Enriched Stable Isotope Labels,* Anal. Chem. 69, 1510–1517
21. Jacobson KB, Arlinghaus HF, 1992. *Development of Resonance Ionization Spectroscopy for DNA Sequencing and Genome Mapping,* Anal. Chem. 64, 315A–328A
22. Jurinke C, van den Boom D, Cantor RC, Köster H, 2002. *The use of MassARRAY technology for high throughput genotyping,* Adv. Biochem. Eng. Biotechnol. 77, 57–74
23. Kirpekar F, Berkenkamp S, Hillenkamp F, 1999. *Detection of double–stranded DNA by IR– and UV–MALDI mass spectrometry,* Anal. Chem. 71, 2334–2339
24. Kirpekar F, Nordhoff E, Larsen LK, Kristiansen K, Roepstorff P, Hillenkamp F, 1998. *DNA sequence analysis by MALDI mass spectrometry,* Nucleic Acid Research 26(11), 2554–2559
25. Little DP, Cornish TJ, O'Donnell MJ, Braun A, Cotter RJ, Köster H, 1997. *MALDI on a Chip: Analysis of Arrays of Low–Femtomole to Subfemtomole Quantities of Synthetic Oligonucleotides and DNA Diagnostic Products Dispensed by a Piezoelectric Pipet.,* Anal. Chem. 69, 4540–4546
26. O'Donnell MJ, Tang K, Köster H, Smith CL, Cantor CR, 1997. *High–Density, Covalent Attachment of DNA to Silicon Wafers for Analysis by MALDI–TOF Mass Spectrometry,* Anal. Chem. 69, 2438–2443
27. Chakrabarti MC, Schwarz FP, 1999. *Thermal stability of PNA/DNA and DNA/DNA dupexes by differential scanning calorimetry,* Nucleic Acids Research 27(24), 4801–4806
28. Hoheisel JD, 1997. *Oligomer–chip technology,* TIBTECH 15, 465–469
29. Matysiak S, Reuthner F, Hoheisel JD, 2001. *Automating parallel peptide synthesis for the production of PNA library arrays,* BioTechniques 31, 896–904
30. Nielsen PE, 1999. Applications of peptide nucleic acids, Current Opinion in Biotechnology 10, 71–75

31. Nielsen PE, 1997. *Peptide nucleic acid (PNA) from DNA recognition to antisense and DNA structure*, Biophysical Chemistry 68, 103–108
32. Ratilainen T, Holmén A, Tuite E, Nielsen PE, Nordén B, 2000. *Thermodynamics of sequence-specific binding of PNA to DNA*. Biochemistry, 39, 7781–7791
33. Ratilainen T, Holmén A, Tuite E, Haaima G, Christensen L, Nielsen PE, Nordén B, 1998. *Hybridization of peptide nucleic acid*, Biochemistry 37, 12331–12342
34. Ray A, Norden B, 2000. *Peptide nucleic acid (PNA): medical and biotechnical applications and promise for the future*, The FASEB Journal 14, 1041–1060
35. see articles in: Peptide Nucleic Acids: Protocols and Applications, 1999, Nielson PE, Egholm M, eds., Horizon Scientific Press, Wymondham, UK
36. Weiler J, Gausepohl H, Hauser N, Jensen ON, Hoheisel JD, 1997. *Hybridization based DNA screening on peptide nucleic acid (PNA) oligomer arrays*, Nucleic Acids Research 25(14), 2792–2799
37. Arlinghaus HF, Ostrop M, Friedrichs O, Feldner J, Gunst U, Lipinsky D, 2002. *DNA sequencing with ToF–SIMS*, Surf. Interf. Anal. 33, 35–39
38. Arlinghaus HF, Höppener C, Drexler J, 2000. *TOF–SIMS Characterization of DNA and PNA Biosensor Chips*, in: Secondary Ion Mass Spectrometry SIMS XII, Benninghoven, A., Bertrand, P., Migeon, H.–N., Werner, H.W., eds., Elsevier, Amsterdam, 951–954
39. Arlinghaus HF, Kwoka MN, Jacobson KB, 1997. *Analysis of Biosensor Chips for Identification of Nucleic Acids*, Anal. Chem. 69, 3747–3753
40. Arlinghaus HF, Ostrop M, Friedrichs O, Feldner J, 2003. *Genome Diagnostic with TOF–SIMS*, Appl. Surf. Sci., 203–204, 689–692
41. Feldner JC, Ostrop M, Friedrichs O, Sohn S, Lipinsky D, Gunst U, Sohn S, Arlinghaus HF, 2003. *TOF–SIMS Investigation of the Immobilization Process of Peptide Nucleic Acids*, Appl. Surf. Sci., 203–204, 722–725
42. Höppener C, Drexler J, Ostrop M, Arlinghaus HF, 2000. *Investigation of the Immobilization Process of Nucleic Acid, in: Secondary Ion Mass Spectrometry SIMS XII*, Benninghoven, A., Bertrand, P., Migeon, H.–N., Werner, H.W., eds., Elsevier, Amsterdam, 915–918
43. Griffin TJ, Smith LM, 2000. *Single–nucleotide polymorphism analysis by MALDI–TOF mass spectrometry*, TIBTECH, 18, 77–84
44. Ross PL, Lee K, Belgrader P, 1997. *Discrimination of single–nucleotide polymorphisms in human DNA using peptide nucleic acid probes detected by MALDI–TOF mass spectrometry*, Anal. Chem. 69, 4197–4202
45. Storm N, Darnhofer B, van den Boom D, Rodi CP, 2003. *MALDI–TOF mass spectrometry–based SNP genotyping*, Methods in Mol. Biol. 212, 241–62
46. Syvänen A, 2001. *Accessing genetic variation: Genotyping single nucleotide polymorphisms*, Nature Reviews 2, 930–942
47. Wang J, Rivas G, Cai X, Chicharro M, Parrado C, Dontha N, Begleiter A, Mowat M, Palecek E, Nielsen PE, 1997. *Detection of point mutation in p53 gene using a peptide nucleic acid biosensor,* Analytica Chimica Acta 344, 111–118
48. Wittung–Stafshede P, Rodahl M, Kasemo B, Nielsen P, Norden B, 2000. *Detection of point mutations in DNA by PNA–based quarz–crystal biosensor*, Colloids and Surfaces A: Physicochem. Eng. Aspects 174, 269–273
49. Berkenkamp S, Kirpekar F, Hillenkamp F, 1998. *Infrared MALDI mass spectrometry of large nucleic acids*, Science 281, 260–262
50. Karas M, Bachman D, Bahr U, Hillenkamp F, 1987. Int. J. Mass Spectrom. Ion Processes, 78, 53–68

51. Arlinghaus HF, 2002, *Static Secondary Ion Mass Spectrometry (SSIMS)* in: Surface and Thin Film Analysis. Principles, Instrumentation, Applications, Bubert H, Jenett H, eds., Wiley–VCH, 86–106
52. Benninghoven A, Rüdennauer FG, Werner HW, 1987. *Secondary Ion Mass Spectrometry*, Wiley, New York
53. see articles in the Proceeding of Secondary Ion Mass Spectrometry SIMS II – SIMS XIII
54. Vickerman JC, Briggs D, 2001. *TOF–SIMS, Surface Analysis by Mass Spectrometry*, IM Publication, Charlton, UK
55. Stapel D, Brox O, Benninghoven A, 1999. Appl. Surf. Sci. 140, 156–67
56. Prewett PD, Jefferies DK, 1980. J. Phys. D 13, 1747–1755
57. Niehuis E, Heller T, Feld H, Benninghoven A, 1987. J. Vac. Sci. Technol. A 5, 1243
58. Karataev VI, Mamyrin BA, Shmikk DV, 1972. Sov. Phys. Techn. Phys. 16, 1177
59. Schueler BW, 1992. Microsc. Microanal. Microstruct. 3, 119
60. Iltgen K, Bendel C, Niehuis E, Benninghoven A, 1997. J. Vac. Sci. Technol. A 15, 460
61. Sakurai T, Matsuo T, Matsuda H, 1985. Int. J. Mass. Spectrom. Ion Phys. 63, 273
62. Schreiber F, 2000. *Structure and growth of self–assembling monolayers*, Progress in Surface Science 65, 151–256
63. Huang E, Zhou F, Deng L, 2000. *Studies of surface coverage and orientation of DNA molecules immobilized onto preformed alkanethiol self–assemled monolayers*, Langmuir 16, 3272–3280
64. Kröger, K., Jung, A., Barzen, C., Gaulitz, G., 2002, *Versatile biosensor surface based on peptide nucleic acid with label free and total internal reflection fluorescence detection for quantification of endocrine disruptors*, Analytical Chimica Acta, 469(1), 37–48

Part II

DNA Microarrays

10

Analysis of DNA Sequence Variation in the Microarray Format

Ulrika Liljedahl, Mona Fredriksson, and Ann-Christine Syvänen

10.1 Introduction

Single nucleotide polymorphisms (SNPs) are sequence positions, where more than one nucleotide is observed when DNA sequences of multiple individuals within a population or between populations are compared. SNPs are the most frequent type of genetic variation in the human genome, and they occur at one out of every thousand to two thousand nucleotides. Following the completion of the draft sequence of the human genome [1, 2], it has become feasible to compare DNA sequences from multiple individuals and populations both experimentally and in silico, to identify large sets of SNPs. Today more than four million SNPs are included in public databases, and a large fraction of these SNPs have been assigned to a defined position in the genome (www.ncbi.nlm.nih.gov/SNP). The number of SNPs with known allele frequencies in various populations is also growing rapidly.

Depending on their genomic locations, the phenotypic consequences of the SNPs differ. SNPs in coding regions of genes may alter the amino acid sequence of the encoded proteins, thus affecting their structure and function, and consequently their physiological role. SNPs located in the regulatory regions of a gene may affect the binding of transcription factors, thereby influencing the expression level of the gene. Most of the SNPs are located in non-coding regions of the genome, where they have no known impact on the phenotype of an individual. These SNPs are useful as genetic markers in forensic identification, in tissue typing, for population genetic studies and evolutionary studies. SNPs (point mutations) causing monogenic disorders have been routinely analyzed for diagnostics and identification of disease carriers for more than a decade. In pharmacogenetics, SNPs in genes for drug metabolizing enzymes are analyzed to assess an individual's response to drug treatment [3]. As molecules other than drug metabolizing enzymes, such as drug receptors or transporters, are becoming targets for pharmacogenetic analysis [4], this field is a rapid growing area of SNP typing today. SNPs in candidate genes are often used as markers in association studies aiming at identifying genes predisposing to multifactorial disorders. The hope that SNPs may be useful as markers in genome–wide association or linkage studies to identify these genes, has stimulated efforts to increase throughput and decrease the cost of methods for SNP genotyping.

Most of the currently used genotyping methods depend on amplification of the genomic region of interest by the polymerase chain reaction (PCR) [5–7] to provide sufficient sensitivity and specificity to detect a SNP among the 3×10^9 base pairs of DNA that constitute the human genome. However, today PCR is the major bottle–neck for high throughput genotyping of previously known SNPs at different locations of the genome due to the difficulty of performing multiplex amplification [8]. In applications where complete genes or exons are resequenced to detect previously unknown mutations, the problem of designing multiplex PCR is avoided to some extent.

The microarray format is attractive for analyzing previously known SNPs as well as for resequencing because of the potential of increasing the throughput of the assay by simultaneous and highly parallel analysis of multiple sequence variants. The cost for the reagents is also reduced owing to the miniaturized format of the microarrays. The microarray format was first designed for expression profiling, where typically very large numbers of mRNA species are analyzed in a relatively small number of samples [9]. The standard microscope slide format used for expression profiling, where one sample is analyzed

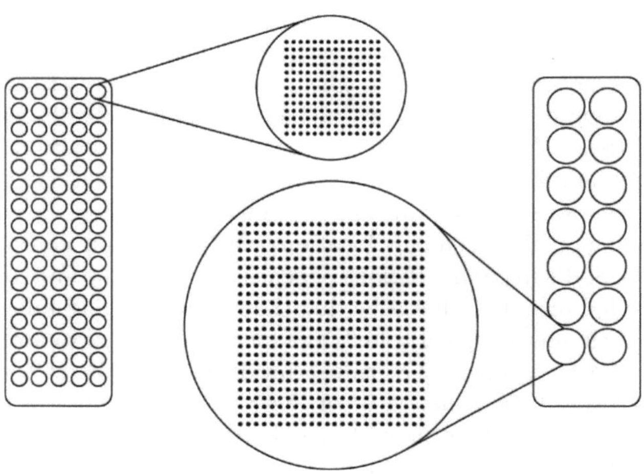

Fig. 10.1. 'Array–of–arrays' conformation. A standard microscope slide is divided into 80 subarrays with a diameter identical to that of a 384–well microtiter plate reaction well (left image). Up to $14 \times 13 = 182$ oligonucleotide spots can be printed per subarray at a center–to–center distance of 200 μm. If more SNPs are to be analyzed the 'array–of–arrays' format can be converted to a format with subarrays with the same diameter as a reaction well of a 96–well microtiter plate. In this case 14 separate subarrays fit per slide and $24 \times 24 = 576$ oligonucleotide spots can be printed in each subarray (right image)

per slide, is not practical for SNP genotyping studies, where a large number of samples are to be analyzed for each set of SNPs.

To circumvent this problem an 'array–of–arrays' conformation, that allows parallel analysis of up to 80 samples for each set of SNPs on a single microscope slide [4, 10, 11], has been devised (Fig. 10.1). Each microarray is divided into multiple separate reaction wells by a silicon rubber grid that is placed on the microscope slide (Fig. 10.2). A similar 'array–of–arrays' concept is also utilized in a 384–well–microtiter plate format instead of using a microscope slide (SNPstream UHT, Orchid Biosciences [12]). The 'array–of–arrays' format was originally devised for genotyping by allele-specific primer extension [10], but the format can equally well be used with all other reaction principles for SNP–typing.

10.2 Principles of Genotyping

Most of the techniques used for analysis of genetic variation are based on either hybridization with short allele specific oligonucleotide (ASO) probes or on the action of DNA modifying enzymes such as DNA–polymerases and ligases to determine the sequence variation.

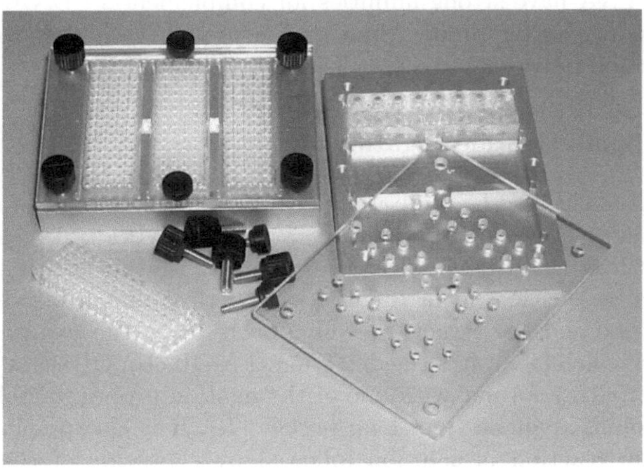

Fig. 10.2. The microarray reaction rack. A custom made aluminium reaction rack that holds three microarray slides is used as an incubation chamber in the microarray based minisequencing reactions. A silicone grid is used to separate the different samples on the microarray. Reusable silicon rubber grids are moulded on an inverted 384– or 96–well microtiter plate using PDMS (polydimethyl siloxane, e.g. Elastosil RT 625A and B, Wacker–Chemie) according to the manufacturer's instruction, followed by cutting the grid to match the size of the slides

10.2.1 Hybridization

In hybridization with ASO–probes, the destabilising effect of a single nucleotide mismatch between an oligonucleotide probe and its target sequence is utilized to distinguish between sequence variants (Fig. 10.3a). The reaction conditions are optimized with respect to ionic strength and temperature to provide maximal discrimination between the two sequence variants. However, the stability of the oligonucleotide–target hybrid is also affected by the sequence flanking the SNP–position, as well as by the secondary structure of the template. Therefore there is no single set of reaction conditions that would provide optimal specificity for all SNPs in multiplexed hybridization assays.

Multiplex analysis using ASOs on microarrays is used in the Affymetrix GeneChip® assay, where the difficulty in assay design is circumvented by using arrays with tens of different allele-specific oligonucleotides for each SNP to be analyzed [13] and by accepting a reduced success rate [14]. Other attempts to circumvent the specificity problem of multiplexed ASO–assays is to employ temperature gradients [15] or electric field gradients (e.g. Nanogen) [16] to the microarrays. In these methods optimal discrimination between match and mismatch is achieved at a specific point of the gradient.

Peptide nucleic acids (PNA) or locked nucleic acids (LNA) can also be used to increase the power of ASO hybridization. Due to their chemical structure, PNA and LNA have strong affinities for complementary DNA, which allows for the use of shorter probes than the natural ASO–probes to improve the discrimination between the SNP alleles [17, 18].

10.2.2 Oligonucleotide Ligation

In the oligonucleotide ligation assay (OLA) [19], the ability of a DNA ligase to discriminate between a match and a mismatch hybridization at the ligation point is utilized. An allele-specific probe and a ligation probe are hybridized to a target sequence, and in the case of a perfect match between the allele-specific probe and the target, the junction between the two probes is closed by ligation which facilitates the detection (Fig. 10.3b). OLA has been adopted to the microarray format with one of the ligation probes immobilized [20] or with immobilized single stem loop probes [21]. It is also possible to perform the ligation reaction in solution followed by capturing of the products on microarrays [22] or microparticles [23] by hybridization to generic tag or zip–code oligonucleotides.

Padlock probes are circularisable oligonucleotide ligation probes with specific target recognition sequences in their 5′ and 3′ ends and a connecting sequence between the target specific regions [24]. When hybridized to its target sequence the two ends of the probe are brought adjacent to each other, and the junction is ligated when there is a perfect match. Proof of principle of highly multiplex padlock probe ligation using ‘molecular inversion probes’ in solution has recently been shown [25]. In this assay the circularized probes are detected by PCR with tagged primers followed by capture on microarrays. Another novel, highly multiplexed ligation assay is used in a bead array format [26].

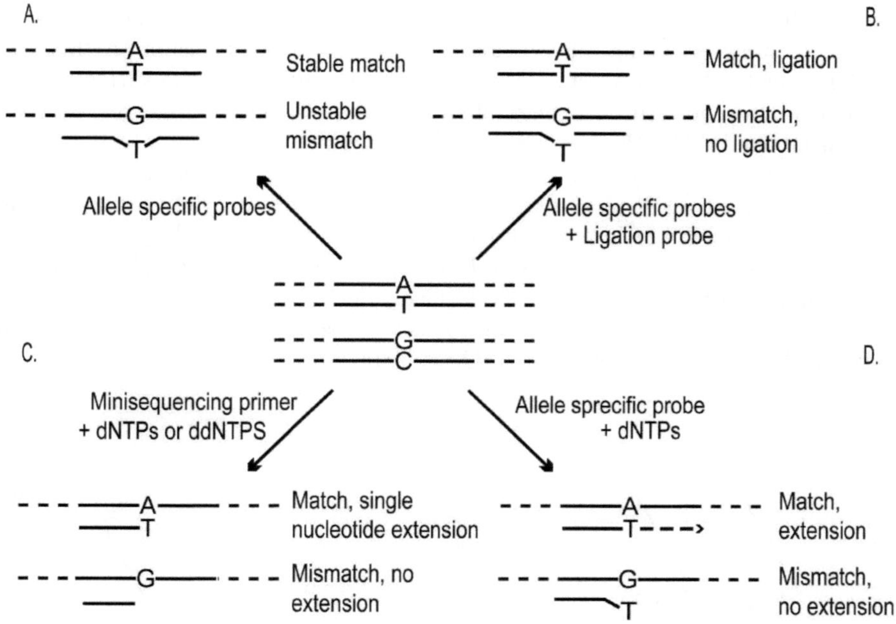

Fig. 10.3. Reaction principles for SNP genotyping. Detection of the A–allele in an A to G transition is shown; the G–allele would be detected analogously. **(a)** Hybridization with allele-specific oligonucleotides (ASO). Two ASO probes are required for each SNP to be analyzed, and a nucleotide near the middle position of the probe is complementary to the allelic variant of the SNP. The reaction conditions are set to allow only perfect matches to be stable and detectable. **(b)** In the oligonucleotide ligation assay (OLA) a ligation probe and an allele-specific probe are used for detection of the allelic variant of the SNP. When there is a perfect match between the allele-specific probe and the target sequence, the junction between the two probes can be closed with a ligase. **(c)** Minisequencing single nucleotide primer extension. A minisequencing primer that anneals immediately adjacent to the SNP–position will be extended with a nucleotide complementary to the nucleotide at the variable site by the action of a DNA polymerase. **(d)** Allele-specific primer extension. A primer with an allele-specific 3′–end anneals to the target sequence. Only in case of a perfect match between the primer and the target sequence, the primer will be extended

10.2.3 DNA Polymerase Assisted Methods

In minisequencing, also denoted single nucleotide primer extension (SNE) and single base extension (SBE), a DNA polymerase is used to extend a detection primer, that anneals immediately adjacent to the site of the SNP, with a labelled nucleotide analogue [27, 28] (Fig. 10.3c). In the microarray format of minisequencing, also denoted arrayed primer extension (APEX), the SNP-specific detection primers are attached covalently to the surface of activated microscope slides through their 5′–end, and their 3′–ends are extended with labelled ddNTPs that are complementary to the nucleotide at the SNP site [4, 29–33] (Fig. 10.4a). The primer extension reaction allows specific genotyping of most SNPs at similar reaction conditions using only a single primer per SNP, which are important features in the multiplexed assays in a microarray format. In a side–by–side comparison with ASO hybridization in the same microarray format, the minisequencing reaction provided ten–fold higher power of discrimination between heterozygous and homozygous genotypes than hybridization with ASO probes [29].

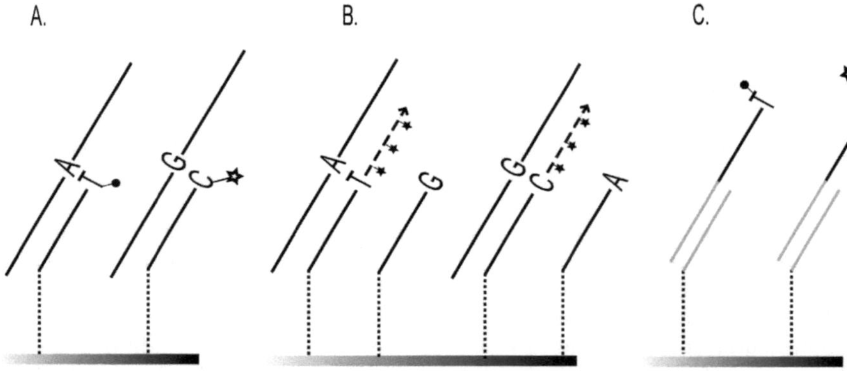

Fig. 10.4. Reaction principles for primer extension on microarrays. Detection of a heterozygous sample is presented. In direct minisequencing on microarrays **(a)** one minisequencing primer for each SNP is immobilised, and multiplex PCR products, fluorescently labelled ddNTPs and a DNA polymerase are added. The primer extension is allowed to proceed on the surface of the array, followed by fluorescence scanning with a laser scanner. For the allele-specific primer extension **(b)** two oligonucleotides with the 3′–nucleotide complementary to the two possible nucleotides of each SNP are immobilized on the array. In the presence of a perfectly matched target sequence the allele-specific oligonucleotide becomes extended by a DNA–polymerase. In the tag array based minisequencing **(c)** cyclic single nucleotide primer extension reactions are carried out in solution in the presence of fluorescently labelled dideoxynucleotides with the minisequencing primers carrying an extra tag–sequence in their 5′–end. Generic arrays of oligonucleotides that are complementary to the tag–sequences are used to capture the product on the microarray after the cyclic minisequencing reactions

DNA polymerases may also be utilized for SNP genotyping by allele-specific primer extension in the microarray format (Fig. 10.3d). In this case two immobilized primers with 3′–ends complementary to either of the nucleotides of the SNP are used [10] (Fig. 10.4b). In this approach, primer extension will only occur when there is a perfect match in the 3′–end of the primer. The allele-specific primer extension reaction is more dependent on the reaction conditions than minisequencing, but its specificity has been increased by analyzing RNA templates in conjunction with reverse transcriptase reactions in the presence of trehalose [10] which has allowed accurate genotyping in a large study where 140,000 genotypes where produced [34]. Another approach for increasing the specificity of allele-specific primer extension is to include apyrase in the reaction to prevent the slower mismatched extension reaction [35].

In an alternative format of the minisequencing system, multiplex cyclic primer extension reactions are performed in solution with primers tailed with 5′–tag sequences. The products of the minisequencing reaction are then captured to complementary tag sequences immobilized on the microarray by hybridization (Fig. 10.4c). This flexible genotyping strategy that was first described for microspheres [36,37], has been used in conjunction with both low-density [38] and high density [39] microarrays. In the latter application, the high density GeneChip® platform was combined with genotyping by single nucleotide primer extension.

The tag–array assays are more flexible to design compared to the minisequencing approach with immobilized extension primers, since the array is generic and thus can be used for many different sets of SNPs. The 'array–of–arrays' format is particularly well suited for genotyping by the flexible tag–array approach [11]. Additionally, the cyclic extension reaction also serves to increase the signal strength. The accuracy of the primer extension reactions in solution allows multiplex quantification of variant alleles present as a small minority (2–5%) of a sample [11].

In the following section two important features of the microarray based assays, namely production of microarrays and labelling strategies will be discussed in more detail.

10.3 Performing the Assays in Practice

10.3.1 Production of Microarrays

The manufacturing of microarrays can be performed through in situ synthesis of oligonucleotides on the surface of the microarray, or by chemical immobilization of presynthesised oligonucleotides. The material used for microarrays must have low autofluorescence and high binding capacity of oligonucleotides. Glass meets these criteria, and in addition it is non-porous, which allow the use of small reaction volumes, and it is durable to both heat and chemicals.

In situ synthesis of oligonucleotides at high density on a glass surface using light directed photolithography has been developed by Affymetrix [40]. These GeneChip® arrays are used for expression analysis and for genotyping using ASO–probes. The photolithographic synthesis proceeds in the 3′–5′ direction, which makes the GeneChip® arrays impossible to use in direct primer extension assays, where a free 3′–end is needed for the polymerase to extend. A proposed strategy for avoiding this limitation is to perform the in situ synthesis in 3′–5′–direction with a subsequent inversion of the oligonucleotide on the surface [41]. Direct in situ synthesis on glass surfaces in the 5′–3′– direction using 5′–phosphoramidites has also been proposed [42]. However, the most frequently used method for producing microarrays for primer extension is to attach presynthesised oligonucleotides on the glass surface. Covalent attachment is preferred over passive adsorption since it can be better controlled than in situ synthesis. Covalent attachment also allows for better accessibility for the oligonucleotide in the proceeding genotyping reaction, and allows the use of more stringent washing protocols than arrays prepared by adsorption [43–45].

We have previously compared six chemical reactions for immobilization of oligonucleotides on a surface for application in the microarray based minisequencing method [32]. Both commercially and in–house coated slides were evaluated to identify the slide with the best binding capacity and most favorable performance in the minisequencing reaction with respect to background fluorescence prior to and after the reaction, as well as signal intensities and power of genotype discrimination. We found the CodeLink™ Activated Slides from Amersham Biosciences (previously denoted CodeLink™ Activated Slides, Motorola and 3DLink™ Activated Slides, SurModics) to have the highest binding capacity of oligonucleotides relative to the in–house coated isothiocyanate slides that served as reference. Although the mercaptosilane slides (Orchid Biosciences [46]) binding disulfide-modified oligonucleotides have lower binding capacity than the CodeLink™ slides, the slides performed equally well in the minisequencing reaction because of their lower background fluorescence.

10.3.2 Labelling Strategies

In principle any detection strategy, such as radioactivity, colorimetry and fluorescence may be used in the microarray format, but fluorescence is the far most frequently used principle today. The Affymetrix GeneChip® system employs an indirect fluorescence detection strategy, in which the target sequence is first labelled using a biotin–conjugated nucleotide, which is visualized in a subsequent staining reaction with a fluorescent streptavidin–phycoerythrin conjugate [13,47]. Many different fluorophores are available for direct labelling, followed by detection using fluorescence microscopes, CCD cameras or fluorescence scanners with photomultiplier tubes. In OLA two allele-specific fluorescently labelled oligonucleotides are required for each SNP [22]. An advantage

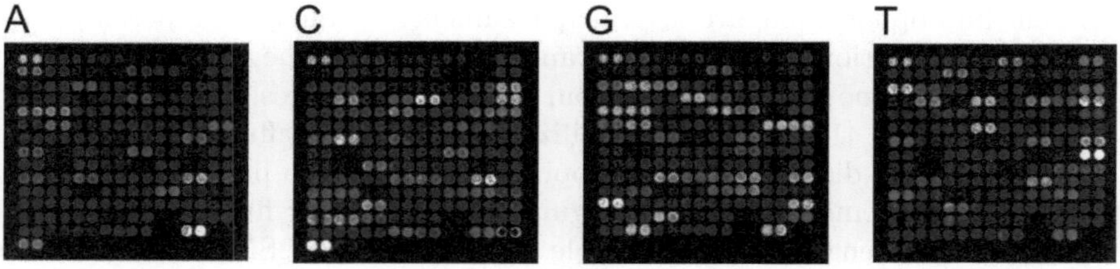

Fig. 10.5. Fluorescence scan image of cyclic minisequencing products captured on a generic microarray carrying complementary tag sequences for detection of 55 SNPs in duplicate. The minisequencing reactions were performed with the four ddNTPs labelled with different fluorophores (Texas Red–ddATP, Tamra–ddCTP, R110–ddGTP, Cy5–ddUTP) and detected with a four color laser scanner (equipped with the excitation lasers: Blue Argon 488 nm, Green HeNe 543.5 nm, Yellow HeNe 594 nm and Red HeNe 632.8 nm) according to the protocol provided in Table 10.3.2 The rainbow color scale corresponds to the different signal intensities with blue as low and white as saturated signal

of primer extension assisted reactions over OLA is that an unlabelled oligonucleotide primer becomes labelled in the actual detection reaction, which reduces the cost of the assay.

In allele-specific primer extension, dNTPs labelled with a single fluorophore are used [10,35], while multiple fluorophores are available and can be used in a variety of minisequencing single nucleotide primer extension assay designs. The same fluorophore may be used on all four nucleotides, in which case four separate reactions are performed for each sample [4,32]. Three dif-

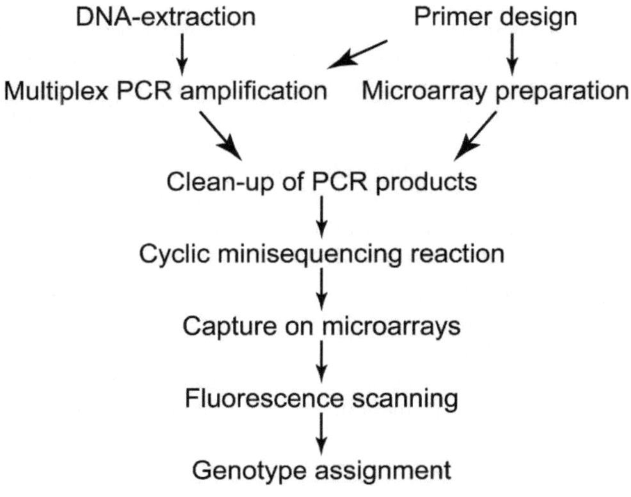

Fig. 10.6. Steps of the tag–array based minisequencing procedure. The steps are explained in detail in the protocol provided in Table 10.3.2

ferent fluorophores [38] are in principle sufficient to analyst all possible SNPs in a single reaction if both DNA strands are utilized. The use of four different fluorophores, one for each of the four dideoxynucleotides, is the most convenient approach [11, 31, 48]. The utilization of multiple fluorophores requires that they have distinct non-overlapping wavelengths to limit the 'cross–talk' between their emission spectra. Figure 10.5 shows four fluorescence scans at different wavelengths for one sample genotyped for 55 SNPs. The efficiency and sequence specificity of the DNA polymerase is affected both by the ddNTP and the fluorophore, but most of all by the sequence context of the SNP [11]. Figure 10.6 outlines the steps of the procedure for performing multiplexed genotyping by minisequencing using tag–arrays. An experimental protocol is provided in Table 10.3.2.

Table 10.1. Protocol for minisequencing in the tag–array format using four fluorophores

Step of the Procedure	**Notes**
1. Design of PCR primers The primers should have similar T_m and low self complementarity to reduce primer dimer formations.	There is no publicly available software for design of PCR–primers for multiplex reactions.
2. Design of minisequencing primers Minisequencing primers are 20–27 nucleotides in length and have similar T_m. In their 5′–end a 20 nucelotide tag sequence (Affymetrix GeneChip® Tag Collection) is incorporated.	A T_m of 55–60°C ensures specificity in the following cyclic primer extension reaction. The tag sequence should be selected not to favor formation of secondary structures (i.e. hairpin loops).
3. Preparation of microarrays The complementary tag sequences contain a 15 T–residue spacer and an amino–group in the 3′–end to enable chemical immobilization. A 25 µM solution of the oligonucleotide in 150 mM sodium phosphate buffer pH 8.5 is printed on CodeLink™ activated slides (Amersham Biosciences) according to the manufacturer's protocol.	Different types of slides with a variety of chemical reaction types are available. We use CodeLink™ activated slides since they performed best in a comparison for our application [32]. The oligonucleotides are printed in duplicate spots on the microarray.
4. Multiplex PCR amplification Typical reaction conditions are U/µl of a thermostable DNA polymerase, 1.5–4 mM $MgCl_2$, 0.2 mM dNTP, 220 ng DNA and 0.14 µM primers in 5–50 µl reaction volumes in 96– or 384–well micro–titer plates.	Multiplex PCR of more than ten fragments has proven difficult to reproduce in multiple samples [10, 33] The pooled PCR products can be used directly or they can be concentrated by ethanol pre–

Continued on next page

Step of the Procedure	Notes
PCR program: 94°C for 10 minutes, then 94°C for 0.5–1 minute, 55–68°C (depending on the T_m of the primers) for 0.5–1 minute, 72°C for 1.5 minutes for 35 cycles and a final extension at 72°C for 7 minutes The multiplex PCR products from each sample are pooled.	cipitation or by spin dialysis with Centricon® devices (Millipore Corporation) to increase the amount of amplicons.
5. Clean–up of PCR products Seven µl of pooled PCR product is treated with 0.5 U/µl Exonuclease I and 0.1 U/µl shrimp alkaline phosphatase (USB Corporation) in 4–8 mM $MgCl_2$, 50 mM Tris–HCl, pH 9.5, in a 10.5 µl volume at 37°C for 30–60 minutes The enzymes are inactivated at 99°C for 15 minutes	Exonuclease I degrades the excess of PCR primers and shrimp alkaline phosphatase inactivates the dNTPs. The $MgCl_2$ concentration has to be optimized and adjusted according to the amount added with the PCR product.
6. Cyclic minisequencing reaction The reaction mixture contains 10.5 µl of enzyme-treated PCR product, 5.0 mM of each tagged minisequencing primers, 0.09–0.27 µM of fluorescent ddNTPs (TexasRed–ddATP, TAMRA–ddCTP, R110–ddGTP, Cy5–ddUTP (Perkin Elmer Life Sciences)), 0.017% Triton–X–100, 50 mM Tris–HCl pH 9.5, 0.07 U/µl of ThermoSequenase™DNA Polymerase (Amersham Biosciences) and 1 nM of control templates in a 15 µl volume. The reaction is repeated for 33 cycles of 95°C and 55°C for 20 seconds each.	Avoid exposing the fluorophores to light to prevent bleaching. The fluorophores should have distinct and non-overlapping emission spectras. It may be advantageous to use Cy5–ddUTP at a higher concentration than the other ddNTPs. The control templates are four synthetic oligonucleotides mimicking a four allelic SNP for which the primer will be extended with A, C, G or T respectively. Up to 99 cycles can be performed.
7. Capture on microarrays The slides are preheated to 42°C in a custom–made aluminium reaction rack (Fig. 10.2). Fifteen µl of minisequencing reaction product, 0.4 nM of TAMRA-labelled control oligonucleotide in 22µl of 6× SSC, are added to each reaction well on the microscope slide. After hybridization for 2–3 hours at 42°C, the slides are briefly rinsed with 4× SSC at room temperature and washed twice for 5 minutes with 2× SSC, 0.1% SDS at 42°C and twice for 1 minute with 0.2× SSC at room temp. The slides are dried by centrifugation for 5 minutes at 500 rpm.	To avoid drying of the reaction wells, which can lead to high background fluorescence, a wet tissue paper is placed on the plexiglass lid and covered with saran–wrap and aluminium foil. The control oligonucleotide is a fluorescently labelled, synthetic sequence that hybridize to its complementary sequence on the slide. 1 × SSC: 150 mM sodium chloride, 15 mM sodium citrate pH 7.0.

Continued on next page

Step of the Procedure	Notes
8. Fluorescence scanning Fluorescence signals on the slides are detected using a four color laser scanner (e.g. ScanArray® 5000, Perkin Elmer LifeSciences). The signal intensities are measured with the analysis software of the scanner (QuantArray®).	Figure 10.5 shows fluorescence images of a microarray scanned at four different wavelengths after hybridization of a cyclic minisequencing product.
9. Genotype assignment The mean value of the signals from the duplicate spots is corrected for the average background in the reaction well. Genotypes are assigned by calculating the ratio between the signal intensity from one of the alleles divided by the sum of the signals from both alleles using a Microsoft Excel™macro.	A software for genotype as–signment for SNPs is the SNPsnapper software, available at: http://www.bioinfo.helsinki.fi/SNPSnapper/

10.4 Conclusion

During the past few years much effort has been targeted at developing technology for analyzing DNA sequence variation in the microarray format. Microarray-based methods have also been applied in a number of clinical, genomic and evolutionary studies. Table 10.2 provides some examples of these applications. So far the studies have been of modest size, but with the possibility of a high level of multiplexing to bring down the costs of the microarray-based assays, we can foresee studies on a much larger scale that will increase our understanding of the role of DNA sequence variation in health and disease.

Table 10.2. Examples of applications of microarray-based analysis of DNA sequence variants

Application	Reaction principle	Comment	Ref.
Comparative sequencing	ASO–hybridization	Introduction of microarray concept	[49]
Cystic fibrosis mutations	Affymetrix GeneChip®	First use of GeneChip® for genotyping	[50]
Recessive disease muta–tions in Finland	Minisequencing primer extension	Proof of principle for primer extension on arrays	[29]
Mutation detection in the ATM gene	ASO–hybridization GeneChip®	Strategy for multiplex PCR design	[13]
Risk factors for myocar–dial infarction	Minisequencing primer extension	^{33}P–detection	[51]

Continued on next page

Application	**Reaction principle**	**Comment**	**Ref.**
Map of 2,200 SNPs	ASO–hybridization GeneChip®	First 'large scale' SNP effort	[52]
Ancestral alleles of human SNPs	ASO–hybridization GeneChip®	Large study 99,000 genotypes	[53]
Detection of minority K–ras mutations	Oligonucleotide ligation	Zip–code approach	[22]
Panel of 142 human SNPs	Tag–array single base extension	High density GeneChip® tag–arrays	[39]
Hemochromotosis and connexin mutations	Single nucleotide primer extension	Two color fluorescence detection	[30]
Panel of 76 human SNPs	Tag–array single base extension	Low density tag–arrays	[38]
Detection of β–thalassemia mutations	Arrayed primer extension	Four color fluorescence detection	[31]
SNPs in the human mu opioid receptor gene	Allele-specific single nucleotide primer extension	Gelpad microchips	[54]
Population frequencies of recessive mutations	Allele-specific primer extension	Large study 140,000 genotypes	[34]
Y–chromosomal SNPs in Finno–Ugric population	Minisequencing primer extension	Detection by ^{33}P and single color fluorescence	[33]
Linkage disequilibrium map of chromosome 22	Arrayed primer extension	Analysis of 900 SNP–markers in 50 Estonian samples	[55]
Quantitative analysis of interferon–related SNPs	Tag–array minise–quencing single nucleotide primer extension	Four color fluorescence detection	[11]
Genome wide mapping of allelic imbalances	ASO–hybridization GeneChip®	Genotyping of 1200 SNPs	[56]
Resequencing exon 7 of the p53 gene	Arrayed primer extension	Four color fluorescence detection	[48]
Pharmacogenetics of hypertension.	Minisequencing primer extension	Single color fluorescence detection. 74 SNPs	[4]

References

1. Lander ES, Linton LM, Birren B, Nusbaum C, Zody MC, Baldwin J et al. (2001) *Initial sequencing and analysis of the human genome* Nature 409: 860–921

2. Venter JC, Adams MD, Myers EW, Li PW, Mural RJ, Sutton GG et al. (2001) *The sequence of the human genome* Science 291: 1304–1351
3. Evans WE and Relling MV (1999) *Pharmacogenomics: translating functional genomics into rational therapeutics* Science 286: 487–491
4. Liljedahl U, Karlsson J, Melhus H, Kurland L, Lindersson M, Kahan T et al. (2003) *A microarray minisequencing system for pharmacogenetic profiling of antihypertensive drug response* Pharmacogenetics 13: 7–17
5. Saiki RK, Bugawan TL, Horn GT, Mullis KB and Erlich HA (1986) *Analysis of enzymatically amplified beta-globin and HLA-DQ alpha DNA with allele-specific oligonucleotide probes* Nature 324: 163–166
6. Mullis KB and Faloona FA (1987) *Specific synthesis of DNA in vitro via a polymerase-catalyzed chain reaction Methods* Enzymol 155: 335–350
7. Saiki RK, Gelfand DH, Stoffel S, Scharf SJ, Higuchi R, Horn GT et al. (1988) *Primer-directed enzymatic amplification of DNA with a thermostable DNA polymerase* Science 239: 487–491
8. Shuber AP, Grondin VJ and Klinger KW (1995) *A simplified procedure for developing multiplex PCRs* Genome Res 5: 488–493
9. Schena M, Shalon D, Davis RW and Brown PO (1995) *Quantitative monitoring of gene expression patterns with a complementary DNA microarray* Science 270: 467–470
10. Pastinen T, Raitio M, Lindroos K, Tainola P, Peltonen L and Syvanen AC (2000) *A system for specific, high-throughput genotyping by allele-specific primer extension on microarrays* Genome Res 10: 1031–1042
11. Lindroos K, Sigurdsson S, Johansson K, Ronnblom L and Syvanen AC (2002) *Multiplex SNP genotyping in pooled DNA samples by a four-colour microarray system* Nucleic Acids Res 30: e70
12. Bell PA, Chaturvedi S, Gelfand CA, Huang CY, Kochersperger M, Kopla R et al. (2002) *SNPstream UHT: ultra-high throughput SNP genotyping for pharmacogenomics and drug discovery* Biotechniques 30: S70–77
13. Hacia JG, Sun B, Hunt N, Edgemon K, Mosbrook D, Robbins C et al. (1998) *Strategies for mutational analysis of the large multiexon ATM gene using high-density oligonucleotide arrays* Genome Res 8: 1245–1258
14. Cho RJ, Mindrinos M, Richards DR, Sapolsky RJ, Anderson M, Drenkard E et al. (1999) *Genome-wide mapping with biallelic markers in Arabidopsis thaliana* Nat Genet 23: 203–207
15. Fotin AV, Drobyshev AL, Proudnikov DY, Perov AN and Mirzabekov AD (1998) *Parallel thermodynamic analysis of duplexes on oligodeoxyribonucleotide microchips* Nucleic Acids Res 26: 1515–1521
16. Radtkey R, Feng L, Muralhidar M, Duhon M, Canter D, DiPierro D et al. (2000) *Rapid, high fidelity analysis of simple sequence repeats on an electronically active DNA microchip* Nucleic Acids Res 28: E17
17. Ross PL, Lee K and Belgrader P (1997) *Discrimination of single-nucleotide polymorphisms in human DNA using peptide nucleic acid probes detected by MALDI-TOF mass spectrometry* Anal Chem 69: 4197–4202
18. Orum H, Jakobsen MH, Koch T, Vuust J and Borre MB (1999) *Detection of the factor V Leiden mutation by direct allele-specific hybridization of PCR amplicons to photoimmobilized locked nucleic acids* Clin Chem 45: 1898–1905
19. Landegren U, Kaiser R, Sanders J and Hood L (1988) *A ligase-mediated gene detection technique* Science 241: 1077–1080

20. Lizardi PM, Huang X, Zhu Z, Bray-Ward P, Thomas DC and Ward DC (1998) *Mutation detection and single-molecule counting using isothermal rolling-circle amplification* Nat Genet 19: 225–232
21. Broude NE, Woodward K, Cavallo R, Cantor CR and Englert D (2001) *DNA microarrays with stem-loop DNA probes: preparation and applications* Nucleic Acids Res 29: E92
22. Gerry NP, Witowski NE, Day J, Hammer RP, Barany G and Barany F (1999) *Universal DNA microarray method for multiplex detection of low abundance point mutations* J Mol Biol 292: 251–262
23. Iannone MA, Taylor JD, Chen J, Li MS, Rivers P, Slentz-Kesler KA et al. (2000) *Multiplexed single nucleotide polymorphism genotyping by oligonucleotide ligation and flow cytometry* Cytometry 39: 131–140
24. Nilsson M, Malmgren H, Samiotaki M, Kwiatkowski M, Chowdhary BP and Landegren U (1994) *Padlock probes: circularizing oligonucleotides for localized DNA detection* Science 265: 2085–2088
25. Hardenbol P, Baner J, Jain M, Nilsson M, Namsaraev EA, Karlin-Neumann GA et al. (2003) *Multiplexed genotyping with sequence-tagged molecular inversion probes* Nat Biotechnol 21: 673–678
26. Oliphant A, Barker DL, Stuelpnagel JR and Chee MS (2002) *BeadArray technology: enabling an accurate, cost-effective approach to high-throughput genotyping* Biotechniques Suppl: 56–58, 60–51
27. Syvanen AC, Aalto-Setala K, Harju L, Kontula K and Soderlund H (1990) *A primer-guided nucleotide incorporation assay in the genotyping of apolipoprotein E* Genomics 8: 684–692
28. Syvanen AC (1999) *From gels to chips:"minisequencing"; primer extension for analysis of point mutations and single nucleotide polymorphisms* Hum Mutat 13: 1–10
29. Pastinen T, Kurg A, Metspalu A, Peltonen L and Syvanen AC (1997) *Minisequencing: a specific tool for DNA analysis and diagnostics on oligonucleotide arrays* Genome Res 7: 606–614
30. Fortina P, Delgrosso K, Sakazume T, Santacroce R, Moutereau S, Su HJ et al. (2000) *Simple two-color array-based approach for mutation detection* Eur J Hum Genet 8: 884–894
31. Kurg A, Tonisson N, Georgiou I, Shumaker J, Tollett J and Metspalu A (2000) *Arrayed primer extension: solid-phase four-color DNA resequencing and mutation detection technology* Genet Test 4: 1–7
32. Lindroos K, Liljedahl U, Raitio M and Syvanen AC (2001) *Minisequencing on oligonucleotide microarrays: comparison of immobilisation chemistries* Nucleic Acids Res 29: e69
33. Raitio M, Lindroos K, Laukkanen M, Pastinen T, Sistonen P, Sajantila A et al. (2001) *Y-chromosomal SNPs in Finno-Ugric-speaking populations analyzed by minisequencing on microarrays* Genome Res 11: 471–482
34. Pastinen T, Perola M, Ignatius J, Sabatti C, Tainola P, Levander M et al. (2001) *Dissecting a population genome for targeted screening of disease mutations* Hum Mol Genet 10: 2961–2972
35. O'Meara D, Ahmadian A, Odeberg J and Lundeberg J (2002) *SNP typing by apyrasemediated allele-specific primer extension on DNA microarrays* Nucleic Acids Res 30: e75

36. Cai H, White PS, Torney D, Deshpande A, Wang Z, Marrone B et al. (2000) *Flow cytometry-based minisequencing: a new platform for high-throughput single-nucleotide polymorphism scoring* Genomics 66: 135–143
37. Chen J, Iannone MA, Li MS, Taylor JD, Rivers P, Nelsen AJ et al. (2000) *A microsphere-based assay for multiplexed single nucleotide polymorphism analysis using single base chain extension* Genome Res 10: 549–557
38. Hirschhorn JN, Sklar P, Lindblad-Toh K, Lim YM, Ruiz-Gutierrez M, Bolk S et al. (2000) *SBE-TAGS: an array-based method for efficient single-nucleotide polymorphism genotyping* Proc Natl Acad Sci U S A 97: 12164–12169
39. Fan JB, Chen X, Halushka MK, Berno A, Huang X, Ryder T et al. (2000) *Parallel genotyping of human SNPs using generic high-density oligonucleotide tag arrays* Genome Res 10: 853–860
40. Pease AC, Solas D, Sullivan EJ, Cronin MT, Holmes CP and Fodor SP (1994) *Light-generated oligonucleotide arrays for rapid DNA sequence analysis* Proc Natl Acad Sci U S A 91: 5022–5026
41. Kwiatkowski M, Fredriksson S, Isaksson A, Nilsson M and Landegren U (1999) *Inversion of in situ synthesized oligonucleotides: improved reagents for hybridization and primer extension in DNA microarrays* Nucleic Acids Res 27: 4710–4714
42. Beier M and Hoheisel JD (2002) *Analysis of DNA-microarrays produced by inverse in situ oligonucleotide synthesis* J Biotechnol 94: 15–22
43. Guo Z, Guilfoyle RA, Thiel AJ, Wang R and Smith LM (1994) *Direct fluorescence analysis of genetic polymorphisms by hybridization with oligonucleotide arrays on glass supports* Nucleic Acids Res 22: 5456–5465
44. Beier M and Hoheisel JD (1999) *Versatile derivatisation of solid support media for covalent bonding on DNA-microchips* Nucleic Acids Res 27: 1970–1977
45. Dolan PL, Wu Y, Ista LK, Metzenberg RL, Nelson MA and Lopez GP (2001) *Robust and efficient synthetic method for forming DNA microarrays* Nucleic Acids Res 29: E107–107
46. Rogers YH, Jiang-Baucom P, Huang ZJ, Bogdanov V, Anderson S and Boyce-Jacino MT (1999) *Immobilization of oligonucleotides onto a glass support via disulfide bonds: A method for preparation of DNA microarrays* Anal Biochem 266: 23–30
47. Chee M, Yang R, Hubbell E, Berno A, Huang XC, Stern D et al. (1996) *Accessing genetic information with high-density DNA arrays* Science 274: 610–614
48. Tonisson N, Zernant J, Kurg A, Pavel H, Slavin G, Roomere H et al. (2002) *Evaluating the arrayed primer extension resequencing assay of TP53 tumor suppressor gene* Proc Natl Acad Sci U S A 99: 5503–5508
49. Southern EM, Maskos U and Elder JK (1992) *Analyzing and comparing nucleic acid sequences by hybridization to arrays of oligonucleotides: evaluation using experimental models* Genomics 13: 1008–1017
50. Cronin MT, Fucini RV, Kim SM, Masino RS, Wespi RM and Miyada CG (1996) *Cystic fibrosis mutation detection by hybridization to light-generated DNA probe arrays* Hum Mutat 7: 244–255
51. Pastinen T, Perola M, Niini P, Terwilliger J, Salomaa V, Vartiainen E et al. (1998) *Array-based multiplex analysis of candidate genes reveals two independent and additive genetic risk factors for myocardial infarction in the Finnish population* Hum Mol Genet 7: 1453–1462
52. Wang DG, Fan JB, Siao CJ, Berno A, Young P, Sapolsky R et al. (1998) *Large-scale identification, mapping, and genotyping of single-nucleotide polymorphisms in the human genome* Science 280: 1077–1082

53. Hacia JG, Fan JB, Ryder O, Jin L, Edgemon K, Ghandour G et al. (1999) *Determination of ancestral alleles for human single-nucleotide polymorphisms using high-density oligonucleotide arrays* Nat Genet 22: 164–167
54. LaForge KS, Shick V, Spangler R, Proudnikov D, Yuferov V, Lysov Y et al. (2000) *Detection of single nucleotide polymorphisms of the human mu opioid receptor gene by hybridization or single nucleotide extension on custom oligonucleotide gelpad microchips: potential in studies of addiction* Am J Med Genet 96: 604–615
55. Dawson E, Abecasis GR, Bumpstead S, Chen Y, Hunt S, Beare DM et al. (2002) *A first-generation linkage disequilibrium map of human chromosome 22* Nature 418: 544–548
56. Primdahl H, Wikman FP, von der Maase H, Zhou XG, Wolf H and Orntoft TF (2002) *Allelic imbalances in human bladder cancer: genome-wide detection with high-density single-nucleotide polymorphism arrays* J Natl Cancer Inst 94: 216–223

11

High Sensitivity Expression Profiling

Ramesh Ramakrishnan, Paul Bao, and Uwe R. Müller

11.1 Introduction

DNA microarrays were originally conceived to provide a new means for rapid sequence analysis [1–3] but it was soon recognized that they presented a powerful new tool to determine the relative transcript abundance of multiple genes [4, 5]. Expression microarrays have been shown to provide valuable insights in the areas of target discovery [6], mechanism of drug action [7–9], genes and pathways involved in various cellular responses and pathophysiologies [10–12], exon mapping [13], chemosensitivity [14, 15] and tumor classification [16, 17]. Clinically, expression microarrays have been used in studies utilizing gene expression signatures to distinguish primary breast cancers from multifocal disease [18] and to predict disease outcome, surpassing currently used clinical and histopathological methods [19–22].

The probes used for expression array fabrication can be made from clones of genes, PCR amplicons, or oligonucleotides [6, 23–26], and various methods for their attachment and linkage to the array surface have emerged (Chaps. 2 and 3). While cDNA probes and PCR amplicons are typically arrayed in a buffer that contains both strands, oligonucleotide probes are single-stranded and complementary to the mRNA or cDNA target sequences, respectively.

The primary target for expression analysis is mRNA, but it is typically converted to cDNA prior to use for two reasons: first, DNA is much more stable and therefore more easily handled and stored; second cDNA synthesis provides a convenient method to produce labelled targets by incorporation of fluorescent or hapten labelled nucleotides during the reverse transcription reaction. When the target material is limiting, various methods can be employed to either enhance the signal, or to amplify the mRNA (see below). As reviewed in Chap. 8, the standard labelling system consists in direct incorporation of fluorescent nucleotides using a two color labelling scheme, but indirect labelling via incorporation of haptens provides for alternate and potentially higher sensitivity detection schemes (Chap. 8 and below). The type of labels used and the exact conditions for labelling and hybridization are critically important and have a profound impact on the sensitivity of the system. For most slide based arrays, hybridization is carried out under a coverslip in a humidity chamber, followed by washing and staining, if indirect labelling systems are used. However, for enhanced hybridization rates and more consistent performance, automated hybridization chambers and complete hybridization systems have been developed.

Although all array based expression systems are based on a determination of relative transcript abundance by comparing the copy number in a specific sample to that in a reference, there are two different approaches involving either labelling both samples with the same 'color' and hybridizing to separate arrays, or labelling both samples with different 'colors', and hybridizing them competitively to the same array. The latter was first pioneered by Kallioniemi et al. by comparing genomic DNA from different sources [27]. The advantage is that differences between the two arrays that can affect either the hybridization (e.g. spot morphologies, probe amounts) or the detection (e.g. shading) are eliminated, typically resulting in improved CVs (coefficients of variation) for color ratios as compared to the CVs of raw hybridization signals [24]. However, the two color approach also has disadvantages including varying incorporation rates of different fluorophors, the need for more reference sample in multiple experiments, spectral overlap between dyes, more expensive imaging systems, and, in case of multiple haptens, more complex signal amplification protocols.

Many methods for improvements and optimizations have emerged during the past decade, most, of course, to improve manufacturability, specificity and/or sensitivity. This is where some significant differences in the type of probe used for manufacturing the array have appeared. In fact, while most data suggest that equivalent results are obtained between oligo and cDNA arrays [24, 25, 28], some data suggest otherwise [29, 30].

Because oligo and cDNA arrays each have a set of advantages and disadvantages, we have combined our experiences to describe in this chapter high sensitivity expression systems achieved with either format, using as examples the Motorola CodeLink™ (now Amersham CodeLink™) oligonucleotide array and the Corning CMT™ cDNA-based expression arrays (no longer commercially available, but see Chap. 5).

11.2 Oligonucleotide Expression Arrays

11.2.1 Array Design

The use of oligonucleotides instead of clones or PCR amplified DNA sequences as probes for expression arrays has significant advantages since they can be designed to hybridize specifically to any gene in the sample, provided sequence information is available. The ease with which oligonucleotides can be synthesized reduces complexity in the manufacturing and quality control areas, since it eliminates the need for clone tracking and handling, PCR amplification and sequence verification. Further, the specificity associated with oligonucleotide arrays enables the study and analysis of splice variants [31] and the ability to differentiate closely related members of gene families.

Typical arrays fabricated with oligo probes of 20–30 nucleotides in length have sensitivity limitations, but this limitation can be minimized by extending the length to 50 bases or more [25]. However, depending on the method of array fabrication, this may result in other disadvantages. For example, for in situ synthesis [32–34] the lower coupling efficiencies on the array can limit purity with significant impact on specificity and sensitivity. This typically limits the probe length to about 25 bases. In addition, in situ synthesis does not allow an independent confirmation of the fidelity of synthesis. On the other hand, synthesis of oligonucleotides prior to deposition on the array incurs a significant cost not just in synthesis, but also in purification and sequence confirmation, which increases proportionally with oligo size. In addition, this approach requires covalent attachment of the oligos to the array surface. However, several innovative solutions in chemistry and systems engineering have been proposed to address these obstacles [35, 36].

Covalent attachment of prefabricated oligonucleotides circumvents some of the constraints imposed by earlier in situ synthesis methods and allows new elements to be added without redesigning the entire microarray. The emphasis here shifts to the ability to reproducibly attach probes. One approach includes fabrication of arrays by photochemical as well as chemical attachment [37]. Incorporation of specific functional moieties at the 5′ end of oligonucleotides can serve as a pseudo–purification step. Since only full-length oligonucleotides will receive the attachment group and attach to the matrix efficiently, non-specific adsorption of the oligonucleotide can be virtually eliminated.

Longer oligonucleotides (60–mer to 70–mer) exhibit chemical characteristics similar to cDNAs in that they can be attached directly (non-covalently) to slide surfaces without the need for any specialized attachment chemistry. However, for maximum attachment, a UV–crosslinking step is advisable and improves sensitivity (personal observations), though probes retained on a glass surface in this manner may not exhibit the same degree of conformational flexibility or accessibility as do those retained via end attachment [38]. Alternatively, oligonucleotides can be modified by incorporation of biotin or haptens at either end, and they can then be anchored efficiently on surfaces coated with streptavidin or anti-hapten antibodes, respectively. The disadvantage of such an attachment scheme is that the biological interaction must remain intact throughout the assay, imposing constraints upon subsequent hybridization and array processing.

For applications in expression analysis, the oligonucleotide probes are generally designed towards the 3′ end of a RNA transcript, primarily to reduce the effects of transcript degradation [26]. Probe design is also guided by priming and amplification schemes (random hexamer versus oligo–dT), which impact the regions of the transcript represented in the cDNA or cRNA sample. Although a set of heuristics has been proposed for probe design [39], there is as yet no definitive algorithm that can select a set of hundreds of probe sequences with maximum hybridization efficiency under isothermal conditions. In order to tackle this problem, rapid probe prototyping as well as the use of multiple probes per transcript in expression arrays offer a pragmatic solution. Basic studies on heteroduplex formation as it pertains to microarrays are now underway [40] and analogies to antisense oligonucleotides (whose efficacy depends on hybridization and transcript cleavage) may provide better solutions in the future [41].

Given that k_2, the second order rate constant for hybridization, is proportional to the square root of the length of the shortest strand participating in duplex formation [42], it is not surprising that larger oligonucleotides show improved hybridization kinetics. A recent study has demonstrated the utility of 60–mer fabricated in situ for expression analysis and has shown good sensitivity under various hybridization conditions. However, two 60–mer have to differ by at least 18 nucleotides in order to achieve sufficient specificity, though this depends very much on the exact location of each mismatch and the relative content of deoxycytidine in the oligo sequence [24]. It appears

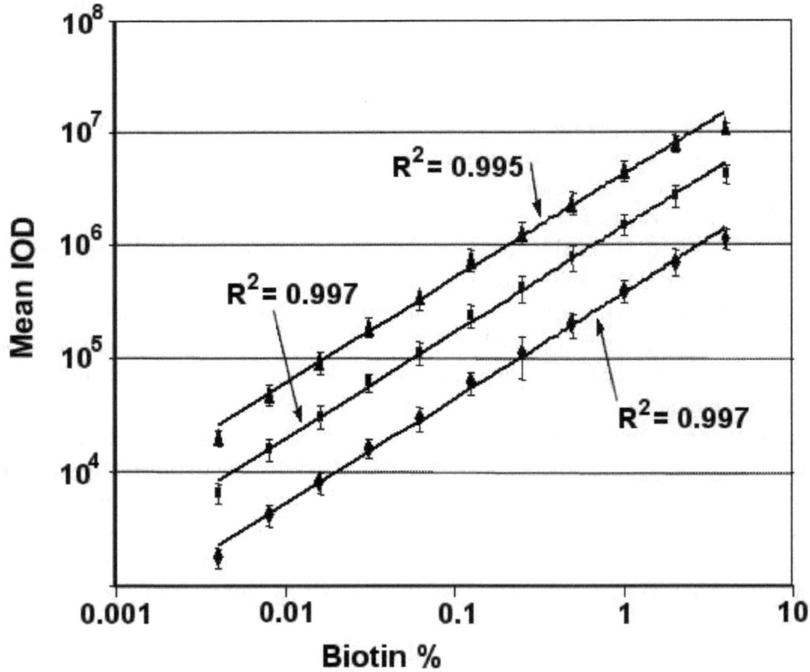

Fig. 11.1. The biotinylated probe chip: linearity of the detection process. Slides were processed with streptavidin–alexa 647 and scanned at PMT voltages of 500 (*diamonds*, $R^2 = 0.997$), 600 (*squares*, $R^2 = 0.997$), and 700 (*triangles*, $R^2 = 0.995$). Each data point represents the average of sixteen replicates per slide

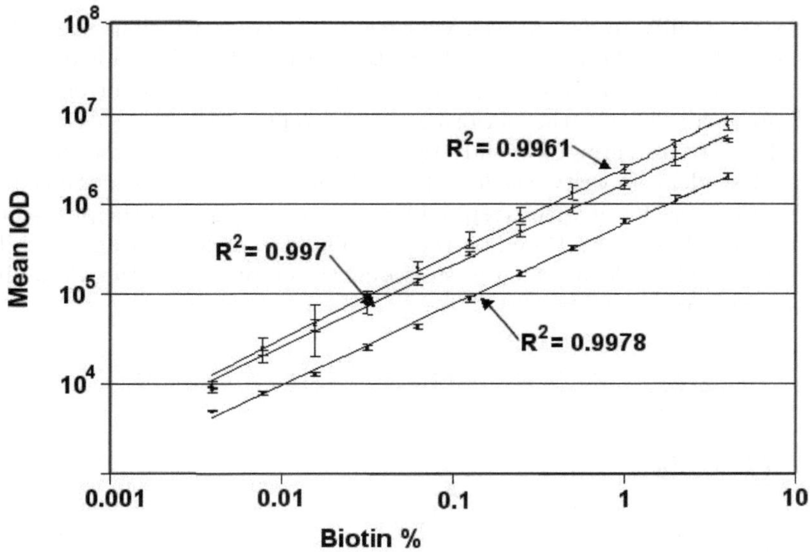

Fig. 11.2. Slides were processed with Streptavidin–Alexa 647 ($R^2 = 0.9961$), Strepta-vidin–Cy3 ($R^2 = 0.997$), or Streptavidin–phycoerythrin ($R^2 = 0.9978$) and scanned at a PMT voltage of 600. IOD = integrated optical density

now that most manufacturers prepare 30 to 75–mer oligonucleotide arrays for expression applications.

Labelling of cDNA and cRNA can be done by several methods, as discussed below, but for the development of the Motorola CodeLink^{TM}array system an indirect labelling system was chosen. The reason is that biotin incorporation has many advantages to direct labelling, including the fact that biotin-labelled nucleotides are efficient substrates for many DNA and RNA polymerases, resulting in better yields than incorporation of cyanine dyes (unpublished observations). Moreover, cDNAs or cRNAs containing biotinylated nucleotides have denaturation, reassociation, and hybridization parameters similar to those of their unlabelled counter–parts [43].

In order to test the biotin labelling system, the biotinylated probe chip was developed by the Motorola group as an analytical tool to monitor the linearity, variability, and sensitivity of the detection process [44]. This chip contains unlabelled oligonucleotide probes mixed with biotinylated probes of the same length in increasing ratios ranging from 0.000122% to 4%, with the final probe concentration per spot kept constant. This enables the assessment of fluorescence detection independent of hybridization, since the presence of biotinylated probes can be simply monitored using Streptavidin–Alexa 647. As shown in Fig. 11.1, the dose response of Streptavidin–Alexa 647 binding is linear ($R^2 > 0.99$) for three logs of biotinylated probe concentrations (0.004% to 4%), without observing saturation, and with very low variability. Furthermore, the ability to detect even very low probe concentrations suggests that this detection method may be sufficiently sensitive to detect low expressers.

Since a variety of indirect detection methods are used in current commercial products [26, 45], the performance of streptavidin–Alexa 647 was compared to streptavidin–Cy3 and streptavidin–phycoerythrin on these biotin chips. Excellent linearity was found for all these methods, although the signal intensities were lower with streptavidin–phycoerythrin with the protocols and slides used (Fig. 11.2). Based on these findings, a single-color approach based on streptavidin–Alexa 647 was subsequently employed in the CodeLink™ assay system.

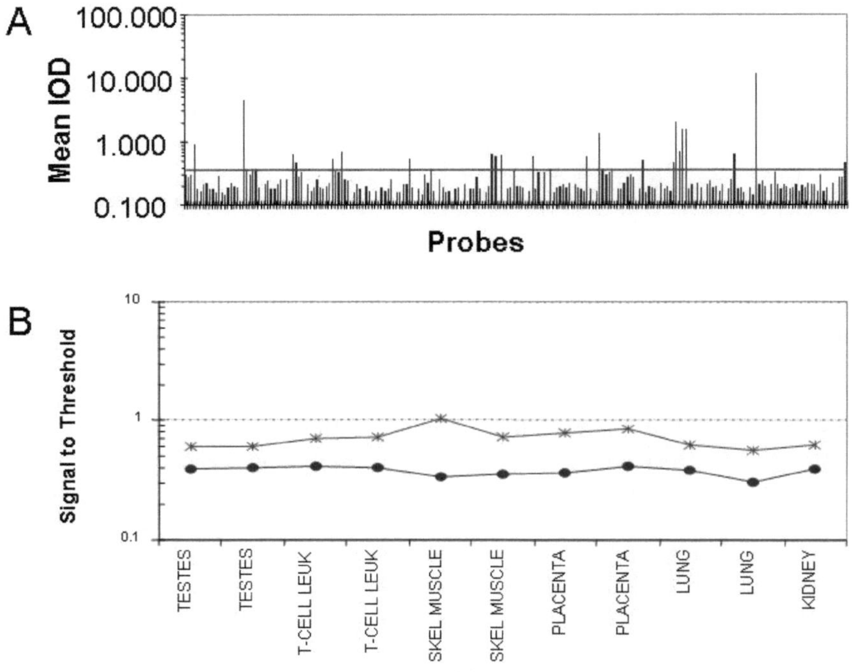

Fig. 11.3. Threshold determination. The negative control threshold can be used to define the lower limits of detection. **(a)** Graph shows mean IOD for negative control probes used to calculate the threshold. Each slide has 216 negative control probes (54 probes in 4× redundancy). Threshold was calculated using 20% trimmed mean for each slide (10% of the highest signals and 10% of the lowest signals were removed from the probes population) and the remaining probes were used to calculate the threshold. 9.44% of the untrimmed population of the negative control probes is above the threshold. The line indicates the threshold as calculated by the mean and three standard deviations. **(b)** The negative control values are constant in six different samples. The mean (*asterisks*) and median (*circles*) negative control values were calculated from six tissues (each tissue was hybridized in duplicate)

11.2.2 Use of a Threshold to Define Lower Limits of Detection and Nonspecific Binding

Specificity during and after the hybridization reaction can be monitored efficiently through the use of negative controls, i.e. probes which do not cross–hybridize to the complex message for which the array was designed. For the CodeLink™ product a negative control probe set was developed, consisting of approximately 55 bacterial sequences that were designed, FASTA verified, and empirically shown not to cross–hybridize to human transcripts. The threshold was determined by calculating the mean negative control value and adding three standard deviations (99.7% confidence). An example is shown in Fig. 11.3a using in vitro synthesized complementary RNA (cRNA) from human liver as target, where 9.44% of the untrimmed population of the negative control probes were found to be above the threshold. Using 6 different tissues in multiple hybridizations, it was shown that the same set of probes was trimmed each time by this process, pointing to some of the potential shortcomings in either oligo design or sequence accuracy. Nevertheless, the data indicate that, if used appropriately, this set of bacterial probes can be universally applied to indicate the cross–hybridization threshold since the mean and median signal intensities do not change significantly between a variety of tissues (Fig. 11.3b).

11.2.3 Sensitivity Measurements Using Oligonucleotide Arrays

One of the most common methods to evaluate sensitivity of an oligonucleotide-based expression microarray is the use of spiking experiments with exogenous bacterial transcripts that are complementary to a set of positive control probes on the array. These elements would have to be different from those which serve as the negative control elements and which are used to generate the negative control threshold. Results from a representative experiment are shown in Fig. 11.4, where defined amounts of 6 different in situ synthesized transcripts were spiked into the complex human message prior to reverse transcription and labelling. After hybridization, the fluorescence was determined at the cognate bacterial probe spots and plotted against the mass of spiked transcript. Each array contains 3 different probes per bacterial control gene, and each probe is represented 4 times across the slide. Figure 11.4 shows the data for one of the six transcripts (*araB*) for each of the 3 different probe spots. The amount of mRNA used for spiking was chosen to represent a mass ratio ranging from 1:6,000 to 1:6,000,000 spiked mRNA:total RNA. Assuming that 5% of the total RNA population is polyA+ RNA [25], this is equivalent to a mass ratio range of 1:300 to 1:300,000 spiked mRNA to polyA+ RNA. As expected, different probes show different signal to threshold ratios (due to different affinities) at the same spike level. However, all probes displayed a signal above threshold at the 1:300,000 spike level. This is equivalent to 1 copy per cell according to [45–47], and exceeds that sensitivity according

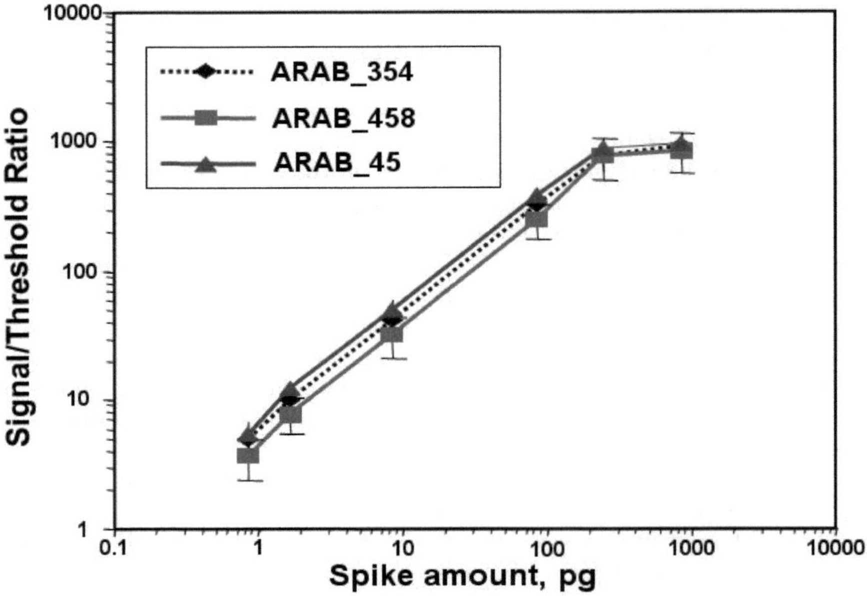

Fig. 11.4. Sensitivity and dynamic range. Nine exogenous bacterial transcripts were spiked into the complex mRNA from human liver, each at increasing concentrations (bacterial RNA: total human liver RNA as 1:6,000; 1:20,000; 1:60,000; 1:600,000; 1:3,000,000; 1:6,000,000. The signal: threshold ratio was determined by dividing the fluorescence for each bacterial positive control probe by the negative control threshold. The data for the *araB* transcripts is shown, with three bacterial control probes designed to hybridize to each transcript

to [24, 48]. On further diluting target to achieve a mass ratio of 1:15,000,000 of transcript per 5 μg of total RNA (Fig. 11.5), the signal–to–threshold ratios of the *araB* probes were still found to be significantly above background.

The relative intensity values obtained in these spiking experiments can be used to normalize and compare the results of different arrays. For that purpose the signal intensity is divided by the negative control threshold previously described.

By spiking at increasing mass ratios, it was possible to determine not only the absolute sensitivity but also the dynamic range of the assays. A linear transcript concentration-dependent dose response was obtained over two orders of magnitude reaching a signal saturation near the highest spike concentration (equivalent to a 1:300 mass ratio). Similar data were obtained for the other bacterial probes and transcripts (not shown).

Once a target is present in the hybridization reaction, its ability to be captured by the microarray is a function of the hybridization efficiency, which in many cases is limited by two–dimensional and three–dimensional diffusion [49]. In order to enhance passive diffusion, the cRNA was already frag-

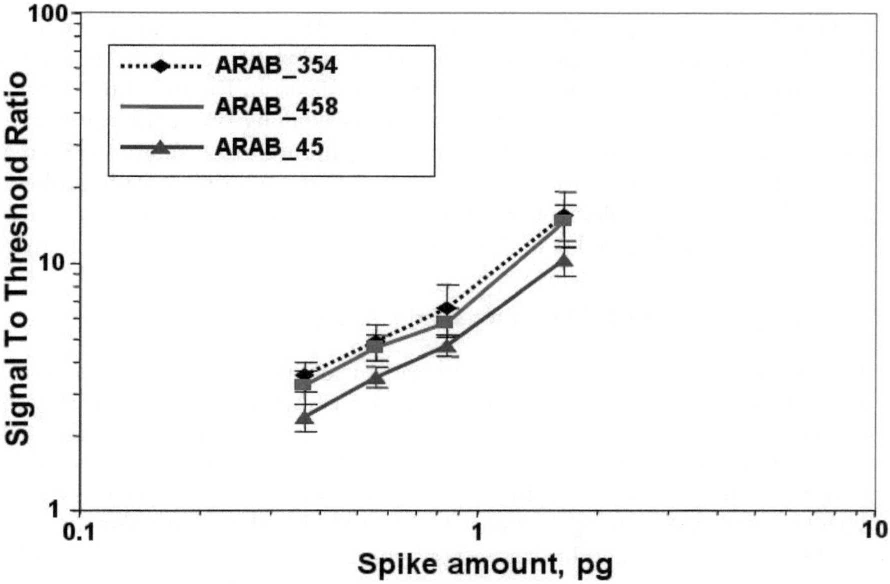

Fig. 11.5. Sensitivity and dynamic range extended. Exogenous bacterial transcripts to the *araB* gene were spiked into the complex mRNA from human liver, at increasing concentrations (bacterial RNA: total human liver RNA as 1:3,000,000, 1:6,000,000, 1:9,000,000, and 1:15,000,000. The signal: threshold ratio was determined by dividing the fluorescence for each bacterial positive control probe by the negative control threshold. Three bacterial control probes were designed to hybridize to each transcript

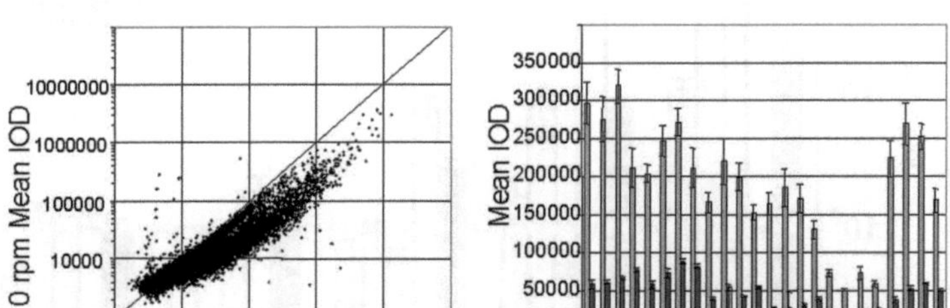

Fig. 11.6. The effect of mixing during hybridization. **(a)** The average signal intensities with (x axis) and without (y axis) mixing were plotted for all probes (~ 9300). The bowing shows the enhanced signal intensities with mixing. **(b)** Signal intensities from 24 positive bacterial control probes (x axis) when their complementary transcripts were spiked into the total RNA at a mass ratio of 1:2,000,000. The first bar in each pair represents the signal intensity obtained with mixing and the second bar in each pair represents the signal intensity obtained without mixing

mented before adding to the hybridization reaction. To examine the effect of diffusion limitation, an experiment was performed to demonstrate the effect of vigorous shaking and mixing during the hybridization. Figure 11.6a shows an average 3.1 $\pm$ 5.6 fold increase in signal intensity along the entire signal range for $\sim$ 9500 probes when mixing was employed. Figure 11.6b shows the increase in signal intensity due to mixing for 24 positive bacterial control probes (x axis) when their complementary transcripts were spiked into the total RNA at a mass ratio of 1:2,000,000. An average increase of 4.2 $\pm$ 1.4 was found.

11.2.4 Specificity of Oligonucleotide Arrays

The specificity of this oligo array hybridization platform was examined by introducing one or more mismatches in the center of each of five different oligonucleotide probes, each designed to hybridize to different human transcripts. One (1×), two (2×), three (3×), or four (4×) adjacent mismatches were introduced into the middle of a probe and the fluorescence associated with the perfect match was compared with that of each type of mismatch. Figure 11.7 shows the data obtained after hybridization with human liver cRNA. Each probe was represented four times per array, with the red horizontal line representing the threshold fluorescence as defined above. As expected, each

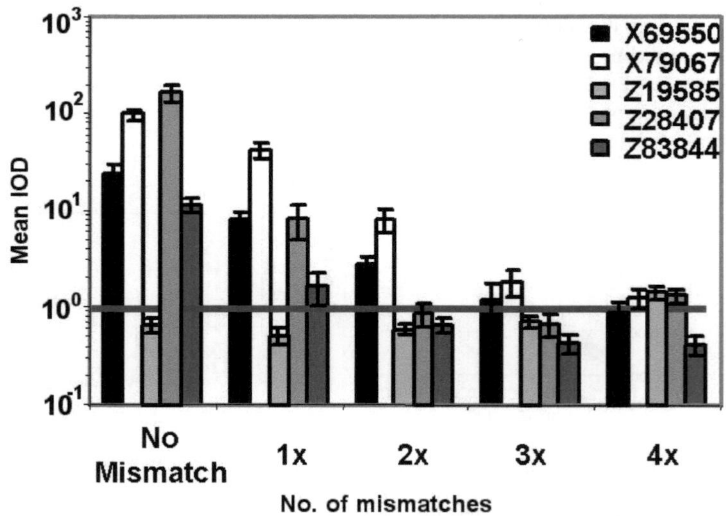

Fig. 11.7. Specificity of the CodeLink^{TM} Expression Bioarray platform. Specificity was determined by introducing one (1×), two (2×), three (3×), or four (4×) adjacent mismatches into the middle of a probe and determining fluorescence associated with the perfect match (no mismatch) and each of its mismatch probes. This analysis was performed for five probe sets after hybridization with cRNA generated from human liver total RNA. Each mismatch probe has four repeats across the slide. The line represents threshold

probe type within a set gave significantly different signal intensities (IOD), demonstrating the effect of probe sequence on hybridization efficiency. In fact, one of these probes (Z19585) did not give a signal above threshold and was eliminated from the analysis. However, the variability between the 4 repeats of the same probe was minimal, as indicated by the relatively small standard deviations. Analysis of the mismatched probes revealed that their signal was reduced. Though each of the 4 hybridization competent single (1×) mismatch probes gave signals above background, their signals were reduced by a factor ranging from about 2 to 22 fold, presumably due to the effects of flanking sequences. Two of the 2× and 3× mismatch probes, respectively, also gave signals at or above threshold, but with significant further reductions. Similar results have been obtained with cRNA target generated from other tissues such as human skeletal muscle and placenta. Since the oligos used here were 30 nucleotides long, these data suggest that under the appropriate conditions and with optimum oligo sequence design, 5% sequence variations can be distinguished. This ability to distinguish a small number of mismatches between highly homologous genes or exons provides an important advantage over cDNA and 60–mer oligonucleotide arrays.

11.2.5 Validation of Relative Transcript Levels with Real Time Quantitative Reverse Transcription PCR Assays

As a preliminary validation that the oligo array platform generated precise but also accurate answers, differential expression ratios from this platform were compared to those obtained using quantitative reverse transcription PCR (Taqman) assays for a set of 54 genes, using the same RNA sample as target source. Although the actual sequence of the oligonucleotide on the array was not identical to that of the Taqman probe, in the majority of cases the probes did overlap. For the entire data set of 54 genes, there was a good correlation (correlation coefficient of 0.76) between the changes reported by each system (Fig. 11.8).

11.3 cDNA-based Expression Arrays

11.3.1 Array Design and Manufacture

Most of the advantages and disadvantages of oligo versus cDNA arrays have already been discussed above and only a few points will be highlighted here. The development of microarrays using PCR amplified cDNAs as probes was enabled by the availability of a significant number of known genes in cloned format from several sources. Most of these were derived either as partial or full-length clones or expressed sequence tags (EST) by reverse transcription from mRNA, but they can also be cloned directly from the genome [50]. Either way, confirmation of clone purity and sequence integrity is critical and

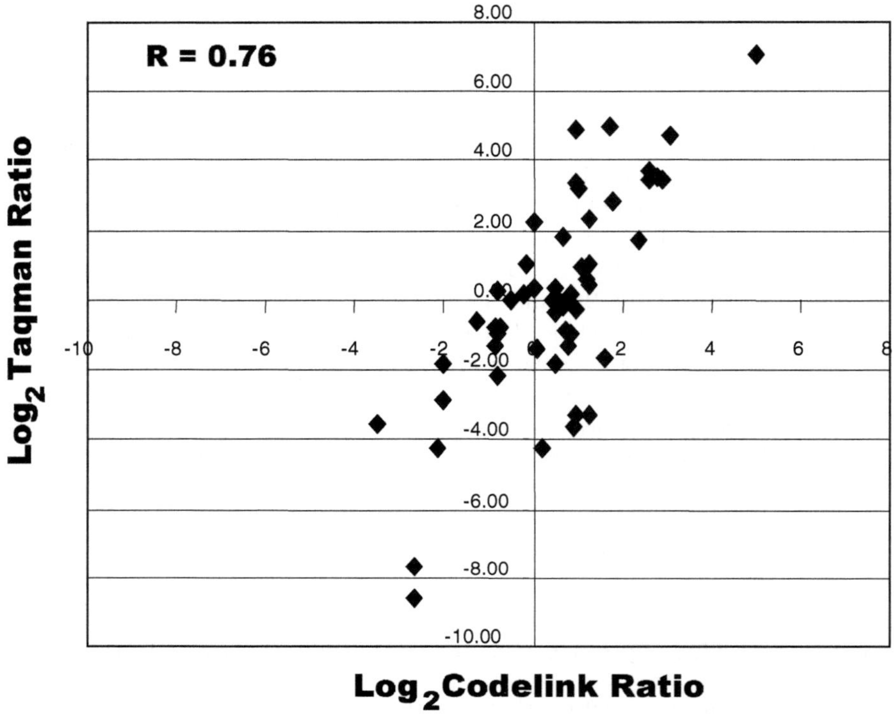

Fig. 11.8. Correlation of differential expression ratios with Taqman. The $\log_2$ of the differential expression ratio obtained with Taqman when heart and brain were compared was plotted on the y-axis versus the $\log_2$ of the ratio obtained with the CodeLink™ Expression Bioarray platform using the same RNAs on the x-axis. The correlation coefficient (R = 0.76) was based on all 54 genes

presents a major cost in the manufacturing process for commercial manufacturers. Equally important are the choice of the array support and the attachment chemistry, as they can have a significant impact on array performance (see also Chaps. 2 and 3). First, the density and conformation of the probe will affect the hybridization kinetics, and while for oligos some of these parameters have been defined [51], there is limited information available for cDNA arrays. Instead, the optimum probe concentration in the printing solution has typically been determined empirically [52], and in our hands printing of double-stranded PCR products at a concentration of 0.25 µg/µl (50% DMSO, 50 mM citrate buffer) resulted in arrays with good sensitivity and dynamic range, as discussed below (see also Chap. 5). A second reason for carefully choosing the right surface is the signal enhancements that can be achieved either through optical effects [53–55] or better hybridization kinetics as a result of a 3D or flow–through chip [56–58]. But perhaps most important is the use of a surface designed for minimum non-specific target binding and minimum autofluorescence (if fluorescently labelled target is used). It was shown by Shena et al. that reducing slide background through modification of the attachment chemistry can improve the sensitivity by a factor of 10 [5]. However, we have noticed that adsorption of organic molecules from the environment upon storage of arrays can dramatically increase autofluorescence and render them useless. In many cases treatment of the array with sodium borohydride before hybridization can offer an easy remedy [59]. This process will do little to improve the autofluorescence of the glass itself, unless the glass has been treated to adsorb or reflect any unwanted photons from within or the underside of the slide [53]. In order to insure that the array has a minimum of background autofluorescence, we recommend using a high quality glass substrate (e.g. GAPS™slides, Corning) and scanning all arrays at a high voltage setting prior to use. Assuming that a quality array (for a description of quality parameters see [52]) has been fabricated, the tasks turn to target preparation and labelling.

11.3.2 Target Preparation and Labelling

Many studies have focused on the preparation and labelling of cDNA from either total RNA or isolated poly(A)+ mRNA, since the effect of target concentration, label type and density on sensitivity is obvious. However, there are various definitions and descriptions of sensitivity. Given the variety of arrays, labels and detection modes, the reported amounts of non-amplified target required per hybridization varies up to 100–fold between different publications, ranging from a few to more than 100 µg total RNA [60,61] or 200 ng to a few µg of purified poly (A)+ mRNA [52,62]. A better way to express sensitivity is to define the minimum amount of a specific target needed per hybridization to give a detectable signal over noise, as originally described by Schena et al. [4]. This can be accomplished by spiking specific amounts of one or more types of a synthetic mRNA into the target RNA before reverse transcription as described above for oligo arrays. By these criteria, arrays with longer probes (cDNA or $>$ 50–mer oligos) have been shown to be more sensitive than short oligo arrays (25–mer) [63], which is not unexpected. Without any target or signal amplification, the sensitivities for cDNA arrays have been reported to be around 2 pg (0.006 fmol) of a unique Cy3 or Cy5 labelled mRNA per spot and per hybridization [52] compared to 20 pg ($\sim$ 0.06 fmol) for 60–mer oligos [24, 63], and $\geq$ 0.3 fmol for 30–mer oligo arrays [63]. However, Cy5 is typically somewhat less sensitive [63, 64].

Assuming approximately 10 pg total RNA [60,65] and 100,000 transcripts per eukaryotic cell [24], mRNA represents approximately 0.5% of the total RNA. Thus, based on the above cited sensitivity limits detection of 1 copy of a given transcript per cell would require on the order of 40 µg of total RNA or 200 ng purified mRNA. Since for many applications, such as fine-needle aspirates, this much material can not be obtained, signal or target amplification procedures may offer a suitable solution. Incorporation of biotinylated nucleotides during the RT step, followed by fluorescently labelled streptavidin is commonly used, as described above. Efficient incorporation of aminated random primers and/or aminoallyl nucleotides during cDNA synthesis, and subsequent chemical conversion of the primary amine groups to fluorescent moieties has provided > 10 fold improvements in signal strength [66] and reduced the required amount of material to as little as 1 µg total RNA [67].

Better amplification can be achieved by incorporating a T7 primer during the cDNA synthesis, followed by transcription using the cDNA as template [68,69]. The resulting amplification is linear, reaching amplification levels of several orders of magnitude without significant distortion of transcript ratios [68, 70, 71]. Our own data suggest a 5800 fold amplification, starting with the total RNA from as few as 10,000 HepG2 cells ($\sim$ 11 pg/cell) and resulting in 312 pg/cell of aRNA (assuming 0.5% of total RNA is mRNA) [72]. An even more powerful target amplification can be achieved by combining reverse transcription with PCR allowing expression analysis of single cells [73]. However, the non-linear PCR step may distort the transcript ratios [74].

11.3.3 Hybridization and Detection

Clearly, any target or signal amplification procedures add some complexity to the assay process, and a high sensitivity 2-color assay format with a simple reverse transcription reaction and direct incorporation of label is very desirable. We have developed a protocol that allows detection of single copy mRNAs starting with 2–5 µg of total RNA (i.e. $\sim$ 500,000 human cells) without the need for any signal or target amplification. This was only possible because of the combination of quality slides and cDNA arrays, reduction of autofluorescence by treatment with sodium borohydride [59], and the optimized labelling and hybridization protocol shown in the Appendix. In fact, prior to discovering the benefits of the sodium borohydride treatment more than 10 times that amount of target was needed to get similar results with the same arrays [75].

When using this protocol to test various tissues on Corning CMT^{TM}4K Cancer arrays (containing $\sim$ 2000 cancer related genes in duplicate) we typically found that virtually all genes represented on the array were expressed in the tissues analyzed. As shown in Fig. 11.9, more than 95% of the probe spots have a net Signal/Noise (S/N) ratio for Cy5 of 5 or larger (slide A), whereby this ratio for the negative control probes (bacterial genes) ranged between 1.4 and 1.8 (not shown). The Cy3 net S/N ratio is somewhat lower and 4% less genes are detectable. This is because of the differences in the average background for these two fluorophore, which was on the order of 176 RFU for Cy5 and 383 RFU for Cy3 in our experimental set–up.

Of course, the amount of target hybridizing to the probe is not only a function of the target concentration, but also of solution stringency. By lowering the salt concentration from 2.25× SSC to 1.25× SSC (slide B), up to 20% fewer genes become detectable, depending on the net S/N ratio that one chooses

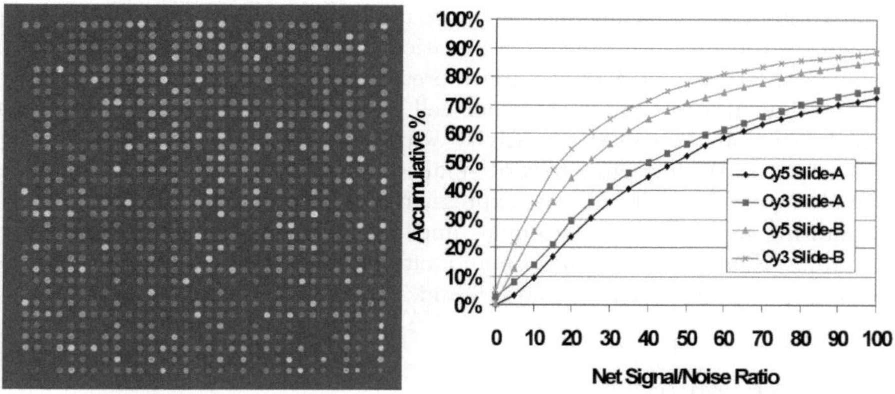

Fig. 11.9. Expression analysis of vitamin D3 treated MCF breast cancer cells. The image on the left shows one of 4 arrays comprising the 4K array after hybridizing a mixture of Cy3 labelled cDNA from untreated cells and Cy5 labelled cDNA from vitamin D3 treated cells (5 μg each). The red spot in the last column reveals a 50-fold upregulation of the gene 24 Hydroxylase. The graph on the right shows the accumulative percentage of probe spots vs their Net Signal/Noise for two separate slides hybridized under different stringencies [779]. Slide A: 2.25× SSC; slide B: 1.25× SSC; Net Signal = total signal of a spot reduced by local background

for identifying a spot. Thus, it is critical to tune the hybridization conditions in order to achieve maximum sensitivity with minimum cross–hybridization. Since the negative control spots in our example show no significant target binding (net S/N $\sim$ 1), the lower stringency is adequate and allows detection of very low abundant transcripts. Of course, transcripts with significant sequence similarity can not be distinguished under these conditions, which is a previously mentioned drawback of cDNA arrays. It is therefore crucial to confirm expression results of a given array system with other methodologies for quantitation of mRNAs. The system we had chosen in this example was the comparison of vitamin D3 treated and untreated MCF breast cancer cells. These had been studied extensively by several laboratories and the 50–fold upregulation of the 24 Hydroxylase gene, for example, was expected. Several of the $\sim$ 70 genes that were found to be either up- or down-regulated by our array analysis were then checked with real-time PCR, with generally good qualitative agreement between these techniques [76]. This in itself does not prove that the positive signals on all spots are real, especially for signals that are barely above background. If sufficient arrays and sample are available, a dose response test can help to differentiate between real and false positive signals, since net signal strength should scale with target concentration [75, 77]. Alternatively, one can boost the net S/N ratio by improving the hybridization kinetics. For most coverslip type hybridizations the reaction is diffusion limited and can be improved 3–5 fold by actively moving the hybridization fluid, as shown above and also in several other reports [77, 78]. This is particularly important if volume displacers are used, such as polydextrans [79], which increase the effective target concentration but also the fluid viscosity [77].

Since at least 50% of the genes present in a genome are expected to be expressed at less than 1 copy per cell and most of the others are present in fewer than 10 copies, sensitivity will remain the key issue for this type of analysis. With a push toward smaller sample sizes and ideally single cell analysis, stochasticity in gene expression will become the ultimate limit [80], requiring multiplexing of samples and arrays to overcome these statistical hurdles. Finally, miniaturization and automation will provide some additional solutions, as discussed in Chaps. 6 and 7.

11.4 Appendix

Assay Protocol for Expression Microarrays The following protocol is based on methods worked out by the Biochemistry research group at Corning, Inc. [50, 59, 75–77] and includes recent improvements.

1. *Reagents*
   - 5× FSS buffer: 250 mM Tris–HCl, 375 mM KCl, 15 mM $MgCl_2$
   - dNTP mix: 10 mM each of dGTP, dATP, and dTTP, 1 mM of dCTP
   - RevT solution: 8 µl 5× FSS buffer, 4 µl 0.1 M DTT, 2 µl dNTP mix and 1 µl of 1 mM dCTP–Cy3 or 1 mM dCTP–Cy5, and 2 µl of reverse transcriptase
   - Universal Hybridization Kit (Cat. No. 40026, Corning Incorporated), consisting in: Universal Wash Reagent A, Universal Wash Reagent B, Universal Pre-Soak Solution, Sodium Borohydride Pre-Soak Tablets, Universal Pre-Hybridization Solution, and Universal Hybridization Buffer
     Wash Soln 1: 50 ml Wash Reagent A, 447.5 ml water, 2.5 ml Wash Reagent B
     Wash Soln 2: 75 ml Wash Reagent A, 1425 ml water
     Wash Soln 3: 300 ml Wash Soln 2, 1200 ml water
2. *Labelling of total human RNA*
   - mix 1–5 µg of purified total human RNA, 3 µg of random hexamers (1 ug/ul) and nuclease free water; final volume 23 µl
   - incubate for 5 minutes at 70°C, quick chill on ice and spin down
   - add 17 µl of RevT solution, mix well and incubate for 2 hours, 42°C
   - add 1 µl (2 U/µl) RNase H and 0.25 µl RNase A (30 µg/µl); incubate 15 minutes, 37°C
   - purify cDNAs with Qiagen's PCR purification kit and reduce the volume by evaporation to about 5–8 µl
3. *Autofluorescence reduction and prewash:*
   - incubate slides in Universal Pre-Soak Solution with 1 tablet of $NaBH_4$ at 42°C, 20–30 minutes, then transfer successively to Wash Solution 2

(RT, 10 sec), Universal Pre-Hybridization Solution (42°C, 15 minutes), Wash Solution 2 (RT, 1 minute), Wash Solution 3 (RT, 30 sec)
- dry slides by low speed spin (1000 rpm) at RT, 1 minute

4. *Hybridization*
   - dissolve labelled cDNA in 60 μl Universal Hybridization buffer
   - denature the target mixture at 95°C for 3 minutes, then spin for 20 sec at RT
   - place onto the array, cover with 24 × 60 mm cover–slip (avoid bubbles!)
   - incubate in a high humidity hybridization chamber at 42°C, 14–20 hours
5. *Post hybridization processing*
   - immerse slides in Wash Solution 1 (2 minutes), remove coverslip, then incubate for 5 minutes, all at 42°C
   - transfer slides successively to Wash Solution 1 (5 minutes, 42°C), Wash Solution 2 (10 minutes, RT), wash solution 3 (2 minutes, RT), wash solution 3 (2 minutes, RT)
   - dry slides by low speed spin (1000 rpm) for 1 minute at RT; store in the dark

References

1. Drmanac R, Drmanac S, Chui G, Diaz R, Hou A, Jin H, Jin P, Kwon S, Lacy S, Moeur B, Shafto J, Swanson D, Ukrainczyk T, Xu C, Little D. 2002. *Sequencing by hybridization (SBH): advantages, achievements, and opportunities.* Adv Biochem Eng Biotechnol 77:75–101
2. Khrapko KR, Lysov YuP, Khorlyn AA, Shick VV, Florentiev VL, Mirzabekov AD. 1989. *An oligonucleotide hybridization approach to DNA sequencing.* FEBS Lett 256:118–22
3. Lysov IuP, Florent'ev VL, Khorlin AA, Khrapko KR, Shik VV. 1988. *Determination of the nucleotide sequence of DNA using hybridization with oligonucleotides.* A new method. Dokl Akad Nauk SSSR 303:1508–11
4. Schena M, Shalon D, Davis RW, Brown PO. 1995. *Quantitative monitoring of gene expression patterns with a complementary DNA microarray.* Science 270:467–70
5. Schena M, Shalon D, Heller R, Chai A, Brown PO, Davis RW. 1996. *Parallel human genome analysis: microarray–based expression monitoring of 1000 genes.* Proc Natl Acad Sci USA 93:10614–9
6. Heller RA, Schena M, Chai A, Shalon D, Bedilion T, Gilmore J, Woolley DE, Davis RW. 1997. *Discovery and analysis of inflammatory disease–related genes using cDNA microarrays.* Proc Natl Acad Sci USA 94:2150–5
7. Huang P, Feng L, Oldham EA, Keating MJ, Plunkett W. 2000. *Superoxide dismutase as a target for the selective killing of cancer cells.* Nature 407:390–5
8. Lam LT, Pickeral OK, Peng AC, Rosenwald A, Hurt EM, Giltnane JM, Averett LM, Zhao H, Davis RE, Sathyamoorthy M, Wahl LM, Harris ED, Mikovits JA, Monks AP, Hollingshead MG, Sausville EA, Staudt LM. 2001. *Genomic–scale measurement of mRNA turnover and the mechanisms of action of the anticancer drug flavopiridol.* Genome Biol 2:RESEARCH0041

9. Marton MJ, DeRisi JL, Bennett HA, Iyer VR, Meyer MR, Roberts CJ, Stoughton R, Burchard J, Slade D, Dai H, Bassett DE Jr, Hartwell LH, Brown PO, Friend SH. 1998. *Drug target validation and identification of secondary drug target effects using DNA microarrays.* Nat Med 4:1293–301
10. Bittner M, Meltzer P, Chen Y, Jiang Y, Seftor E, Hendrix M, Radmacher M, Simon R, Yakhini Z, Ben-Dor A, Sampas N, Dougherty E, Wang E, Marincola F, Gooden C, Lueders J, Glatfelter A, Pollock P, Carpten J, Gillanders E, Leja D, Dietrich K, Beaudry C, Berens M, Alberts D, Sondak V. 2000. *Molecular classification of cutaneous malignant melanoma by gene expression profiling.* Nature 406:536–40
11. Iyer VR, Eisen MB, Ross DT, Schuler G, Moore T, Lee JC, Trent JM, Staudt LM, Hudson J Jr, Boguski MS, Lashkari D, Shalon D, Botstein D, Brown PO. 1999. *The transcriptional program in the response of human fibroblasts to serum.* Science 283:83–7
12. Perou CM, Sorlie T, Eisen MB, van de Rijn M, Jeffrey SS, Rees CA, Pollack JR, Ross DT, Johnsen H, Akslen LA, Fluge O, Pergamenschikov A, Williams C, Zhu SX, Lonning PE, Borresen-Dale AL, Brown PO, Botstein D. 2000. *Molecular portraits of human breast tumours.* Nature 406:747–52
13. Shoemaker DD, Schadt EE, Armour CD, He YD, Garrett-Engele P, McDonagh PD, Loerch PM, Leonardson A, Lum PY, Cavet G, Wu LF, Altschuler SJ, Edwards S, King J, Tsang JS, Schimmack G, Schelter JM, Koch J, Ziman M, Marton MJ, Li B, Cundiff P, Ward T, Castle J, Krolewski M, Meyer MR, Mao M, Burchard J, Kidd MJ, Dai H, Phillips JW, Linsley PS, Stoughton R, Scherer S, Boguski MS. 2001. *Experimental annotation of the human genome using microarray technology.* Nature 409:922–7
14. Scherf U, Ross DT, Waltham M, Smith LH, Lee JK, Tanabe L, Kohn KW, Reinhold WC, Myers TG, Andrews DT, Scudiero DA, Eisen MB, Sausville EA, Pommier Y, Botstein D, Brown PO, Weinstein JN. 2000. *A gene expression database for the molecular pharmacology of cancer.* Nat Genet 24:236–44
15. Staunton JE, Slonim DK, Coller HA, Tamayo P, Angelo MJ, Park J, Scherf U, Lee JK, Reinhold WO, Weinstein JN, Mesirov JP, Lander ES, Golub TR. 2001. *Chemosensitivity prediction by transcriptional profiling.* Proc Natl Acad Sci USA 98:10787–92
16. Alizadeh AA, Eisen MB, Davis RE, Ma C, Lossos IS, Rosenwald A, Boldrick JC, Sabet H, Tran T, Yu X, Powell JI, Yang L, Marti GE, Moore T, Hudson J Jr, Lu L, Lewis DB, Tibshirani R, Sherlock G, Chan WC, Greiner TC, Weisenburger DD, Armitage JO, Warnke R, Staudt LM, et al. 2000. *Distinct types of diffuse large B–cell lymphoma identified by gene expression profiling.* Nature 403:503–11
17. Golub TR, Slonim DK, Tamayo P, Huard C, Gaasenbeek M, Mesirov JP, Coller H, Loh ML, Downing JR, Caligiuri MA, Bloomfield CD, Lander ES. 1999. *Molecular classification of cancer: class discovery and class prediction by gene expression monitoring.* Science 286:531–7
18. Unger MA, Rishi M, Clemmer VB, Hartman JL, Keiper EA, Greshock JD, Chodosh LA, Liebman MN, Weber BL. 2001. *Characterization of adjacent breast tumors using oligonucleotide microarrays.* Breast Cancer Res 3:336–41
19. Bhattacharjee A, Richards WG, Staunton J, Li C, Monti S, Vasa P, Ladd C, Beheshti J, Bueno R, Gillette M, Loda M, Weber G, Mark EJ, Lander ES, Wong W, Johnson BE, Golub TR, Sugarbaker DJ, Meyerson M. 2001. *Classification of human lung carcinomas by mRNA expression profiling reveals distinct adenocarcinoma subclasses.* Proc Natl Acad Sci USA 98:13790–5

20. Garber ME, Troyanskaya OG, Schluens K, Petersen S, Thaesler Z, Pacyna-Gengelbach M, van de Rijn M, Rosen GD, Perou CM, Whyte RI, Altman RB, Brown PO, Botstein D, Petersen I. 2001. *Diversity of gene expression in adenocarcinoma of the lung*. Proc Natl Acad Sci USA 98:13784–9
21. Pomeroy SL, Tamayo P, Gaasenbeek M, Sturla LM, Angelo M, McLaughlin ME, Kim JY, Goumnerova LC, Black PM, Lau C, Allen JC, Zagzag D, Olson JM, Curran T, Wetmore C, Biegel JA, Poggio T, Mukherjee S, Rifkin R, Califano A, Stolovitzky G, Louis DN, Mesirov JP, Lander ES, Golub TR. 2002. *Prediction of central nervous system embryonal tumour outcome based on gene expression*. Nature 415:436–42
22. van 't Veer LJ, Dai H, van de Vijver MJ, He YD, Hart AA, Mao M, Peterse HL, van der Kooy K, Marton MJ, Witteveen AT, Schreiber GJ, Kerkhoven RM, Roberts C, Linsley PS, Bernards R, Friend SH. 2002. *Gene expression profiling predicts clinical outcome of breast cancer*. Nature 415:530–6
23. Brown PO, Botstein D. 1999. *Exploring the new world of the genome with DNA microarrays*. Nat Genet 21:33–7
24. Hughes TR, Mao M, Jones AR, Burchard J, Marton MJ, Shannon KW, Lefkowitz SM, Ziman M, Schelter JM, Meyer MR, Kobayashi S, Davis C, Dai H, He YD, Stephaniants SB, Cavet G, Walker WL, West A, Coffey E, Shoemaker DD, Stoughton R, Blanchard AP, Friend SH, Linsley PS. 2001. *Expression profiling using microarrays fabricated by an ink–jet oligonucleotide synthesizer*. Nat Biotechnol 19:342–7
25. Kane MD, Jatkoe TA, Stumpf CR, Lu J, Thomas JD, Madore SJ. 2000. *Assessment of the sensitivity and specificity of oligonucleotide (50mer) microarrays*. Nucleic Acids Res 28:4552–7
26. Lipshutz RJ, Fodor SP, Gingeras TR, Lockhart DJ. 1999. *High density synthetic oligonucleotide arrays*. Nat Genet 21:20–4
27. Kallioniemi A, Kallioniemi OP, Sudar D, Rutovitz D, Gray JW, Waldman F, Pinkel D. 1992. *Comparative genomic hybridization for molecular cytogenetic analysis of solid tumors*. Science 258:818–21
28. Yuen T, Wurmbach E, Pfeffer RL, Ebersole BJ, Sealfon SC. 2002. *Accuracy and calibration of commercial oligonucleotide and custom cDNA microarrays*. Nucleic Acids Res 30:e48
29. Kuo WP, Jenssen TK, Butte AJ, Ohno-Machado L, Kohane IS. 2002. *Analysis of matched mRNA measurements from two different microarray technologies*. Bioinformatics 18:405–12
30. Li J, Pankratz M, Johnson JA. 2002. *Differential gene expression patterns revealed by oligonucleotide versus long cDNA arrays*. Toxicol Sci 69:383–90
31. Hu GK, Madore SJ, Moldover B, Jatkoe T, Balaban D, Thomas J, Wang Y. 2001. *Predicting splice variant from DNA chip expression data*. Genome Res 11:1237–45
32. Fodor SP, Read JL, Pirrung MC, Stryer L, Lu AT, Solas D. 1991. *Light–directed, spatially addressable parallel chemical synthesis*. Science 251:767–73
33. Maskos U, Southern EM. 1992. *Oligonucleotide hybridizations on glass supports: a novel linker for oligonucleotide synthesis and hybridization properties of oligonucleotides synthesised in situ*. Nucleic Acids Res 20:1679–84
34. Nanthakumar A, Pon RT, Mazumder A, Yu S, Watson A. 2000. *Solid–phase oligonucleotide synthesis and flow cytometric analysis with microspheres encoded with covalently attached fluorophores*. Bioconjug Chem 11:282–8

35. Graves DJ. 1999. *Powerful tools for genetic analysis come of age.* Trends Biotechnol 17:127–34
36. Marshall WS, Boymel JL. 1999. Drug Discovery Today. 2:34–42
37. Elghanian R, Xu Y, McGowen J, Siethoff M, Liu CG, Winick J, Fuller N, Ramakrishnan R, Beuhler A, Johnson T, Mazumder A, Brush C. 2001. Nucleosides, Nucleotides and Nucleic Acids. 20:1371–1375
38. Lemeshko SV, Powdrill T, Belosludtsev YY, Hogan M. 2001. *Oligonucleotides form a duplex with non-helical properties on a positively charged surface.* Nucleic Acids Res 29:3051–8
39. Lockhart DJ, Dong H, Byrne MC, Follettie MT, Gallo MV, Chee MS, Mittmann M, Wang C, Kobayashi M, Horton H, Brown EL. 1996. *Expression monitoring by hybridization to high–density oligonucleotide arrays.* Nat Biotechnol 14:1675–80
40. Mir KU, Southern EM. 1999. *Determining the influence of structure on hybridization using oligonucleotide arrays.* Nat Biotechnol 17:788–92
41. Matveeva OV, Tsodikov AD, Giddings M, Freier SM, Wyatt JR, Spiridonov AN, Shabalina SA, Gesteland RF, Atkins JF. 2000. *Identification of sequence motifs in oligonucleotides whose presence is correlated with antisense activity.* Nucleic Acids Res 28:2862–5
42. Wetmur JG. 1991. *DNA probes: applications of the principles of nucleic acid hybridization.* Crit Rev Biochem Mol Biol 26:227–59
43. Langer PR, Waldrop AA, Ward DC. 1981. *Enzymatic synthesis of biotin–labeled polynucleotides: novel nucleic acid affinity probes.* Proc Natl Acad Sci USA 78:6633–7
44. Ramakrishnan R, Dorris D, Lublinsky A, Nguyen A, Domanus M, Prokhorova A, Gieser L, Touma E, Lockner R, Tata M, Zhu X, Patterson M, Shippy R, Sendera TJ, Mazumder A. 2002. *An assessment of Motorola CodeLink microarray performance for gene expression profiling applications.* Nucleic Acids Res 30:e30
45. Lockhart DJ, Winzeler EA. 2000. *Genomics, gene expression and DNA arrays.* Nature 405:827–36
46. Wang X, Ghosh S, Guo SW. 2001. *Quantitative quality control in microarray image processing and data acquisition.* Nucleic Acids Res 29:E75–5
47. Wodicka L, Dong H, Mittmann M, Ho MH, Lockhart DJ. 1997. *Genome–wide expression monitoring in Saccharomyces cerevisiae.* Nat Biotechnol 15:1359–67
48. Hastie ND, Bishop JO. 1976. *The expression of three abundance classes of messenger RNA in mouse tissues.* Cell 9:761–74
49. Chan V, Graves DJ, McKenzie SE. 1995. *The biophysics of DNA hybridization with immobilized oligonucleotide probes.* Biophys J 69:2243–55
50. Hong Y, Muller UR, Lai F. 2003. Toxicology in Vitro 17:85–92
51. Riccelli PV, Merante F, Leung KT, Bortolin S, Zastawny RL, Janeczko R, Benight AS. 2001. *Hybridization of single–stranded DNA targets to immobilized complementary DNA probes: comparison of hairpin versus linear capture probes.* Nucleic Acids Res 29:996–1004
52. Yue H, Eastman PS, Wang BB, Minor J, Doctolero MH, Nuttall RL, Stack R, Becker JW, Montgomery JR, Vainer M, Johnston R. 2001. *An evaluation of the performance of cDNA microarrays for detecting changes in global mRNA expression.* Nucleic Acids Res 29:E41–1
53. Che D, Bao Y, Muller UR. 2001. *Novel surface and multicolor charge coupled device–based fluorescent imaging system for DNA microarrays.* J Biomed Opt 6:450–6

54. Neuschaefer D, Budach W, Wanke C. 2002. SPIE's BIOS 2002. Proceeding Abstract 4626–115:189
55. Stimpson DI, Hoijer JV, Hsieh WT, Jou C, Gordon J, Theriault T, Gamble R, Baldeschwieler JD. 1995. *Real–time detection of DNA hybridization and melting on oligonucleotide arrays by using optical wave guides.* Proc Natl Acad Sci USA 92:6379–83
56. Broude NE, Woodward K, Cavallo R, Cantor CR, Englert D. 2001. *DNA microarrays with stem–loop DNA probes: preparation and applications.* Nucleic Acids Res 29:E92
57. Cheek BJ, Steel AB, Torres MP, Yu YY, Yang H. 2001. *Chemiluminescence detection for hybridization assays on the flow–thru chip, a three–dimensional microchannel bio–chip.* Anal Chem 73:5777–83
58. Fredrickson HL, Perkins EJ, Bridges TS, Tonucci RJ, Fleming JK, Nagel A, Diedrich K, Mendez-Tenorio A, Doktycz MJ, Beattie KL. 2001. *Towards environmental toxicogenomics – development of a flow–through, high–density DNA hybridization array and its application to ecotoxicity assessment.* Sci Total Environ 274:137–49
59. Raghavachari N, Bao Y, Li G, Xie X, Müller UR. 2003. *Reduction of autofluorescence on DNA microarrays and slide surfaces by treatment with sodium borohydride.* Anal Biochem 312:102–5
60. Duggan DJ, Bittner M, Chen Y, Meltzer P, Trent JM. 1999. *Expression profiling using cDNA microarrays.* Nat Genet 21:10–4
61. Mayanil CS, George D, Freilich L, Miljan EJ, Mania-Farnell B, McLone DG, Bremer EG. 2001. *Microarray analysis detects novel Pax3 downstream target genes.* J Biol Chem 276:49299–309
62. DeRisi J, Penland L, Brown PO, Bittner ML, Meltzer PS, Ray M, Chen Y, Su YA, Trent JM. 1996. *Use of a cDNA microarray to analyse gene expression patterns in human cancer.* Nat Genet 14:457–60
63. Relogio A, Schwager C, Richter A, Ansorge W, Valcarcel J. 2002. *Optimization of oligonucleotide–based DNA microarrays.* Nucleic Acids Res 30:e51
64. Storhoff JJ, Marla SS, Hagenow S, Mehta H, Lucas A, Garimella V, Patno TJ, Buckingham W, Cork WH, Müller UR. 2002. SPIE Proceedings. SPIE Proceedings 4937:1–7
65. Sussman H. 2002. The Scientist. 16(13):37–38
66. Schroeder BG, Peterson LM, Fleischmann RD. 2002. *Improved quantitation and reproducibility in Mycobacterium tuberculosis DNA microarrays.* J Mol Microbiol Biotechnol 4:123–6
67. Xiang CC, Kozhich OA, Chen M, Inman JM, Phan QN, Chen Y, Brownstein MJ. 2002. *Amine–modified random primers to label probes for DNA microarrays.* Nat Biotechnol 20:738–42
68. Pabon C, Modrusan Z, Ruvolo MV, Coleman IM, Daniel S, Yue H, Arnold LJ Jr. 2001. *Optimized T7 amplification system for microarray analysis.* Biotechniques 31:874–9
69. Van Gelder RN, von Zastrow ME, Yool A, Dement WC, Barchas JD, Eberwine JH. 1990. *Amplified RNA synthesized from limited quantities of heterogeneous cDNA.* Proc Natl Acad Sci USA 87:1663–7
70. Vernon SD, Unger ER, Rajeevan M, Dimulescu IM, Nisenbaum R, Campbell CE. 2000. *Reproducibility of alternative probe synthesis approaches for gene expression profiling with arrays.* J Mol Diagn 2:124–7

71. Wang E, Miller LD, Ohnmacht GA, Liu ET, Marincola FM. 2000. *High–fidelity mRNA amplification for gene profiling.* Nat Biotechnol 18:457–9
72. Raghavachari et al. Ambion TechNotes. 9((3)): http://www.ambion.com-/techlib/tn/93–9313.html
73. Theilgaard-Monch K, Cowland J, Borregaard N. 2001. *Profiling of gene expression in individual hematopoietic cells by global mRNA amplification and slot blot analysis.* J Immunol Methods 252:175–89
74. Brail LH, Jang A, Billia F, Iscove NN, Klamut HJ, Hill RP. 1999. *Gene expression in individual cells: analysis using global single cell reverse transcription polymerase chain reaction (GSC RT–PCR).* Mutat Res 406:45–54
75. Bao P, Frutos AG, Greef C, Lahiri J, Muller U, Peterson TC, Warden L, Xie X. 2002. *High–sensitivity detection of DNA hybridization on microarrays using resonance light scattering.* Anal Chem 74:1792–7
76. Srilatha S, Raghavachari N, Bao YP, Muller UR, Feldman D. 2003. *Breast Cancer Research and Treatment.* in press
77. Yuen PK, Li G, Bao P, Müller UR. 2003. Lab on a Chip. 3:46–50
78. Adey NB, Lei M, Howard MT, Jensen JD, Mayo DA, Butel DL, Coffin SC, Moyer TC, Slade DE, Spute MK, Hancock AM, Eisenhoffer GT, Dalley BK, McNeely MR. 2002. *Gains in sensitivity with a device that mixes microarray hybridization solution in a 25–microm–thick chamber.* Anal Chem 74:6413–7
79. Wahl GM, Stern M, Stark GR. 1979. *Efficient transfer of large DNA fragments from agarose gels to diazobenzyloxymethyl–paper and rapid hybridization by using dextran sulfate.* Proc Natl Acad Sci USA 76:3683–7
80. Elowitz MB, Levine AJ, Siggia ED, Swain PS. 2002. *Stochastic gene expression in a single cell.* Science 297:1183–6

12

Applications of Matrix-CGH (Array-CGH) for Genomic Research and Clinical Diagnostics

Carsten Schwaenena, Michelle Nesslinga, Bernhard Radlwimmera,
Swen Wessendorf, and Peter Lichtera

12.1 Introduction

One of the major scientific achievements of the past decade was the rapid development of genomic research, resulting in the comprehensive sequence information of the human genome. This information has provided the basis for the identification of a vast number of novel genomic aberrations in tumors and hereditary diseases.

Cancer etiology and development is associated with hereditary or acquired genomic alterations. Among these, genomic imbalances play a prominent role with deletions indicating the localization of tumor suppressor genes (e.g. NF2, P53 or ATM) or amplifications frequently affecting protooncogenes (e.g. MYC). Such aberrations may lead to an inactivation or, by a so–called "dosage effect", activation of genes relevant to the initiation and progression of tumor cells. Genomic imbalances also play an important role in the field of clinical genetics. Many human mental retardation syndromes, congenital malformations and miscarriages are caused by defined copy number gains or losses of various chromosomal regions, whole chromosomes, or by small subtelomeric chromosome rearrangements [1–5]. Besides the most frequent aneuploidies of human chromosomes, such as Patau syndrome (trisomy of chromosome 13), Edward syndrome (trisomy of chromosome 18) or Down syndrome (trisomy of chromosome 21), a number of congenital diseases are associated with smaller imbalances, mostly microdeletions: e.g. Prader–Willi syndrome (15q12), Angelman syndrome (15q12), Williams syndrome (7q11.2), or DiGeorge syndrome (22q11.21). Identification of chromosomal imbalances has significantly contributed to the detection of genes playing a pathogenic role and the elucidation of molecular mechanisms responsible for defined phenotypes in malignant or congenital diseases.

Our current understanding of chromosomal alterations is mainly based on chromosome banding analysis, visualization of targeted genomic regions by fluorescence in situ hybridization (FISH) to metaphase chromosomes or interphase cell nuclei, or traditional comparative genomic hybridization (CGH).

In the last decade, CGH was extensively applied to the investigation of recurrent imbalances in hematological malignancies and solid tumors. CGH to metaphase chromosomes is a molecular cytogenetic technique that allows genome–wide screening for imbalanced chromosomal regions independent of the need to prepare metaphase chromosomes from the specimen to be investigated. This is of particular importance in the analysis of tumor cells, as in many instances dividing cells are difficult or even impossible to obtain. Thus, CGH circumvents the limitations of conventional karyotype analysis including mandatory short-term culturing of the tumor cells which might induce additional chromosomal aberrations. In principle, equal amounts of differentially labelled genomic test (e.g. tumor) and control DNA are used as probes for fluorescence in situ hybridization (FISH) onto chromosomes of normal metaphase cells immobilized on glass slides. The comparison of the obtained hybridization–signal intensities of both DNA probes represents an average of all imbalances present in the sample genome [6–10].

CGH allows genome–wide localization of chromosomal imbalances without prior knowledge of specific regions of interest, and has been used to study a large variety of solid tumors and hematological malignancies [11–16]. Such studies have revealed a wealth of novel genomic aberrations, contributed to the identification of novel genes, and provided the basis for improved cancer classification schemes. Conventional CGH, however, has not become a diagnostic tool in clinical settings, since the method is technically demanding, difficult to automate and has limited spatial resolution. Due to the degree of DNA condensation of metaphase chromosomes, resolution of CGH is restricted to approximately 3–10 Mb for low copy number gains and losses [17–19] and 2 Mb (a product of the degree of repetition and the size of the amplicon) for high level amplifications [8, 20].

To circumvent these problems, it was mandatory to replace the metaphase chromosomes as targets for comparative in situ hybridization. A chip-based technique, termed "matrix–CGH" or "array–CGH", was developed [21] allowing analyses at a much higher resolution and providing a basis for extensive automation. For this approach, the chromosome targets are substituted by well-defined genomic DNA fragments (e.g. specific for chromosomal regions or genes) cloned in various types of vectors (e.g. BACs, PACs, cosmids, plasmids). These fragments are immobilized on glass surfaces in order to generate a microarray where each clone is represented on a distinct position of the matrix. When the technique was first reported [21] most of the steps of the procedure were performed manually. This approach has been extended and modified with regard to automation and array size [22–24]. Rapid and reproducible positioning of large numbers of DNA fragments is achieved using ink jet, split pin or capillary-based robotic printing systems. This allows high resolution genomic screening of thousands of defined DNA targets immobilized on glass slides in a single experiment. The spatial resolution of matrix–CGH is highly superior to that of chromosomal CGH. Resolution is limited mainly by the size of the spotted DNA fragments, and by the fact that hybridization signal strength decreases with decreasing fragment size and complexity. Using BAC or PAC clones as targets, single–copy number changes can be detected with a resolution similar to that of interphase FISH in a single hybridization experiment. Analysis of multiple genomic regions by inter-phase FISH, in contrast, would require either multiple hybridization experiments or a complex multicolor FISH approach applied to a series of cell nuclei. High level amplifications down to a size of several kb can be detected by using cDNA arrays as hybridization targets [25, 26].

Matrix–CGH analysis is based on a co-hybridization of differentially fluorescent-labelled genomic test and control DNA. Following hybridization, the signal intensities of both fluorochromes are measured for each target sequence and the respective normalized signal ratios are calculated. The obtained genomic profile indicates gains or losses of chromosomal regions like low copy number losses, such as deletions, low copy number gains, such as trisomies, or high level amplifications. A representative example is shown in Figs. 12.1 and 12.2. One should keep in mind that comparative genomic hybridization does not allow the identification of balanced chromosome aberrations such as balanced translocations or inversions. While matrix–CGH is dedicated to the detection of net genomic imbalances, genomic DNA arrays could also contribute to the fine mapping of breakpoints in rearranged chromosomes. Provided a marker chromosome is prepared by flow sorting or micromanipulation techniques and the chromosome-derived labelled DNA is hybridized to a comprehensive genomic array, breakpoints could be pinpointed to a single fragment immobilized on the chip depending on the chip design [27].

12.2 Technical Aspects

Different DNA targets have been tested for use in matrix–CGH. The most common are genomic DNA fragments cloned in different vector types (BAC, PAC, cosmid) [21, 22, 24]. These spanning DNA inserts (up to 300 kbp) are prepared from bacterial cultures and sheared by sonification to a fragment length of 500–5000 kbp [22, 24], or generated by applying PCR-based amplification procedures [23, 28]. One disadvantage of the preparation from large bacterial cultures is that it is laborious and expensive and has to be repeated whenever the DNA supply has been exhausted. To overcome these problems, methods of whole–genome amplification, such as degenerate–oligonucleotide–primed–PCR (DOP–PCR, [29]) PCR and single-cell comparative genomic hybridization (SCOMP); [30]), have been applied to PCR–amplify BAC and PAC clone DNA [23, 27]. For these methods, DNA is prepared on a small scale, and the PCR products, once obtained, can be repeatedly re-amplified producing a large supply of target DNA. Furthermore, use of PCR-amplified BAC and PAC sequence pools has the advantage of allowing the simultaneous preparation of thousands of DNA fragments ready to be immobilized on a

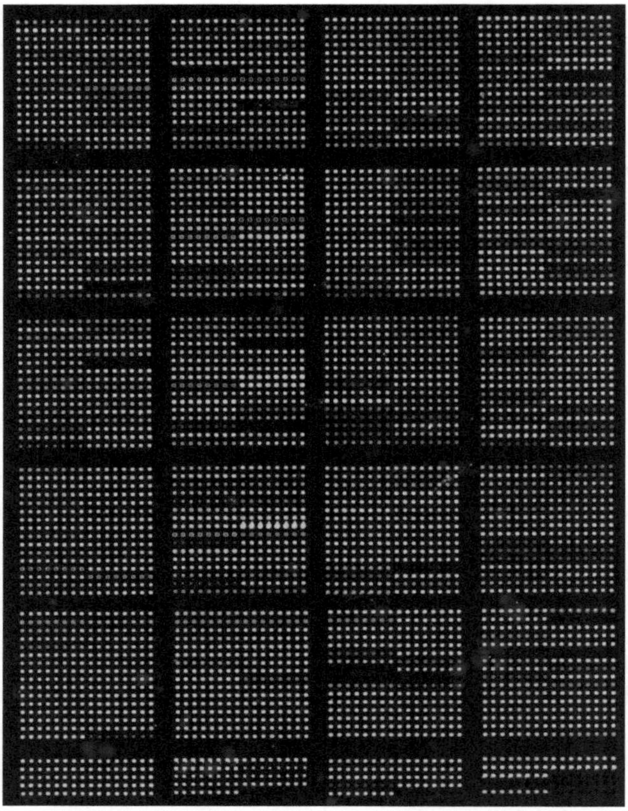

Fig. 12.1. Matrix–CGH chip co-hybridized with Cy3-labelled HL60–tumor DNA and Cy5-labelled male–control DNA. Clones are spotted in replicas of eight. Red and green spot color indicates losses and gains in HL60 relative to control DNA, respectively

microarray [27]. Alternatively, cDNA arrays have been used in some genomic profiling studies [25, 26].

These protocols are compatible with commercially available printing devices equipped with split–pins, capillaries or ink jet systems. Printing of e.g. one nanoliter of DNA solution yields homogeneous spots of 70–150 μm in diameter. For large-scale microarrays the split–pin or capillary technology is superior due to a much higher printing speed. Technical parameters affecting spot density, spot quality (temperature, humidity) and immobilization efficiency (glass surface, chemical fixation) are equivalent to those used in other current DNA microarray protocols.

Concerning the source of DNA to be investigated, best matrix–CGH results are obtained with fresh or frozen tissue or cell samples. However, many clinical specimens are formalin-fixed and paraffin embedded. From such samples only partially degraded genomic DNA can be prepared. While it is possible to perform matrix–CGH from such archived collections, the success of these studies is highly dependent on the quality of the fixative (buffered formalin is required) and the duration of the fixation. As this information is often not available, the success rate of studies on archived material can be quite limited.

For labelling of genomic DNA probes, a number of different protocols are used, e.g. incorporation of dye-labelled nucleotides by nick–translation, universal PCR or random primer extension. Apart from labelling of very small DNA amounts, e.g. from microdissected tumor samples with universal PCR reactions, random labelling currently is the most widely used protocol for matrix–CGH. In comparison to expression profiling by DNA microarrays, matrix–CGH has a much higher demand with respect to sensitivity and signal linearity. This is due to the necessity to measure subtle ratio differences for the detection of monoallelic gains or losses of < 0.5 (0.5 for a deletion and 1.5 for a trisomy compared to 1.0 for a balanced state). Therefore tiny variations in signal intensities have to be detected with high accuracy. Additionally, in contrast to constitutional diseases, ratio differences in primary tumor specimens are diminished according to the sample's content of 'contaminating' stromal tissue (fibrocytes, leukocytes, vessel endothelial cells) or adjacent non-malignant cells exhibiting a normal diploid karyotype. For reliable genomic profiling, the proportion of cells to be analyzed, e.g. the content of tumor cells, should be at least 50%.

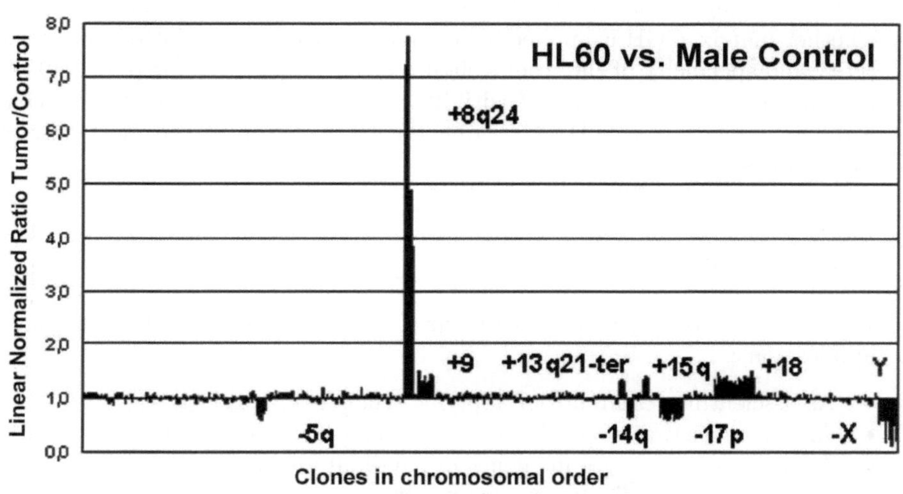

Fig. 12.2. Matrix–CGH profile of the tumor cell line HL60. BAC and PAC clones are arranged in chromosomal order along the X-axis. Linear normalized ratios of HL60 and male–control DNA fluorescence signals are shown on the Y-axis. All the genomic imbalances characteristic of HL60 are detected

As in expression analyses, the raw fluorescence ratios of matrix–CGH images have to be normalized to compensate for unequal incorporation rates of the fluorescence dyes and other biases. Normalization can either be performed globally, using all clones of the array, or by selecting clones corresponding to genomic regions that likely are in a diploid or 'balanced' state. During the development of matrix–CGH it became clear that a robust normalization procedure is needed, since the ratio values of some target fragments seem to depend on parameters which are not fully understood. With the development of new types of arrays, this problem has become more evident. Especially when screening tumor cell genomes this becomes an issue, since it is a priori not known whether a specific sequence used for normalization is actually part of a genomic imbalance. To overcome this problem, normalization of an experiment should be based on the median ratio of a large number of DNA clones (> 100), which are more or less linearly distributed across the whole human genome.

Technical issues that still need to be resolved include the development of protocols for quantitative amplification of small amounts of DNA extracted from small numbers of cells. Efforts have been made to use DOP–PCR for matrix–CGH of microdissected paraffin-embedded cells [31], and SCOMP has been successfully used for chromosomal CGH. Recently, a new method, hyperbranched strand displacement amplification [32], was tested for matrix–CGH. The authors found that using 1000 or more cells of starting material, gene–dosage alterations of threefold or more could be detected [33]. Technical demands also depend on whether the experiments are part of a research study or whether matrix–CGH is applied as a diagnostic tool. In the latter case, sensitivity and specificity of the results should be as high as possible (e.g. $\gg 95\%$). As a means of increasing the sensitivity, we have established a protocol using 8 replica spots for each DNA fragment with exclusion of the most extreme ratio values [34].

12.3 Applications

Automated genomic profiling by matrix–CGH can be envisioned for two major applications:

- As a scientific research tool applying whole genome chips
- Arrays consisting of contiguously mapping DNA fragments
- Arrays testing for specific genes or candidate regions
- In clinical applications allowing rapid and automated diagnosis based on arrays dedicated to the detection of disease specific imbalances

High resolution genomic DNA chips covering the genome will allow high resolution screenings aimed at the detection of previously unknown quantitative genomic alterations. Currently, such chips consist of arrays of fragments mapping at defined intervals [23], but in the near future whole genome chips, carrying, e.g., a complete tiling path of fragments of the human genome, will become available. The identified aberrations will allow the identification of critical chromosomal regions or might even pinpoint critical genes, e.g. tumor suppressor genes or oncogenes. Aberrant chromosomal regions can be further narrowed to microdeletions or single imbalanced DNA fragments by subsequent molecular analysis with specialized arrays consisting of contiguously mapping genomic DNA fragments ('contigs'). Depending on the size of the imbalanced chromosomal region and the desired physical resolution, BACs, PACs, cosmids or sets of cDNAs of adjacent genes are used. Whenever disease-relevant chromosomal regions have been identified by any method, contig chips are a suitable starting point for studies aiming at the identification of disease genes. The feasibility of this approach has been recently demonstrated in 116 patients with hereditary neurofibromatosis type 2. In this study, the chromosomal region 22q12 was analyzed with defined contigs uncovering small deletions as small as 40 kbp in size [35]. Thus, matrix–CGH allows bridging of the gap between imbalances approximately 10 Mbp in size, assessed, e.g. by cytogenetic methods, and smaller imbalances only some 100 kbp in length.

A further approach utilizes various designs of so-called onco chips, which test for the copy number of genes, selected on the basis of their function (e.g. carcino–genesis) or location (e.g. in tumor-associated imbalanced regions). In a recent study, such a chip was used to identify previously undetected, hidden chromosomal amplifications in high grade non-Hodgkin lymphoma that correlate with the mRNA expression level of candidate genes located in the respective amplicons [36]. Based on this and other studies [37, 38], and in analogy to the novel findings that had been detected with traditional CGH, it

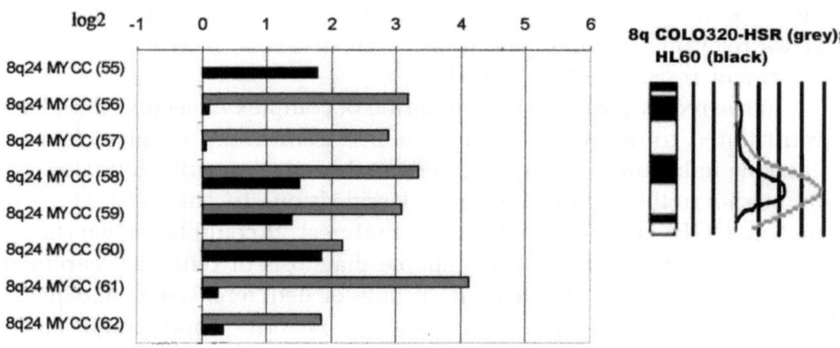

Fig. 12.3. Comparison of matrix–CGH and chromosomal CGH values. Adapted from Wessendorf et al. 2002. Examples are shown for tumor cell lines COLO320–HSR (continuous amplification of 8q24, grey bars) and HL60 (discontinuous amplification of 8q24, black bars). Although the scoring of the ratio values of the two methods is highly concordant, the absolute ratio values of the amplified regions are distributed over a much higher range for array signals, indicating the superior dynamics of matrix–CGH

can be assumed that matrix–CGH will help to uncover numerous amplicons too small to be seen when using conventional methods. Furthermore, due to the method's high resolution, detection of discontinuous amplifications will even become possible (Fig. 12.3). Once the number of candidate genes has been limited by fine–mapping of amplicons, subsequent molecular analyses will lead to the identification of new disease-related genes, in particular onco genes. The significance of the identified amplicon can easily be verified by FISH to tissue microarrays [39], which are a convenient tool to rapidly assess the frequency of the respective amplification in a large series of tumors. Pathogenically significant gene amplifications can also provide interference points for new therapeutic targets. The Her2/neu amplification in breast carcinomas and other tumors serves as a paradigmatic example. Amplification of this gene, which codes for a membrane-bound receptor, is associated with tumor progression. Patients carrying this amplification benefit from treatment with a modified antibody (Herceptin) directed against the receptor. Efforts needed to prove the pathogenic role of a candidate gene, however, should not be underestimated. Even with today's advanced technologies, comprehensive functional characterization, including analysis of DNA sequence, RNA and protein expression levels, posttranslational modifications, molecular interactions in biological pathways, and more, remains a formidable challenge.

Another interesting application of matrix–CGH is the detailed comparison of related tumor samples from an individual patient. This approach is of particular interest to the monitoring of tumor development and progression at different time points including comparison between primary tumors and derived metastases, transformation of tumors towards higher malignancies or analysis of relapse. Information about when and where chromosomal imbalances occur or recur in one individual will potentially aid the discovery of genes relevant to tumor initiation, aggressiveness, metastatic potential, and treatment resistance (e.g. [40, 41]).

In the recent past, the accumulation of complex molecular data has greatly contributed to improvements in tumor classification schemes. Assessment of genomic imbalances at an unprecedented resolution will likely also contribute to further refinements in tumor classification. In this context, two recent matrix–CGH studies are of particular interest. It could be shown that profiling of genomic imbalances allows reliable diagnosis of renal cell carcinoma [42]. A study comparing the genomic profile of dedifferentiated and pleomorphic liposarcomas uncovered a highly distinct pattern of both tumor entities [43]. Interestingly, this distinction was unequivocal using the genomic profile, but was less apparent from the expression profiles obtained from the same tumor series. Thus, depending on the diagnostic question and the tumor type, expression studies are not always superior in their diagnostic potential. This is an important consideration for practical reasons as well, since DNA typically is much more stable than RNA and therefore much better suited for application in routine diagnostics.

While matrix–CGH is now broadly accepted as a research tool, it is often debated to what degree it will really become part of routine diagnostics. Matrix–CGH could be used for the diagnosis of well-characterized recurrent chromosomal aberrations, which predict a homogeneous clinical course. In such a setting, the diagnosis could support treatment decisions and contribute to a further individualization of anticancer therapies. The identification of distinct clinical subgroups with different prognostic outcome is certainly most advanced in hematological malignancies. For example, cytogenetic methods, such as chromosome banding and FISH, are used to define subgroups in patients with chronic lymphocytic or acute leukemias according to their survival probabilities. The impact of such diagnostic data is evident in acute myeloid leukemias, where patients are already treated with either risk-adapted conventional chemotherapy or myeloablative peripheral stem cell transplantation according to their cytogenetic risk profile. Thus, genomic profiles provide important information for a tailored treatment, i.e. each patient receiving the best therapy available when comparing treatment tolerance and individual genetic risk.

Finally, it will be important to reliably diagnose the pattern of genomic alterations with prognostic impact. Disease specific matrix–CGH chips will simplify the identification of relevant chromosomal aberrations, since they allow an automated diagnostic procedure. We have developed such a chip optimized to detect alterations in chronic lymphocytic leukemia of B–cell type. This disease is characterized by a highly variable clinical course. Recurrent chromosomal imbalances were shown to provide strongly significant prognostic markers with deletions including chromosome arms 17p or 11q being associated with rapid disease progression and shorter overall survival of patients [44]. As therapy decisions based on these biological findings appear to become an option that is being tested in clinical trials, we developed a DNA microarray dedicated to meet the clinical needs. Testing of this chip revealed an unprecedented diagnostic specificity and sensitivity [34].

Besides oncological applications, CGH to microarrays will also become an important tool in clinical genetics. In families with unexplained mental retardation or dysmorphic features, as yet unknown microdeletions or cryptic rearrangements associated with small imbalances of genomic material seem to play a major role. Thus, prenatal and postnatal diagnostics in these families would greatly benefit from a method screening for such alterations at a high resolution, such as matrix–CGH with a genome–wide chip design. In case of mental retardation with negative cytogenetic results, a specialized chip covering all telomeric regions could be of particular importance [45], since more than 5% of cases with mental retardation seem to be associated with cryptic telomeric translocations including small genomic imbalances [46].

In addition to the many applications that are currently pursued, genomic microarrays will likely find new uses in other areas as well. For example, it can be envisioned that genomic DNA chips will be applied to study the extent of genomic duplications and deletions that have occurred during evolution and that seem to exist as polymorphisms within populations [47]. Moreover, new technical developments are likely to further increase the potential of genomic DNA arrays. For instance, a combination of immunoprecipitation and hybridization to genomic microarrays could facilitate the assignment of DNA/protein binding sites [48]. Certainly, CGH to microarrays is still in its early phase and the spectrum of applications will likely increase further in the future.

References

1. Knight SJ, Flint J. *Perfect endings: a review of subtelomeric probes and their use in clinical diagnosis.* J Med Genet. 2000 Jun;37(6):401–9
2. Battaglia A, Carey JC. *Diagnostic evaluation of developmental delay/mental retardation: An overview.* Am J Med Genet. 2003 Feb 15;117C(1):3–14
3. Cassidy SB, Dykens E, Williams CA. *Prader-Willi and Angelman syndromes: sister imprinted disorders.* Am J Med Genet. 2000 ;97(2):136–46
4. Donnai D, Karmiloff-Smith A. *Williams syndrome: from genotype through to the cognitive phenotype.* Am J Med Genet. 2000;97(2):164–71
5. Scambler PJ. *The 22q11 deletion syndromes.* Hum Mol Genet. 2000 Oct; 9(16):2421–6
6. Kallioniemi A, Kallioniemi OP, Sudar D, Rutovitz D, Gray JW, Waldman F, Pinkel D. *Comparative genomic hybridization for molecular cytogenetic analysis of solid tumors.* Science. 1992 Oct 30;258(5083):818–21
7. Du Manoir S, Speicher MR, Joos S, Schrock E, Popp S, Dohner H, Kovacs G, Robert-Nicoud M, Lichter P, Cremer T. *Detection of complete and partial chromosome gains and losses by comparative genomic in situ hybridization.* Hum Genet. 1993 Feb;90(6):590–610
8. Joos S, Scherthan H, Speicher MR, Schlegel J, Cremer T, Lichter P. *Detection of amplified DNA sequences by reverse chromosome painting using genomic tumor DNA as probe.* Hum Genet. 1993 Feb;90(6):584–9
9. Kallioniemi OP, Kallioniemi A, Sudar D, Rutovitz D, Gray JW, Waldman F, Pinkel D. *Comparative genomic hybridization: a rapid new method for detecting and mapping DNA amplification in tumors.* Semin Cancer Biol. 1993 Feb;4(1):41–6
10. Speicher MR, du Manoir S, Schrock E, Holtgreve-Grez H, Schoell B, Lengauer C, Cremer T, Ried T. *Molecular cytogenetic analysis of formalin–fixed, paraffin–embedded solid tumors by comparative genomic hybridization after universal DNA–amplification.* Hum Mol Genet. 1993 Nov;2(11):1907–14
11. Forozan F, Karhu R, Kononen J, Kallioniemi A, Kallioniemi OP. *Genome screening by comparative genomic hybridization.* Trends Genet. 1997 Oct;13(10):405–9
12. Zitzelsberger H, Lehmann L, Werner M, Bauchinger M. *Comparative genomic hybridisation for the analysis of chromosomal imbalances in solid tumours and haematological malignancies.* Histochem Cell Biol. 1997 Oct–Nov;108(4–5):403–17
13. Knuutila S, Armengol G, Bjorkqvist AM, el-Rifai W, Larramendy ML, Monni O, Szymanska J. *Comparative genomic hybridization study on pooled DNAs from*

tumors of one clinical–pathological entity. Cancer Genet Cytogenet. 1998 Jan 1;100(1):25–30

14. Bentz M, Werner CA, Dohner H, Joos S, Barth TF, Siebert R, Schroder M, Stilgenbauer S, Fischer K, Moller P, Lichter P. *High incidence of chromosomal imbalances and gene amplifications in the classical follicular variant of follicle center lymphoma.* Blood. 1996 Aug 15;88(4):1437–44
15. Bentz M, Plesch A, Bullinger L, Stilgenbauer S, Ott G, Muller-Hermelink HK, Baudis M, Barth TF, Moller P, Lichter P, Dohner H. *t(11;14)-positive mantle cell lymphomas exhibit complex karyotypes and share similarities with B-cell chronic lymphocytic leukemia.* Genes Chromosomes Cancer. 2000 Mar;27(3):285–94
16. Knuutila S, Aalto Y, Autio K, Bjorkqvist AM, El-Rifai W, Hemmer S, Huhta T, Kettunen E, Kiuru-Kuhlefelt S, Larramendy ML, Lushnikova T, Monni O, Pere H, Tapper J, Tarkkanen M, Varis A, Wasenius VM, Wolf M, Zhu Y. *DNA copy number losses in human neoplasms.* Am J Pathol. 1999 Sep;155(3):683–94
17. Kallioniemi A, Kallioniemi OP, Piper J, Tanner M, Stokke T, Chen L, Smith HS, Pinkel D, Gray JW, Waldman FM. *Detection and mapping of amplified DNA sequences in breast cancer by comparative genomic hybridization.* Proc Natl Acad Sci USA. 1994 Mar 15;91(6):2156–60
18. Bentz M, Plesch A, Stilgenbauer S, Dohner H, Lichter P. *Minimal sizes of deletions detected by comparative genomic hybridization.* Genes Chromosomes Cancer. 1998 Feb;21(2):172–5
19. Kirchhoff M, Gerdes T, Maahr J, Rose H, Bentz M, Dohner H, Lundsteen C. *Deletions below 10 megabasepairs are detected in comparative genomic hybridization by standard reference intervals.* Genes Chromosomes Cancer. 1999 Aug;25(4):410–3
20. Piper J, Rutovitz D, Sudar D, Kallioniemi A, Kallioniemi OP, Waldman FM, Gray JW, Pinkel D. *Computer image analysis of comparative genomic hybridization.* Cytometry. 1995 Jan 1;19(1):10–26
21. Solinas-Toldo S, Lampel S, Stilgenbauer S, Nickolenko J, Benner A, Dohner H, Cremer T, Lichter P. *Matrix-based comparative genomic hybridization: biochips to screen for genomic imbalances.* Genes Chromosomes Cancer. 1997 Dec;20(4):399–407
22. Pinkel D, Segraves R, Sudar D, Clark S, Poole I, Kowbel D, Collins C, Kuo WL, Chen C, Zhai Y, Dairkee SH, Ljung BM, Gray JW, Albertson DG. *High resolution analysis of DNA copy number variation using comparative genomic hybridization to microarrays.* Nat Genet. 1998 Oct;20(2):207–11
23. Snijders AM, Nowak N, Segraves R, Blackwood S, Brown N, Conroy J, Hamilton G, Hindle AK, Huey B, Kimura K, Law S, Myambo K, Palmer J, Ylstra B, Yue JP, Gray JW, Jain AN, Pinkel D, Albertson DG. *Assembly of microarrays for genome-wide measurement of DNA copy number.* Nat Genet. 2001 Nov;29(3):263–4
24. Wessendorf S, Fritz B, Wrobel G, Nessling M, Lampel S, Goettel D, Kuepper M, Joos S, Hopman T, Kokocinski F, Dohner H, Bentz M, Schwaenen C, Lichter P. *Automated screening for genomic imbalances using matrix-based comparative genomic hybridization.* Lab Invest. 2002 Jan;82(1):47–60
25. Pollack JR, Perou CM, Alizadeh AA, Eisen MB, Pergamenschikov A, Williams CF, Jeffrey SS, Botstein D, Brown PO. *Genome-wide analysis of DNA copy-number changes using cDNA microarrays.* Nat Genet. 1999 Sep;23(1):41–6

26. Hyman E, Kauraniemi P, Hautaniemi S, Wolf M, Mousses S, Rozenblum E, Ringner M, Sauter G, Monni O, Elkahloun A, Kallioniemi OP, Kallioniemi A. *Impact of DNA amplification on gene expression patterns in breast cancer.* Cancer Res. 2002 Nov 1;62(21):6240–5
27. Carter NP, Fiegler H, Piper J. *Comparative analysis of comparative genomic hybridization microarray technologies: report of a workshop sponsored by the Wellcome Trust.* Cytometry. 2002 Oct 1;49(2):43–8
28. Carter NP. *Analysis of chromosome rearrengements and replication timing using DNA microarrays.* Abstract, Perigueux; 2003
29. Telenius H, Carter NP, Bebb CE, Nordenskjold M, Ponder BA, Tunnacliffe A. *Degenerate oligonucleotide–primed PCR: general amplification of target DNA by a single degenerate primer.* Genomics. 1992 Jul;13(3):718–25
30. Klein CA, Schmidt-Kittler O, Schardt JA, Pantel K, Speicher MR, Riethmuller G. *Comparative genomic hybridization, loss of heterozygosity, and DNA sequence analysis of single cells.* Proc Natl Acad Sci USA. 1999 Apr 13;96(8):4494–9
31. Daigo Y, Chin SF, Gorringe KL, Bobrow LG, Ponder BA, Pharoah PD, Caldas C. *Degenerate oligonucleotide primed–polymerase chain reaction–based array comparative genomic hybridization for extensive amplicon profiling of breast cancers : a new approach for the molecular analysis of paraffin–embedded cancer tissue.* Am J Pathol. 2001 May;158(5):1623–31
32. Dean FB, Hosono S, Fang L, Wu X, Faruqi AF, Bray-Ward P, Sun Z, Zong Q, Du Y, Du J, Driscoll M, Song W, Kingsmore SF, Egholm M, Lasken RS. *Comprehensive human genome amplification using multiple displacement amplification.* Proc Natl Acad Sci USA. 2002 Apr 16;99(8):5261–6
33. Lage JM, Leamon JH, Pejovic T, Hamann S, Lacey M, Dillon D, Segraves R, Vossbrinck B, Gonzalez A, Pinkel D, Albertson DG, Costa J, Lizardi PM. *Whole Genome Analysis of Genetic Alterations in Small DNA Samples Using Hyperbranched Strand Displacement Amplification and Array–CGH.* Genome Res. 2003 Feb;13(2):294–307
34. Schwaenen C, Nessling M, Wessendorf S, et al. *Automated array-based genomic profiling in chronic lymphocytic leukemia: Development of a clinical tool and discovery of recurrent genomic alterations* Proc Natl Acad Sci USA 2004 Jan;101(4):1039-1044
35. Bruder CE, Hirvela C, Tapia-Paez I, Fransson I, Segraves R, Hamilton G, Zhang XX, Evans DG, Wallace AJ, Baser ME, Zucman-Rossi J, Hergersberg M, Boltshauser E, Papi L, Rouleau GA, Poptodorov G, Jordanova A, Rask-Andersen H, Kluwe L, Mautner V, Sainio M, Hung G, Mathiesen T, Moller C, Pulst SM, Harder H, Heiberg A, Honda M, Niimura M, Sahlen S, Blennow E, Albertson DG, Pinkel D, Dumanski JP. *High resolution deletion analysis of constitutional DNA from neurofibromatosis type 2 (NF2) patients using microarray–CGH.* Hum Mol Genet. 2001 Feb 1;10(3):271–82
36. Wessendorf S, Schwaenen C, Kohlhammer H et al. *Hidden gene amplifications in aggressive B-cell non-Hodgkin lymphomas detected by microarray-based comparative genomic hybridization* Oncogene 2003 Mar;22(9):1425-1429
37. Ishizuka T, Tanabe C, Sakamoto H, Aoyagi K, Maekawa M, Matsukura N, Tokunaga A, Tajiri T, Yoshida T, Terada M, Sasaki H. *Gene amplification profiling of esophageal squamous cell carcinomas by DNA array CGH.* Biochem Biophys Res Commun. 2002 Aug 9;296(1):152–5

38. Zhao J, Roth J, Bode-Lesniewska B, Pfaltz M, Heitz PU, Komminoth P. *Combined comparative genomic hybridization and genomic microarray for detection of gene amplifications in pulmonary artery intimal sarcomas and adrenocortical tumors.* Genes Chromosomes Cancer. 2002 May;34(1):48–57
39. Kononen J, Bubendorf L, Kallioniemi A, Barlund M, Schraml P, Leighton S, Torhorst J, Mihatsch MJ, Sauter G, Kallioniemi OP. *Tissue microarrays for high–throughput molecular profiling of tumor specimens.* Nat Med. 1998 Jul;4(7):844–7
40. Martinez-Climent JA, Alizadeh AA, Segraves R, Blesa D, Rubio-Moscardo F, Albertson DG, Garcia-Conde J, Dyer MJ, Levy R, Pinkel D, Lossos IS. *Transformation of follicular lymphoma to diffuse large cell lymphoma is associated with a heterogeneous set of DNA copy number and gene expression alterations.* Blood. 2002 Oct 24
41. Kraus J, Pantel K, Pinkel D, Albertson DG, Speicher MR. *High–resolution genomic profiling of occult micrometastatic tumor cells.* Genes Chromosomes Cancer. 2003 Feb;36(2):159–66
42. Wilhelm M, Veltman JA, Olshen AB, Jain AN, Moore DH, Presti JC Jr, Kovacs G, Waldman FM. *Array–based comparative genomic hybridization for the differential diagnosis of renal cell cancer.* Cancer Res. 2002 Feb 15;62(4):957–60
43. Fritz B, Schubert F, Wrobel G, Schwaenen C, Wessendorf S, Nessling M, Korz C, Rieker RJ, Montgomery K, Kucherlapati R, Mechtersheimer G, Eils R, Joos S, Lichter P. *Microarray–based copy number and expression profiling in dedifferentiated and pleomorphic liposarcoma.* Cancer Res. 2002 Jun 1;62(11):2993–8
44. Dohner H, Stilgenbauer S, Benner A, Leupolt E, Krober A, Bullinger L, Dohner K, Bentz M, Lichter P. *Genomic aberrations and survival in chronic lymphocytic leukemia.* N Engl J Med. 2000 Dec 28;343(26):1910–6
45. Veltman JA, Schoenmakers EF, Eussen BH, Janssen I, Merkx G, van Cleef B, van Raven-swaaij CM, Brunner HG, Smeets D, van Kessel AG. *High–throughput analysis of sub-telomeric chromosome rearrangements by use of array–based comparative genomic hybridization.* Am J Hum Genet. 2002 May;70(5):1269–76
46. Flint J, Wilkie AO, Buckle VJ, Winter RM, Holland AJ, McDermid HE. *The detection of subtelomeric chromosomal rearrangements in idiopathic mental retardation.* Nat Genet. 1995 Feb;9(2):132–40
47. Lichter P, Joos S, Bentz M, Lampel S. *Comparative genomic hybridization: uses and limitations.* Semin Hematol. 2000 Oct;37(4):348–57
48. Lockhart DJ, Winzeler EA. *Genomics, gene expression and DNA arrays.* Nature. 2000 Jun 15;405(6788):827–36
49. Johansson B, Mertens F, Mitelman F. *Cytogenetic evolution patterns in non-Hodgkin's lymphoma.* Blood. 1995 Nov 15;86(10):3905–14

13

Analysis of Gene Regulatory Circuits

Zirong Li

13.1 Introduction

The gene regulatory circuitry controls the gene expression programs and permits a cell to grow, differentiate, and maintain normal functions within the tissues and organs [1]. It consists of two components: the transcription factors that bind to DNA and regulate expression of neighboring genes, and the cis-regulatory elements that are bound by transcription factors. Typically, a gene has a promoter that can be recognized by multiple transcription factors, and specific expression of the gene is determined by a combination of these factors that bind to the promoter [2, 3]. Simultaneous binding of multiple transcription factors to the promoter is usually required to turn the gene on or off. Once bound to the target genes, the transcription factors recruit chromatin modification complexes or the transcription machinery to activate or repress gene expression [4].

Malfunction of the gene regulatory circuitry is a major cause of human diseases. More than 50 transcription factors have now been linked to genetic lesions that occur in human cancers. In order to understand the molecular basis of cancer, it is necessary to identify the set of genes directly controlled by these regulators. The analysis of the gene regulatory network is not only of substantial medical importance, but also a central problem in biology. Identification of the complete set of target genes for a transcription factor is essential to decode the gene expression programs that produce living cells.

With the availability of complete genome sequences for many organisms and advances in DNA microarray technologies, a method has recently been developed to directly examine the interactions between transcription factors and their target sites in the genome [5, 6]. This technique, known as genome wide location analysis, combines a conventional chromatin immunoprecipitation protocol with microarray technologies to determine the genomic regions that a DNA binding protein recognizes in vivo. It contains four steps: chromatin immunoprecipitation (ChIP), ligation-mediated PCR (LM–PCR), hybridization and microarray analysis (Fig. 13.1).

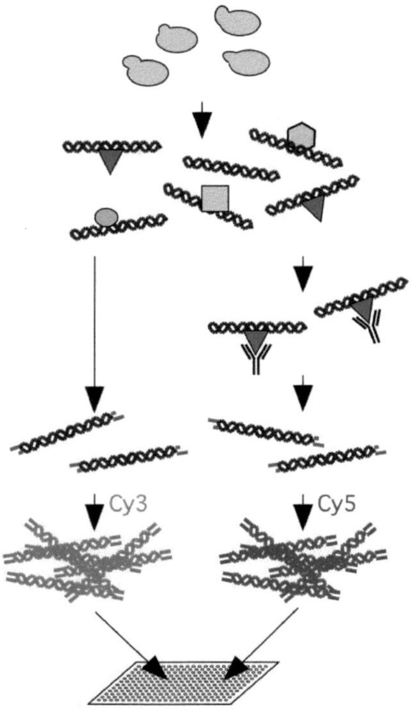

Fig. 13.1. A schematic diagram of the genome wide location analysis

Chromatin immunoprecipitation (ChIP) is a method widely used to study in vivo protein–DNA interactions [7, 8]. Traditionally, this approach has been used to confirm whether a transcription factor is binding to a particular DNA sequence in vivo. Using this method, living cells are first treated with formaldehyde, and then broken apart. The chromosomes are sheared by sonication, and the cross-linked chromatin DNA fragments are immunoprecipitated using a specific antibody against the transcription factor. The enrichment of a particular sequence in the immunoprecipitates is tested by PCR with a pair of gene-specific primers and visualized using gel electrophoresis.

To identify the genomic regions enriched through the ChIP procedure, the immunoprecipitated DNA is amplified through ligation-mediated PCR. Then the DNA is labelled with fluorescent dyes and hybridized to DNA microarrays representing genomic regions of an organism. As a control, the genomic DNA prior to immunoprecipitation is processed in parallel, labelled using a different fluorescent dye and hybridized to the same array. The spots that show a significantly stronger signal in the IP-enriched DNA channel would indicate that the corresponding genomic regions are bound by the transcription factor in vivo.

The genome wide location analysis is emerging as a powerful approach to analyze the genetic regulatory network in cells. It has been successfully used to identify target genes for a number of yeast and mammalian transcription factors [5, 6, 9–11]. For example, the method was first used to characterize the yeast Gal4 protein, a transcription regulator of the galactose metabolism pathway. All of the previously known Gal4 targets were identified, and three novel targets were found and confirmed by independent methods [5]. In another study, Simon [10] used the genome wide location analysis to investigate nine transcription factors that play a role during the yeast cell cycle progression. The results revealed a genetic regulatory network that appears to control the sequential activation of cyclins and other cell cycle regulators. Interestingly, each of these nine transcription factors was found to be a transcriptional target for this network.

Most recently, more than 100 known yeast transcription regulators were characterized using the genome wide location analysis and their targets identified [12]. The target genes for these regulators, which account for nearly all the known yeast transcription factors, were experimentally mapped. The information led to the discovery of six types of regulatory circuitry motifs, which appear to be the basic unit of genetic regulatory networks. This work represents the first comprehensive description of a genetic regulatory network in an organism [12].

The genome wide location analysis has also proved useful to study mammalian transcription factors [11, 13]. One of the main challenges in applying genome wide location analysis to mammalian cells is the availability of DNA microarrays that represent the whole genome. Because the human cells contain more than three billion base pairs per haploid genome, the cost to manufacture DNA microarrays to cover the entire genome is currently very high. Alternatives to the whole genome arrays have been developed. For example, Ren [11] developed DNA microarrays that represent human gene promoters, based on the assumption that these are the most important regulatory regions in the genome [11]. These arrays have been used to identify the target genes for E2F, regulators of mammalian DNA replication and cell cycle [11]. Most known E2F targets were identified in this study. In another approach, genomic DNA libraries enriched for CpG islands were used to make DNA microarrays [13]. Since most human genes have CpG islands in their promoters, such arrays can also be used to identify potential target genes for human transcription factors.

In this chapter, a genome wide location analysis protocol is described. The application of this protocol to the human E2F factor is also demonstrated.

13.2 An Experimental Protocol for Genome Wide Location Analysis

This section describes a detailed protocol for genome wide location analysis. The protocol has been used to analyze the in vivo DNA binding sites for human transcription factors. With minor modifications, this protocol can also be used to study DNA binding proteins in other cell types or organisms.

13.2.1 Materials

- Distilled water (dH2O)
- 5×10^8 WI38 cells, of human lung fibroblast origin
- DNA microarrays containing DNA fragments corresponding to human gene promoters
- Rabbit polyclonal antibodies against the transcription factor of interest
- Sheep anti-rabbit IgG conjugated dynabeads (Dynal, Cat# 112.04)
- Crosslinking solution (11% formaldehyde, 0.1 M NaCl, 1 mM Na–EDTA, 0.5 mM Na–EGTA, 50 mM Hepes, pH 8.0)
- 2.5 M glycine solution
- PBS (137 mM NaCl, 2.7 mM KCl, 10 mM Na_2HPO_4, 2 mM KH_2PO_4)
- Lysis Buffer 1 (0.05 M Hepes–KOH, pH 7.5, 0.14 M NaCl, 1 μM EDTA, 10% glycerol, 0.5% NP–40, 0.25%, Triton X–100, protease inhibitor cocktail (Roche Applied Science, CAT# 1836170) added prior to use)
- Lysis Buffer 2 (0.2 M NaCl, 1 μM EDTA, 0.5 μM EGTA, 10 μM Tris, pH 8, protease inhibitor cocktail (Roche Applied Science, CAT# 1836170) added just prior to use)
- Lysis Buffer 3 (1 μM EDTA, 0.5 μM EGTA, 10 μM Tris–HCl, pH 8, protease inhibitor cocktail (Roche Applied Science, CAT# 1836170) added just prior to use)
- RIPA buffer (50 mM Hepes, pH 7.6, 1 mM EDTA, 0.7% DOC, 1% NP–40, 0.5 M LiCl, protease inhibitor cocktail (Roche Applied Science, CAT# 1836170) added prior to use)
- Elution buffer (50 mM Tris, pH 8, 10 mM EDTA, 1% SDS)
- Proteinase K stock solution (20 mg/ml proteinase K (Sigma), 50 mM Tris–HCl, pH 8.0, 1.5 mM Calcium Acetate)
- TE (10 mM Tris–HCl, pH 8.0, 1 mM EDTA, pH 8.0)
- Proteinase K solution (2% glycogen, 5% proteinase K stock solution, TE)
- Linker oligo (oJW102: GCGGTGACCCGGGAGATCTGAATTC; oJW103: GAATTCAGATC; these two oligos are dissolved in dH2O and annealed to make a 15 μM solution in 0.25 M Tris–HCl, pH 8.0)
- Hybridization buffer 1 (2.2 × SSC, 0.22% SDS)
- Hybridization buffer 2 (70% formamide, 3 × SSC, 14.3% dextran sulfate)
- Pre-hybridization buffer (2 × SSC, 0.05% SDS, 0.2% BSA)
- Wash buffer 1 (2 × SSC, 0.1% SDS)
- Wash buffer 2 (0.2 × SSC, 0.1%SDS)
- Wash buffer 3 (0.2 × SSC)

13.2.2 Procedures

Chromatin Immunoprecipitation

Formaldehyde Cross–linking of Cells. The cells grown in plastic dishes are first re-suspended and transferred as 40 ml aliquots into 50 ml tubes. The tubes are placed on ice for 10 minutes, then 1/10 volume, i.e. 4 ml, crosslinking solution is added directly to each tube. The cross–linking reaction is allowed to continue for 10 minutes before being stopped by the addition of 1/20 volume, i.e. 2.2 ml, of 2.5 M glycine solution to each tube. The fixed cells in each tube are harvested by centrifugation at 2000 g for 10 minutes at 4°C. The cell pellets are re-suspended and pooled together with a total of 50 ml cold PBS. These cells are centrifuged again at 2000 g for 5 minutes at 4°C, and the supernatant is removed. After repeating the washing cycle once more, the final cell pellet is snap frozen in liquid nitrogen and stored at –80°C.

Extraction and Fragmentation of Chromatin. The frozen cell pellet from the previous step is re-suspended in 30 ml of Lysis Buffer 1 and incubated for 10 minutes at 4°C on a rocking platform. The cell mixture is then centrifuged at 2000 g for 10 minutes at 4°C. After removing the supernatant, the cell pellet is re-suspended in 24 ml Lysis Buffer 2 and mixed gently at room temperature for 10 minutes on a rocking platform. The cells are then centrifuged at 2000 g for 10 minutes at 4°C. The cell pellet is finally re-suspended in 10 ml of Lysis Buffer 3.

To obtain small chromatin fragments from the above cell extracts, physical shearing forces generated by a sonicator are used. The cell mixture from the previous step is divided into 5 ml aliquots and placed in 15 ml tubes. These tubes are then placed on ice. Cells are continuously sonicated for 25 seconds using a Branson Sonifier 450 with power setting at 5. The sonication is followed by at least 1 minute of incubation on ice to avoid accumulation of heat. The cell mixture is sonicated and chilled for a total of eight cycles (Note that the number of sonication cycles varies with different cell types and cross–linking conditions). Efficiency of sonication can be checked by taking 10 μl of cell extract out for gel analysis after each cycle, with the optimal chromatin DNA around 500–1000 bp. After sonication, the chromatin samples are pooled together, adjusted to 0.5% Sarkosyl (sodium lauryl sarcosine) and gently mixed for 10 minutes at room temperature on a rocking platform. The chromatin solution is then transferred to a centrifuge tube and spun for 10 minutes at 10,000 g to remove cell debris. The supernatant is collected for chromatin immunoprecipitation, or stored at –80°C as 1 ml aliquots.

Immunoprecipitation of Chromatin. The chromatin immunoprecipitation is performed using anti-rabbit IgG-conjugated magnetic beads (Dynal) that are coupled to the polyclonal antibodies. To prepare this material, magnetic beads (100 μl) are centrifuged at 2000 g for 3 minutes at 4°C. After removing the supernatant, the beads are re-suspended in 5 ml cold PBS containing 5 mg/ml Bovin Serum Albumin (BSA, Sigma Cat# A–7906) made immediately before use. This washing cycle is repeated a total of 3 times, and the magnetic beads are re-suspended in 5 ml of cold PBS with BSA. 10 μg rabbit polyclonal antibody is added to the beads mixture and mixed overnight on a rotating platform at 4°C. The following day, the magnetic beads are collected by centrifugation at 2000 g for 5 minutes, washed 3 times with 5 ml cold PBS with 5 mg/ml BSA and re-suspended in 100 μl cold PBS with 5 mg/ml BSA.

The soluble chromatin solution from Step 2 is first adjusted to 0.1% Triton X–100, 0.1% sodium deoxycholate, and 1 mM PMSF. To 1 ml of this mixture, 100 μl of magnetic beads pre-coupled with the antibody are then added. The mixture is incubated at 4°C overnight in a rotating platform. The following day, the magnetic beads are collected using a magnet MPC–E from Dynal, and the supernatant removed by aspiration. The beads are re-suspended in 1 ml RIPA buffer. After incubation on a rotating platform at 4°C for 3 minutes, the magnetic beads are collected with MPC–E again. This washing process is repeated 5 times followed by a wash with 1 ml TE. The beads are collected by centrifugation at 2000 g for 3 minutes and re-suspended in 50 μl elution buffer. To elute the precipitated chromatin, the beads are incubated at 65°C for 10 minutes with constant agitation, then 40 μl of supernatant are collected after a 30 second centrifugation at 2000 g. The eluted chromatin is mixed with 120 μl of TE (1% SDS) and incubated at 65°C overnight to reverse the cross–links.

Purification of Immunoprecipitated DNA. To purify the immunoprecipitated DNA, 120 μl Proteinase K solution is added to the chromatin solution. The mixture is incubated for 2 hours at 37°C to allow digestion of proteins in the precipitates. The sample is then extracted twice with phenol (Sigma, cat# P–4557), once with 24:1 chloroform/isoamyl alcohol (Sigma cat# C–0549). The sample is adjusted to 200 mM NaCl. 2 volumes of ethanol are added to the mixture, which is then incubated for 15 minutes at –80°C or on dry ice. The DNA is then precipitated by centrifugation at 14,000 rpm at 4°C in a micro–centrifuge. The DNA pellet is washed with 70% ethanol and re-suspended in 30 μl TE containing 10 μg DNase-free RNase A (Sigma, cat# 6513) and incubated for 2 hours at 37°C. After the incubation, the DNA is purified with Qiagen PCR kit (Qiagen, cat# 28106) and re-suspended in 50 μl elution buffer included with the kit.

As a control, DNA from an aliquot of chromatin solution is reverse crosslinked and purified in a similar fashion. At this step, PCR reactions using specific primers to amplify certain known target regions can be performed to check whether the chromatin immunoprecipitation is successful. A detailed description of such tests can be found in other publications [8].

Ligation-mediated PCR

Blunting Reaction. The immunoprecipitated DNA obtained from the previous steps usually needs to be amplified and labelled for DNA microarray analysis. To achieve this, a ligation-mediated PCR (LM–PCR) method is used. First, the DNA is treated with T4 DNA polymerase to form blunt ends. The reaction is assembled as follows:

40 µl immunoprecipitated DNA (or 20 ng of control input DNA)
11 µl (10×) T4 DNA pol buffer (NE Biolabs cat # 007–203)
0.5 µl BSA (10 mg/ml) (NE Biolabs cat # 007–BSA)
0.5 µl dNTP mix (20 mM each)
0.2 µl T4 DNA pol (3 U/µl) (NE Biolabs cat # 203L)
add dH2O to a total 112 µl.

The reaction is carried out for 20 minutes at 12°C. Afterwards, the sample is adjusted with 1/10 volume of 3 M sodium acetate (pH 5.2), 1 µg of glycogen (Roche Applied Sciences, cat# 0901393) and is extracted with phenol:chloroform:isoamyl alcohol (25:24:1) (Sigma, cat# P–3803) once, followed by ethanol precipitation (see above). The final DNA pellet is dissolved in 25 µl dH2O.

Ligation Reaction. Assemble the following reaction:

25 µl of DNA
8 µl dH20
10 µl 5× ligase buffer (Invitrogne, cat# 46300–018)
6.7 µl annealed linkers (15 µM)
0.5 µl T4 DNA ligase (New England Biolabs, cat# 202L)
50.2 µl Total

The ligation reaction is allowed to continue for over night at 16°C. On the next day, the DNA is purified by ethanol precipitation and dissolved in 25 µl dH2O.

PCR. The ligated DNA sample is used as template in the following polymerase chain reaction:

25 µl DNA
4 µl 10× ThermoPol reaction buffer (New England Biolabs, cat# B9004S)
4.75 µl ddH2O
5 µl 10× dNTP mix (2.5 mM each dATP, dTTP, dGTP, dCTP)
1.25 µl oligo oJW102 (40 µM stock)
add dH2O to final volume of 40 µl.

The sample is first incubated at 55°C for 2 minutes, then 10 µl of an enzyme mix [8 µl dH2O, 1 µl Taq DNA polymerase (5 U/µl), 1 µl ThermalPol reaction buffer, and 0.025 unit of Pfu polymerase (Stratagene, cat #

600250–51)] is added to the sample. Subsequently, the following PCR cycle is performed:

step 1: 72°C for 5 minutes;
step 2: 95°C for 2 minutes;
step 3: 95°C for 1 minute;
step 4: 60°C for 1 minute;
step 5: 72°C for 1 minute;
step 6: go to step 3 for 22 times;
step 7: 72°C for 5 minutes;
step 8: 4°C forever;

Afterwards, the DNA is purified using the Qiaquick PCR purification kit (Qiagen, cat# 28106) and eluted in 60 µl elution buffer provided with the kit.

DNA Microarray Hybridization

Labelling Immunoprecipitated DNA. To 200 ng of DNA from the previous step, 20 µl of 2.5× random primer solution (from the BioPrime kit, Invitrogen, Cat# 18094–011) and dH2O are added to a final volume of 42.5 µl. The mixture is boiled for 5 minutes and then immediately placed on ice. To initiate the labelling reaction, 5 µl of 10× low dCTP mixture (2.5 mM each for dATP, dTTP and dGTP, and 0.6 mM for dCTP), 1.5 µl of Cy5–dCTP (Amersham, Cat# PA55021) or Cy3–dCTP (Amersham, Cat# PA53021), 40 unit of Klenow DNA polymerase are added to the mixture. The reaction is carried out at 37°C for 2 hours. Finally, the labelled DNA is purified using the Qiagen PCR kit (Qiagen, Cat# 28106).

DNA Microarray Hybridization. 2.5 µg of Cy5-labelled ChIP DNA, 2.5 µg of Cy3-labelled genomic DNA and 36 µg human Cot–1 DNA (Invitrogen, Cat# 15279–011) are mixed together and concentrated by ethanol precipitation. The DNA pellet is dissolved in 22.4 µl of hybridization buffer 1. Then 20 µl of hybridization buffer 2 is added to the mixture, and the sample is incubated first at 95°C for 5 minutes then 42°C for 2 minutes. Subsequently, 4 µl of yeast tRNA (Sigma, cat# R9001 at 10 µg/µl) and 3 µl of 2% BSA are used to adjust the hybridization reaction to 50 µl. This mixture is added to a DNA microarray slide that has been incubated with the pre-hybridization solution for 40 minutes at 42°C. A 25 mm × 60 mm cover slip is then gently placed on top of the sample, and the hybridization is carried out in a hybridization chamber (Corning, cat# 07–200–271) at 60°C overnight in a water bath.

Washing Microarrays. After the hybridization, the microarray slide is washed once with washing buffer 1 at 60°C for 5 minutes, once with washing buffer 2 for 10 minutes at room temperature, and three times with washing buffer 3 at room temperature.

Microarray Analysis and Identification of in vivo DNA Binding Sites

To collect the microarray data, a microarray scanner (GenePix 4000B, Axon Instrument) is used to scan the microarray slide. The microarray image is first analyzed with the image analysis software GenePix pro 3.0 to derive the Cy3 and Cy5 fluorescent intensity and background noise at each spot. Then background intensity is subtracted from the fluorescent intensity at the spot for both Cy3 and Cy5. Normally, the signal from Cy3 is normalized to the Cy5 based on median spot intensities for the entire image. The ratio of Cy5 intensity (usually corresponding to ChIP DNA) over Cy3 intensity (corresponding to input genomic DNA) is calculated, and a P value is calculated using an error model [14]. The genomic regions that have at least 2 fold Cy5/Cy3 ratio with P values less than 0.001 are usually considered as significant binding sites.

13.3 Example: Identifying the Target Genes of Human E2F4

The E2F4 transcription factor plays an important role in cell cycle progression. E2F4 is thought to function by regulating genes involved in G1/S transition, and chromatin immunoprecipitation (ChIP) experiments have shown that E2F4 binds to genes that are activated at the G^1/S boundary [15]. Entry of E2F4 into the nucleus is restricted in G^0 and early G^1, and binding of E2F4 to promoters in quiescent cells coincides with recruitment of p130, diminished acetylation of histone at the promoters, and gene repression. The human promoter microarray we developed recently allows us to systematically identify the direct E2F4 targets.

13.3.1 Experimental Procedures

Primary human fibroblast (WI38) is synchronized to G^0 through serum starvation. These G^0 cells were fixed by formaldehyde, harvested, and disrupted by sonication. E2F4 bond chromatin was enriched by chromatin immunoprecipitation with E2F4 specific antibody SC–1082 (Santa Cruz Biotechnology). E2F4 bond DNA was then purified after proteinase K and RNase A treatment, and amplified by ligation-mediated PCR (LM–PCR). Amplified DNA was subsequently labelled with Cy5–dCTP using the BioPrime Kit (Invitrogen). In the mean time, input DNA that has not been enriched by chromatin immunoprecipitation was labelled with Cy3–dCTP. Cy5 and Cy3 labelled DNA were mixed and hybridized to the human 5K–promoter array in the presence of human Cot–1 DNA under stringent conditions overnight. The DNA microarray was washed and scanned using a GenePix 4000 scanner.

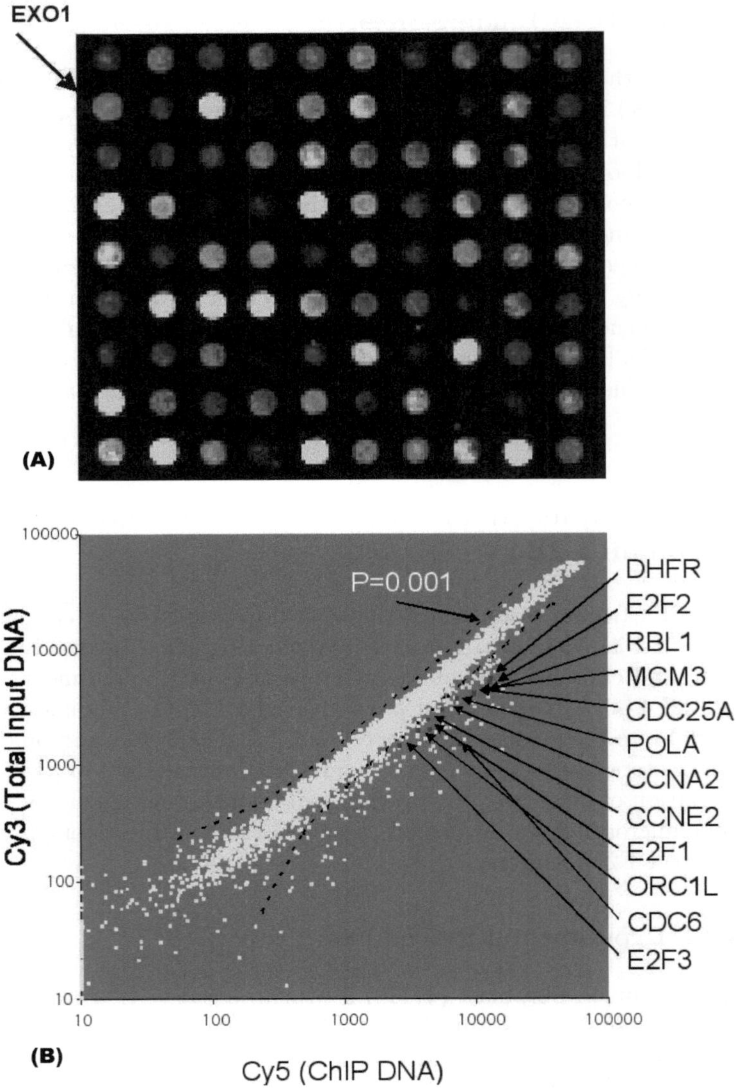

Fig. 13.2. Identification of the E2F4 target genes in WI38 cell. **(a)** A close-up of a scanned image of the human DNA array that contains promoter regions of ∼ 5000 human genes. The arrow points to a spot where the red intensity is significantly higher than the green intensity, indicating that the promoter region of that gene (EXO1) is bound in vivo by E2F4. **(b)** Scatter plot of Cy3-labelled total genomic DNA versus Cy5-labelled E2F4 CHIP-enriched DNA. A P-value cutoff of 0.001 is shown. The red spots represent some previously confirmed E2F4 targets, listed next to the plot

13.3.2 Results and Discussion

The result of the E2F4 location analysis experiment is shown in Fig. 13.2. Our results suggest that the genome wide location analysis procedure is a powerful method to identify in vivo targets of transcription factors. When using the criteria: P–value ≤ 0.001, channel intensity ≥ 200 and ratio ≥ 2, we found 143 genes whose promoters were occupied by E2F4 in physiological condition, indicating that they are putative E2F4 targets. Most of these 143 genes were confirmed earlier by either chromatin immunopreciptation or Affymetrix cDNA expression arrays [11, 16]. We also identified some novel E2F4 targets that fall into several function groups related to cell cycle regulation, DNA replication, DNA repair, G2/M checkpoints and mitotic regulation.

13.4 Summary

Genome wide location analysis is a general method to identify the in vivo binding sites for transcription regulators. The recent use of this method to map the genetic regulatory network in yeast demonstrated that this method is an essential tool for us to understand the mechanisms of gene regulation in cells [12]. Applying this approach to mammalian transcription factors is expected to yield important information about the mechanisms of animal development and pathology of human diseases.

References

1. Davidson, E. H. (2001). *Genomic Regulatory Systems:development and evolution* (San Diego, Academic Press)
2. Tjian, R., and Maniatis, T. (1994). *Transcriptional activation: a complex puzzle with few easy pieces*, Cell 77, 5–8
3. Ptashne, M., and Gann, A. (1997). *Transcriptional activation by recruitment*, Nature 386, 569–77
4. Orphanides, G., and Reinberg, D. (2002). *A Unified Theory of Gene Expression*, Cell 108, 439–451
5. Ren, B., Robert, F., Wyrick, J. J., Aparicio, O., Jennings, E. G., Simon, I., Zeitlinger, J., Schreiber, J., Hannett, N., Kanin, E., et al. (2000). *Genome–wide location and function of DNA binding proteins*, Science 290, 2306–9
6. Iyer, V. R., Horak, C. E., Scafe, C. S., Botstein, D., Snyder, M., and Brown, P. O. (2001). *Genomic binding sites of the yeast cell–cycle transcription factors SBF and MBF*, Nature 409, 533–8
7. Solomon, M. J., Larsen, P. L., and Varshavsky, A. (1988). *Mapping protein–DNA interactions in vivo with formaldehyde: evidence that histone H4 is retained on a highly transcribed gene*, Cell 53, 937–47
8. Orlando, V. (2000). *Mapping chromosomal proteins in vivo by formaldehyde–crosslinked–chromatin immunoprecipitation*, Trends Biochem Sci 25, 99–104

9. Lieb, J. D., Liu, X., Botstein, D., and Brown, P. O. (2001). *Promoter–specific binding of Rap1 revealed by genome–wide maps of protein–DNA association*, Nat Genet 28, 327–34
10. Simon, I., Barnett, J., Hannett, N., Harbison, C. T., Rinaldi, N. J., Volkert, T. L., Wyrick, J. J., Zeitlinger, J., Gifford, D. K., Jaakkola, T. S., and Young, R. A. (2001). *Serial regulation of transcriptional regulators in the yeast cell cycle*, Cell 106, 697–708
11. Ren, B., Cam, H., Takahashi, Y., Volkert, T., Terragni, J., Young, R. A., and Dynlacht, B. D. (2002). *E2F integrates cell cycle progression with DNA repair, replication, and G(2)/M checkpoints*, Genes Dev 16, 245–56
12. Lee, T. I., Rinaldi, N. J., Robert, F., Odom, D. T., Bar-Joseph, Z., Gerber, G. K., Hannett, N. M., Harbison, C. T., Thompson, C. M., Simon, I., et al. (2002). *Transcriptional regulatory networks in Saccharomyces cerevisiae*, Science 298, 799–804
13. Weinmann, A. S., Yan, P. S., Oberley, M. J., Huang, T. H., and Farnham, P. J. (2002). *Isolating human transcription factor targets by coupling chromatin immunoprecipitation and CpG island microarray analysis*, Genes Dev 16, 235–44
14. Roberts, C. J., Nelson, B., Marton, M. J., Stoughton, R., Meyer, M. R., Bennett, H. A., He, Y. D., Dai, H., Walker, W. L., Hughes, T. R., et al. (2000). *Signaling and circuitry of multiple MAPK pathways revealed by a matrix of global gene expression profiles*, Science 287, 873–80
15. Takahashi, Y., Rayman, J. B., and Dynlacht, B. D. (2000). *Analysis of promoter binding by the E2F and pRB families in vivo: distinct E2F proteins mediate activation and repression*, Genes Dev 14, 804–16
16. Iyer, V. R., Eisen, M. B., Ross, D. T., Schuler, G., Moore, T., Lee, J. C., Trent, J. M., Staudt, L. M., Hudson, J., Jr., Boguski, M. S., et al. (1999). *The transcriptional program in the response of human fibroblasts to serum*, Science 283, 83–7

Part III

Protein Microarrays

14

Protein, Antibody and Small Molecule Microarrays

Hendrik Weiner, Jörn Glökler, Claus Hultschig, Konrad Büssow, and Gerald Walter

14.1 Introduction

New and rapidly spreading infectious and lifestyle diseases, together with known killers like cancer and heart disease, particularly threaten older populations and put enormous pressure on our medical capabilities. Today's drug arsenal attacks about 400 targets, while the human genome sequence revealed at least 30,000 genes. The expression of these genes creates a complex puzzle of millions of products and points of interaction between them. Every one of these products is a potential drug or target, provided that the corresponding drug can be shown to be specific and safe in a patient's organism. And as individual patients are different, tests need to be extended to whole populations. Clearly, this can only be handled using high throughput approaches, looking at large numbers of genes and their products simultaneously.

The array format enables miniaturized and parallel analysis of large numbers of diagnostic markers in complex samples [1,2]. The concept of the arrayed library [3] allows gene expression analysis and protein interaction screening on a whole–genome scale. Using automated colony picking and gridding, cDNA or antibody libraries can be expressed and screened as clone arrays [4, 5]. As discussed in this chapter, protein microarrays are constructed from recombinantly expressed and purified proteins, using a range of expression systems. Gene product action can be studied directly if the proteins' structure and functionality is maintained. This requires novel systems for high throughput protein expression that produce sufficient amounts of properly modified and folded molecules. Large numbers of proteins must be arrayed at high density, keeping them intact and biologically active. That is most easily achieved if molecules of the same general structure (e.g. antibodies) are arrayed. Antibody arrays are now becoming an important screening tool for a wide range of molecules in complex mixtures and a robust format for expression profiling of whole genomes. Alternative systems such as nucleotide aptamers should be able to mimic certain protein functions, and as nano- and microfluidic arrays, can make very robust array formats in the future. Differential protein profiles have been used as molecular diagnostics for cancer [6] and might soon be applied to screen high risk populations for tumor markers. In the format of high throughput arrays, differential protein profiles may eventually arrive at the doctor's office and as over–the–counter devices.

14.2 Protein Microarrays

14.2.1 Introduction

A protein microarray is a highly ordered pattern of proteins immobilized on a pre-treated surface of a small and planar metal, plastic, or glass support [7–9]. Microarrays, like microprocessors, use parallelism, miniaturization and automation as three conceptual cornerstones [10]. However, unlike microprocessors, microarrays are not designed to take input signals and, using preprogrammed instructions, convert them into meaningful output. Protein microarray technology enables high throughput analysis of protein functions, such as interactions between proteins, catalysis, binding to drugs and other biochemical reactions [11]. The speed, precision, affordability and efficiency of microarray analysis offer a tremendous experimental advantage over traditional, rather cumbersome, analytical tools using columns, gels, filters and microplates. Microarrays lend themselves to a plethora of applications in biomedical research, clinical diagnostics and in the pharmaceutical industry. This can be inferred from more than 100 protein array-oriented scientific publications in the past two years [12,13]. Ultimately, a single microarray containing the complete set (not taking into account covalently modified isoforms) of 20,000–40,000 proteins expressed in human cells would allow comprehensive assessment of a given protein function. However, as outlined below, such a proteome–wide microarray is not yet on the horizon.

14.2.2 Protein Production, Purity and Printing

Putting diverse protein repertoires on a microarray requires the simultaneous and quality-assured production of many recombinant proteins of high purity. This is a non-trivial exercise that requires an appropriate infrastructure plus expertise, both of which often do not exist in regular and otherwise well-equipped molecular laboratories. Usually, recombinant proteins are produced in a soluble form in *Escherichia coli*, yeast, mammalian or insect cells. In vitro translation is an alternative option since most microarray-based applications require less than 100 μg protein. Current practice in our laboratory involves a long list of quality control steps for the production and isolation of recombinant proteins to assure their purity and fidelity. This list includes the PCR product, vector design, entry clone, expression clone, DNA sequencing of cloned insert and, eventually, the solubility, size and electrophoretic homogeneity of the purified protein product. Such attention to quality is most often very critical for interpretable results from microarray-based binding experiments. Even if induction can significantly increase the abundance of a recombinant protein over background, binding to impurities in the protein preparation, when placed onto an array, can heavily contaminate a true signal.

After being standardized with respect to homogeneity, purity and concentration, the recombinant proteins are ready to be immobilized onto an array support, most often a standard microscope format. Different support designs and surface chemistries have been described in preceding Chaps. 1–3. Established spotting technologies, including needle printing, piezo or solenoid dispensing have been discussed in Chaps. 4 and 5. Contact printing robots allow for up to 50,000 different elements on a slide. Nanotechnology and non-contact printing techniques can further increase the number of elements on an array (Chap. 6). If robotic spotting is unavailable or if only a small subset of proteins are to be analyzed, manual spotting can be an appropriate alternative, e.g. with the apparatus from Schleicher & Schuell or Greiner. The former allows for almost 800 elements on the slide and is originally designed for the company's proprietary slides containing a thin nitrocellulose layer. However, after minor adjustments, this apparatus is also applicable for printing non-layered microscope slides. Whenever an isolated protein is being immobilized it might alter its binding properties with respect to in-solution conditions. This can be minimized through the use of random immobilization as opposed to site-specific immobilization that leaves only a certain part of the molecules accessible for binders. Alternatively, one can immobilize the proteins on a slide that carries a highly hydrophilic layer of nitrocellulose ('Fast Slides', Schleicher & Schuell Bioscience) or polyacrylamide ('Hydrogel', Perkin Elmer Life Sciences). The latter are thought to nicely emulate solution like properties.

In summary, the virtues of functional assays with well constructed protein microarrays include ensured purity, standardized protein amounts and accessibility, on array replicates, ranking of signals possible and an inclusion of both positive and negative controls.

14.2.3 Detection of Small or Large Ligands

Numerous detection strategies have evolved over the years to detect and amplify signals associated with intermolecular binding events. These will not be reviewed here. The advent of fluorescent detection in combination with perfectly flat supports has greatly contributed to the popularity of DNA microarrays. Because of the almost ubiquitous nature of fluorescent detection systems, and because many molecular laboratories are already equipped with the infrastructure for the detection of fluorescent dyes on DNA microarrays, one might choose to concentrate on labelling proteins with the same or similar dyes to those employed for differential analysis on DNA microarrays, namely Cy3/Cy5 dyes (Amersham Biosciences) or Alexa 488/530 dyes (Molecular Probes). Fluorophore labels on proteins can be detected with a sensitivity superior to many other labels. In addition, fluorophore labelled proteins can easily be quantified, e.g. one can easily detect as low as 1 attomol of a fluorophore labelled antigenic protein bound on an antibody immobilized on a microarray (H. Weiner and K. Büssow, unpubl.). However, fluorophore labelling of peptides or small molecules is often not practical due to steric hindrance by labels as large or bigger than the molecule being analyzed. Small molecule binding usually becomes accessible through radiolabels, that can now be detected with suitable microarray-based readers, e.g. from Fuji or Zinsser. Steric hindrance in the microenvironment around a binding site can also be problematic for protein–protein interactions. As a solution, proteins can be radio-labelled metabolically [14, 15], ex vivo [16] or at a single site after purification [17]. Such radiolabelling usually prevents the problems associated with multisite–labelling (biotinylation, fluorophorylation) or secondary detection (antibodies). To radiolabel a protein site–specifically, the protein probe can be constructed as a gluthatione–S–transferase (GST) fusion in that a phosphorylation site for protein kinase A (PKA) is inserted between the GST and the protein part of interest. Vectors for the expression of affinity-tagged fusion proteins that contain a PKA–site are commercially available (Novagen, Amersham Biosciences). The fusion protein has to be phosphorylated by PKA [17] and can then be used as a probe to decorate the microarray. Label-free approaches including mass spectrometry [18, 19] or surface plasmon resonance [20, 21] should be attractive alternatives to detect small or large molecule binding events, as discussed in Chap. 9. Unfortunately, none of these approaches are currently applicable to the detection of binding events on microarrays containing a large set of different proteins.

14.2.4 Caveats

The main challenge for all recombinant techniques is to synthesise properly folded and conformationally correct recombinant proteins, i.e. to emulate the structural integrity of the native protein [22]. This can often not be fully warranted, even if one tries to incorporate co- and post-translational modifications during the production of the protein, e.g., through its expression in insect or mammalian cells. Another problematic aspect is surface denaturation upon spotting, immobilization, storage and assay [18]. Surface denaturation, at least to some extent, always occurs and is often difficult to control, in particular if a variety of proteins is to be treated in parallel and under identical conditions, while each protein requires a particular environment to be fully active; for details see Chaps. 2 and 3. As a result, a given protein function detected on a microarray may be a false positive and not physiologically relevant. Any such result should therefore be confirmed using an in-solution assay, preferably in vivo in an appropriate cellular system.

14.2.5 Conclusions

Almost every cellular process depends on protein activities that are probably controlled by highly specific interactions between proteins and between proteins and other molecules [23,24]. It is therefore not surprising that proteomics is currently being hailed as the next phase of genomic activity [25] and that therapeutic molecules most often are directed to proteins [26]. Appropriately designed protein microarrays are likely to find immediate applications in analytical protein biochemistry and can complement or even replace traditional technologies employed in protein characterization. One of the most promising features of protein microarrays is their potential to serve as a reliable 'early catch' format to fish out a given protein function that can then be characterized more deeply using classical non-array-based protein techniques. This feature is reminiscent to the recently developed high density protein arrays that are constructed from cDNA expression libraries and that are printed on large membranes [27]. Although very useful for certain functional studies [28], such protein arrays are often not acceptable because they carry a redundant set of only unpurified and at least partially denatured recombinant proteins produced in *E. coli*. Clearly, the construction of properly designed protein microarrays often requires hundreds or thousands of different recombinant proteins, non-denatured, of sufficient purity and in workable amounts. As outlined above, the cloning, expression and isolation of such proteins represents the biggest obstacle in the production of a protein microarray, even if only a small set of recombinant proteins is to be arrayed.

14.3 Antibody Microarrays

14.3.1 Introduction

What Are Antibody Microarrays?

Antibody arrays constitute a subset of protein arrays, displaying a certain type of protein in terms of structure and function. Antibodies are here defined as immunoglobulins or their different fragments, such as Fab's, or (reduced to their antigen binding domains) single-chain (sc)Fv's. It is essential that the immobilized antibodies retain their native structure in order to bind their cognate antigen specifically.

Applications

DNA–arrays and PCR have been widely applied to study the transcriptional level of gene expression and correlate patterns to certain phenotypes. However many features of gene function can only be assessed after translation, including modification and intracellular localization of proteins. Even the level of translation may differ from the transcription level of a gene [29]. If we take a look at the diversity of human gene transcripts of currently more than 37,000 [30], the number of possible post-translational modifications on the resulting proteins may increase this complexity beyond a million [31]. Functionally, phosphorylation states can indicate the status of a protein in the signal transduction pathway. Glycosylation of extracellular proteins is decreased or altered in certain types of cancer [32]. Antibodies can detect the three–dimensional conformation of a protein, which is most important for the screening of prions in TSEs (transmissible spongiform diseases) [33]. Antibodies can assess a multitude of other post-translational modifications, emphasizing the demand for antibody arrays to analyze complex protein samples in an efficient manner similar to DNA microarrays.

14.3.2 Current Technology

Originally, antibody arrays have been developed in 96 well–microtitre plates, based on the classical ELISA format. Miniaturization has increased the number of simultaneously detectable antigens, while still using wells to provide for separate incubation chambers. To further integrate the complexity of ELISA experiments, a multiplicity of different antibodies was immobilized in defined spots on the bottom of these wells, hence creating a micro–ELISA format [34]. In order to apply greater amounts of different antibodies to a surface, membrane filters were used as support for recombinant scFv's [5] or antibodies to detect cytokines in patient sera [35].

Microarrays

Early approaches to generate antibody arrays for high throughput screening used either expensive new materials such as specialized ELISA plates and machinery adapted to this format, or a relatively high amount of antibodies and analyte consumed by filter assays. As a consequence, a new format was introduced for microarrays based upon the already well-established microscope glass slide as a basis. Such slides have been extensively used for cDNA microarrays, but then adapted to protein microarrays by Mirzabekov, using gel–pads for the immobilization of protein samples [36]. The robotic equipment developed for cDNA microarray technology was adapted to the production of protein arrays, using glass surfaces to covalently anchor proteins. This enabled the spotting of proteins at a density of 10,000 different samples [11]. Early antibody microarrays were created using poly–L–lysine surfaces as adopted from DNA array technology [37]. However, it became apparent that of the 115 antibody–antigen pairs in these experiments, only half of the immobilized antigens and 20% of the immobilized antibodies remained active.

As antibodies constitute the active part in an immunoassay, special care must be taken to keep these in a native state on the microarrays. Several studies have been focused on finding optimal storage conditions and appropriate surfaces [38, 39]. Of the materials tested, those which covalently immobilism antibodies via epoxy–groups in combination with a surface gave the best results with respect to detection limits and signal to noise ratio. Before such antibody microarrays are created, it is advisable to check the functionality of each antibody individually [38]. Indirect immobilization by biotinylation and streptavidin may improve the performance of antibody arrays up to 10–fold [40]. However to introduce this modification to all antibodies individually would make this approach more costly and time–consuming.

Labelling and Detection

Starting from classical radioactive and enzymatic labelling techniques, covalent fluorescent labels have become standard for the detection of analytes in microarray technology, but see Chaps. 8 and 9 for a detailed review of this and other labelling or label-free techniques. Isothermal rolling–circle amplification has been developed to further increase the sensitivity of fluorescent detection [41]. Preferably, N–succinimide-activated esters of fluorophores such as Cy3 and Cy5 are used in combination, allowing for easy comparison by internal control. For antibody microarrays, either the analyte or a secondary antibody (sandwich assays) must be labelled. However, complex analyte samples are difficult to label homogeneously, preserving epitopes recognized by the immobilized antibodies. Even properties like solubility of the modified proteins might be affected. Alternatively, the application of secondary antibodies matching the primary antibodies on the chip is limited to a small number of different molecules to be screened before the background exceeds the signal. Therefore, sandwich assays could not so far be applied to complexities beyond 38 different sets of antibodies [42].

High sensitivity of detection and minute amounts of sample required are main advantages of microarrays as compared to the classical ELISA. Nanoliter amounts of sample can be applied and immobilized on the support. Putting a cover slide on top of the chip surface during incubation can reduce the amount of analyte. The absolute detection level is dependent on the binding properties of the applied antibody and the complexity of the analyzed sample, but may well reach down to 1 pg/mL using the rolling circle amplification detection [42].

Microwells and Microfluidic Chips

While conventional microarrays only allow the simultaneous screening of two samples at a time, efforts have been made to introduce true multiplexing (as in microtitre ELISA) to the microarray technology. This was achieved partly as described above, by printing small arrays in microtitre wells. However, a true multiplexing is only achieved if all samples are kept in separate compartments, which can be achieved by the synthesis of microchip surfaces bearing microwells, or microfluidic chips that have channels etched on the surface by which all points on the chip can be addressed individually [43].

14.3.3 Current Deficiencies

Source of Antibodies

A major problem of antibody microarrays is the standardized production of many different antibodies. As commercially available monoclonal and polyclonal antibodies can make an array exceedingly expensive, attempts have been made to isolate recombinant antibodies by phage display [44], ribosomal display [45] or even aptamers from nucleic acid libraries [46].

Antibody Performance on Microarrays

Previous studies have demonstrated that there is a widely varying performance of antibodies on microarrays. Many do not show any activity, decreased specificity or a lowered affinity [37,38]. Optimizing the surface and applying indirect immobilization can increase performance. However it would be advantageous to determine and include additional information regarding the suitability of a commercially available antibody in a similar manner as currently available for the application in immunoblotting, indirect ELISA or dot blot. As for antibody fragments in single–chain format derived from phage display libraries, we have found that stability is often impaired by immobilization. While Fab fragments are often found to be more stable than scFv's [47], it remains to be demonstrated that these are better suited for the microarray format.

Surfaces and Hardware

Although a large portion of the hardware equipment was adopted from cDNA microarray technology, such as the microscope slide format, fluorescent detection, microspotting devices and scanners, many of these will have to be optimized to meet the requirements of antibody arrays. Keeping the immobilized antibodies hydrated and reducing the denaturing contact with the surface seems to be necessary to retain these in an active state. Introducing microwells to reduce evaporation may be helpful, but also requires alignment of the handling robots with the surface grid. The same holds true for the microfluidic chips that need a greater extent of additional hardware and protocols to be applied.

14.3.4 Conclusions

Despite the technology of antibody microarrays still being in its infancy, rapid progress has been made. Depending on the application, the diversity and dimension of such microarrays will be ranging from 100 to 10,000 different binders. It will be interesting to see whether the recombinant molecules derived from combinatorial libraries are going to replace the currently favored antibodies in the future. New detection techniques may obviate the need to label the analyte or secondary antibody. Direct *in vitro* synthesis of the binding molecules on the chip may solve storage and activity problems faced today [48]. In summary, the impact antibody microarrays will have on diagnostics and drug discovery is yet to be conceived.

14.4 Peptide and Other Synthetic Arrays

14.4.1 Combinatorial Peptide and Non-Peptide Libraries

Structure determination is a powerful approach to molecular interaction analysis. Techniques such as X–ray crystallography and nuclear magnetic resonance (NMR) offer insights into the spatial arrangement of macromolecules and their complexes. However, since structure determination of biological macromolecules is time consuming and cumbersome, empirical combinatorial methods were developed in parallel to the structure determination methods to address the important topic of structure/activity relationship [49]. These methods mimic natural selection, the driving force behind evolution. They rely on the creation of many different variants of one molecule of interest and the selection of those variants by certain functional criteria. Both combinatorial chemistry and combinatorial biology provide suitable strategies for the creation of and selection from large libraries of diverse but comparable molecules. In these approaches, a library consisting of many different molecules is created and those members with an anticipated property are selected. A variety of different methods for the creation of and the selection from combinatorial libraries have been reviewed exhaustively [50–52].

In combinatorial chemistry, combining different building blocks with suitable chemical reactions creates large numbers of variants. The resulting individual compounds are used to study structure activity relationships of one target molecule systematically. However, the number of compounds that can be individually synthesized is limited. Progress in solid phase synthesis, originally introduced by Merrifield [53–55], gave fast and automated access to individual oligomeric compounds. For the creation of large numbers of individual sequences of monomeric building blocks, various techniques of chemical synthesis have been developed. These fall into two groups, *multiple synthesis* and *parallel synthesis*. A good overview of the different building blocks used for combinatorial chemistry has been provided by Hogan [56]. In *multiple synthesis*, mobile support elements are employed. After each reaction cycle, the segments are separated and regrouped for the next coupling. Examples are the Tea Bag method [57], the use of segmented cellulose filters [58, 59] and the one–bead–one–compound approach [60], combined with the mix–and–split approach [61]. *Parallel synthesis* uses arrays of fixed reactors. Today, several thousand syntheses can be run in parallel due to miniaturization and rapid reagent application. The pin method [62] demonstrated the success of this approach in a convincing way. Geyen et al. performed their reactions on a replicating gadget that was dipped into a microtitre plate filled with reagents for peptide synthesis according to the anticipated sequence. Parallel synthesis on flat supports is another elegant and fast strategy of generating microarrays of biological macromolecules. Its most prominent examples [62] are the macroscopic DNA arrays on glass support, first described by Maskos & Southern [63], the photolithographic Affymax (later Affymetrix) technique [64], and the SPOT method [65–67].

The resulting libraries of natural or artificial building blocks can be screened for active compounds in hybridization or western blotting experiments, while still bound to the solid support used for their synthesis. Their respective position of synthesis is used for the identification of each binding partner. Alternatively, library members are transferred into solution, followed by testing them individually or as pools.

The techniques described above can either be used for synthesis of individual compounds or pools, by using mixtures of building blocks for the coupling reactions. This results in libraries of potential ligands in one reactor. An example for such a pooling strategy is the 'mimotope' approach [68] in which hexameric peptide sequences binding to a certain target structure are determined ab initio. This approach involves iterative testing of pools of peptides at randomized positions and leads to a hexameric peptide sequence with maximal binding strength to the target protein. Frank et al. [69] have proposed a modified version of the 'mimotope' approach that circumvents the iterative screening but allows for direct access to the optimal peptide sequence.

14.4.2 Peptide Libraries to Study Protein–Protein Interactions

Protein–protein interactions are generally believed to be conformationally defined. The contact area between proteins in a complex is often only small and comprises only a short sequence motif. Typical examples include SH3, WW, EBVH1, PDZ and armadillo repeat domains of signalling and structural proteins [70–72]. All these domains bind to short sequence motivs of certain target proteins. Such binding can be mimicked with short synthetic peptides that, however, have a much larger conformational freedom than the respective sequence motive of the target protein.

14.4.3 SPOT Method for the Creation of Peptide Arrays

Among the positionally addressable solid phase synthesis methods, the SPOT synthesis, developed by Ronald Frank [63], is an easy and flexible method for simultaneous, parallel chemical synthesis on membrane supports [66, 67]. SPOT synthesis is used for synthesis of different peptides or peptide mixtures at clearly defined positions on a modified cellulose membrane. These peptide arrays were used to study protein–protein and protein–peptide interactions [73]. In a western blot-like manner, the analyte is incubated with the array on which potential binding partners were synthesized. The positions of binding of the analyte are detected with methods adapted from western blots, and signals can be directly translated into the sequence of the respective peptides.

Epitope mapping of antibodies [63, 74–76] was the first application of this technique. In addition, three different proline-rich repeats of Acta (actin assembly inducing protein A) were identified to be the ligands of VASP (vasodilator stimulated phosphoprotein) and other cellular proteins by Niebuhr et al. [77]. Furthermore, this technique was used to determine the peptide binding motifs of streptavidin [78], which eventually led to the development of the StrepTag [79, 80]. Protein–DNA [81] and protein–metal interactions [82] were studied using peptide arrays prepared by the SPOT method. An investigation of the CaM-regulated activity of the STOP protein in tubulin stabilization has been described recently [83]. A comprehensive review of applications of the SPOT method was published by Frank and Schneider-Mergener [84].

For manufacturing peptide arrays using the SPOT method, N–terminally and side chain-protected amino acids are dissolved in a solvent of low volatility. This solution is distributed by pipetting to defined positions on a modified cellulose membrane. Arrays of ninety–six spots of the size of a standard microplate can be generated manually. For the generation of arrays with more spots, up to 2,000 on a membrane of 20×20 cm, automated SPOT synthesizers have been developed in cooperation between Ronald Frank and Abimed GmbH Langenfeld, Germany. This robot is currently distributed by Intavis (http://www.intavis.com). In the original approach, the entire cellulose membrane was modified by coupling β–alanine (Fmoc–β–alanine) and removing the Fmoc protection group after completion of the coupling reaction. Today, more robust supports suitable for SPOT–synthesis are commercially available (e.g., AIMS Scientific, http://www.aims-scientific-products.de). A kit for the SPOT synthesis is available from Sigma Genosys (http://www.sigma-genosys.com/spot.asp). Technical details of the SPOT synthesis have been reviewed elsewhere [63, 85–87].

14.4.4 Alternative Peptide Array Technology

The throughput of the SPOT synthesis was increased with the introduction of the BioDisk Synthesizer [88]. In this approach, a rotating disk, made of a non-porous polymer, is used as support for the synthesis. Inkjet technology is employed for the delivery of activated protected amino acids and the deprotection reagents. Centrifugal force is used for the removal of the different reagents.

Photolithographic synthesis of peptide arrays was first described by Fodor et al. [64]. The application of this technology to the deprotection of oligonucleotide monomers bound to a suitable solid support resulted in the well-established Affymetrix oligonucleotide arrays. For the synthesis of such arrays, defined photomasks are used, limiting the flexibility of the approach.

Pellois et al. [89] described recently the synthesis of peptide arrays relying on the highly flexible digital micromirror array [90] and conventional peptide chemistry with in-solution removal of acid–labile protecting groups using photogenerated reagents [91–94]. These arrays were used for mapping an antibody with natural and non-natural amino–acids.

Alternative arraying technologies are currently developed aiming at an increased spotting density and production rate of ligands. Various nanodispensing devices for microarrays have been developed recently (e.g., [95]). Laser printer technology has been used as an alternative approach to prepare peptide arrays on paper [96, 97]. Twenty toners are being developed containing Fmoc protected amino acids in a solvent that is solid at room temperature. During standard laser printing, the particles are heated on the paper and the amino acids are coupled to the paper support. The paper is washed to remove uncoupled monomers and subsequently N–terminal protection groups. The next amino acids are coupled to free amino groups of the first immobilized amino acids in the next printing cycle. Laser printing relies on the induction of positive charge by laser or LCD light. Negatively charged toner particles are attracted onto the paper by the positive charges underneath it. Therefore, it should be possible to replace the paper with a computer chip, while charged spots on such chips can be electronically 'switched'.

References

1. Walter G, Büssow K, Cahill D, Lueking A, Lehrach H (2000) *Protein arrays for gene expression and molecular interaction screening.* Current Opinion in Microbiology 3(3): 298–302
2. Walter G, Büssow K, Lueking A, Glokler J (2002) *High–throughput protein arrays: prospects for molecular diagnostics.* Trends in Molecular Medicine 8(6): 250–253
3. Lennon GG, Lehrach H (1991) *Hybridization analyses of arrayed cDNA libraries.* Trends Genet 7(10): 314–317
4. Büssow K, Cahill D, Nietfeld W, Bancroft D, Scherzinger E, Lehrach H, Walter G (1998) *A method for global protein expression and antibody screening on high–density filters of an arrayed cDNA library.* Nucleic Acids Res 26(21): 5007–5008
5. de Wildt RM, Mundy CR, Gorick BD, Tomlinson IM (2000) *Antibody arrays for high–throughput screening of antibody–antigen interactions.* Nat Biotechnol 18(9): 989–994
6. Petricoin EF, et al. (2002) *Use of proteomic patterns in serum to identify ovarian cancer.* Lancet 359(9306): 572–577
7. Jenkins RE, Pennington SR (2001) *Arrays for protein expression profiling: towards a viable alternative to two–dimensional gel electrophoresis?* Proteomics 1(1): 13–29
8. Templin MF, Stoll D, Schrenk M, Traub PC, Vohringer CF, Joos TO (2002) *Protein microarray technology.* Trends Biotechnol 20(4): 160–166
9. Zhu H, Snyder M (2001) *Protein arrays and microarrays.* Curr Opin Chem Biol 5(1): 40–45

10. Schena M, Shalon D, Davis RW, Brown PO (1995) *Quantitative monitoring of gene–expression patterns with a complementary–DNA microarray.* Science 270(N5235): 467–470
11. MacBeath G, Schreiber SL (2000) *Printing proteins as microarrays for high–throughput function determination.* Science 289(5485): 1760–1763
12. BioChipNet [http://www.biochipnet.de/]
13. Microarray Electronic Library [http://www.arrayit.com/e–library/]
14. Conrads TP, Issaq HJ, Veenstra TD (2002) *New tools for quantitative phosphoproteome analysis.* Biochem Biophys Res Comm 290(3): 885–890
15. Ong SE, Blagoev B, Kratchmarova I, Kristensen DB, Steen H, Pandey A, Mann M (2002) *Stable isotope labeling by amino acids in cell culture, SILAC, as a simple and accurate approach to expression proteomics.* Mol Cell Proteomics 1(5): 376–386
16. Glover JF, Wilson TM (1982) *Efficient translation of the coat protein cistron of tobacco mosaic virus in a cell–free system from Escherichia coli.* Eur J Biochem 122(3): 485–492
17. Blanar MA, Rutter WJ (1992) *Interaction cloning: identification of a helix–loop–helix zipper protein that interacts with c–Fos.* Science 256(5059): 1014–1018
18. Aebersold R, Mann M (2003) *Mass spectrometry–based proteomics.* Nature 422(6928): 198–207
19. Borrebaeck CA, Ekstrom S, Hager AC, Nilsson J, Laurell T, Marko-Varga G (2001) *Protein chips based on recombinant antibody fragments: a highly sensitive approach as detected by mass spectrometry.* Biotechniques 30(5): 1126–1130, 1132
20. Karlsson R, Kullman-Magnusson M, Hamalainen MD, Remaeus A, Andersson K, Borg P, Gyzander E, Deinum J (2000) *Biosensor analysis of drug–target interactions: direct and competitive binding assays for investigation of interactions between thrombin and thrombin inhibitors.* Anal Biochem 278(1): 1–13
21. Myszka D, Rich R (2000) *Implementing surface plasmon resonance biosensors in drug discovery.* Pharm Sci Technol Today 3: 310–317
22. Sali A, Glaeser R, Earnest T, Baumeister W (2003) *From words to literature in structural proteomics.* Nature 422(6928): 216–225
23. Gavin AC, et al. (2002) *Functional organization of the yeast proteome by systematic analysis of protein complexes.* Nature 415(6868): 141–147
24. Ho Y, et al. (2002) *Systematic identification of protein complexes in Saccharomyces cerevisiae by mass spectrometry.* Nature 415(6868): 180–183
25. Tyers M, Mann M (2003) *From genomics to proteomics.* Nature 422(6928): 193–197
26. Mitchell P (2002) *A perspective on protein microarrays.* Nat Biotechnol 20(3): 225–229
27. Büssow K, Nordhoff E, Lübbert C, Lehrach H, Walter G (2000) *A human cDNA library for high–throughput protein expression screening.* Genomics 65(1): 1–8
28. Weiner H, Faupel T, Büssow K (2004) Protein arrays for cDNA expression libraries. In: *Protein Arrays.* Humana Press Inc
29. Anderson L, Seilhamer J (1997) *A comparison of selected mRNA and protein abundances in human liver.* Electrophoresis 18(3–4): 533–537
30. Ensembl Database [http://www.ensembl.org/Homo_sapiens]
31. Cahill DJ (2001) P*rotein and antibody arrays and their medical applications.* J Immunol Methods 250(1–2): 81–91

32. Burchell JM, Mungul A, Taylor-Papadimitriou J (2001) *O–linked glycosylation in the mammary gland: changes that occur during malignancy.* J Mammary Gland Biol Neoplasia 6(3): 355–364
33. Safar JG, et al. (2002) *Measuring prions causing bovine spongiform encephalopathy or chronic wasting disease by immunoassays and transgenic mice.* Nat Biotechnol 20(11): 1147–1150
34. Mendoza LG, McQuary P, Mongan A, Gangadharan R, Brignac S, Eggers M (1999) *High–throughput microarray–based enzyme–linked immunosorbent assay (ELISA).* Biotechniques 27(4): 778–780, 782–776, 788
35. Huang RP, Huang R, Fan Y, Lin Y (2001) *Simultaneous detection of multiple cytokines from conditioned media and patient's sera by an antibody–based protein array system.* Anal Biochem 294(1): 55–62
36. Arenkov P, Kukhtin A, Gemmell A, Voloshchuk S, Chupeeva V, Mirzabekov A (2000) *Protein microchips: use for immunoassay and enzymatic reactions.* Anal Biochem 278(2): 123–131
37. Haab BB, Dunham MJ, Brown PO (2001) *Protein microarrays for highly parallel detection and quantitation of specific proteins and antibodies in complex solutions.* Genome Biology 2(2): RESEARCH0004
38. Angenendt P, Glökler J (2003) submitted
39. Angenendt P, Glökler J, Murphy D, Lehrach H, Cahill DJ (2002) *Toward optimized antibody microarrays: a comparison of current microarray support materials.* Anal Biochem 309(2): 253–260
40. Peluso P, et al. (2003) *Optimizing antibody immobilization strategies for the construction of protein microarrays.* Anal Biochem 312(2): 113–124
41. Schweitzer B, Wiltshire S, Lambert J, O'Malley S, Kukanskis K, Zhu Z, Kingsmore SF, Lizardi PM, Ward DC (2000) *Inaugural article: immunoassays with rolling circle DNA amplification: a versatile platform for ultrasensitive antigen detection.* Proc Natl Acad Sci USA 97(18): 10113–10119
42. Schweitzer B, Kingsmore SF (2002) *Measuring proteins on microarrays.* Curr Opin Biotechnol 13(1): 14–19
43. Bernard A, Michel B, Delamarche E (2001) *Micromosaic immunoassays.* Anal Chem 73(1): 8–12
44. Hallborn J, Carlsson R (2002) *Automated screening procedure for high–throughput generation of antibody fragments.* Biotechniques Suppl: 30–37
45. Hanes J, Schaffitzel C, Knappik A, Pluckthun A (2000) *Picomolar affinity antibodies from a fully synthetic naive library selected and evolved by ribosome display.* Nat Biotechnol 18(12): 1287–1292
46. Smith D, Collins BD, Heil J, Koch TH (2003) *Sensitivity and specificity of photoaptamer probes.* Mol Cell Proteomics 2(1): 11–18
47. Kramer K, Fiedler M, Skerra A, Hock B (2002) *A generic strategy for subcloning antibody variable regions from the scFv phage display vector pCANTAB 5 E into pASK85 permits the economical production of F(ab) fragments and leads to improved recombinant immunoglobulin stability.* Biosens Bioelectron 17(4): 305–313
48. He M, Taussig M (2001) *Single step generation of protein arrays from DNA by cell–free expression and in situ immobilisation (PISA method).* Nucleic Acids Res 29(15): e73
49. Petsko GA (1996) *For medicinal purposes.* Nature 384(6604 Suppl): 7–9
50. Abelson JN, ed (1996) *Combinatorial Chemistry. Methods in Enzymology.* Vol. 267, Academic Press: San Diego

51. Cortese R, ed (1996) *Combinatorial Libraries; Synthesis, Screening and Application Potential.* , Walter de Gruyter & Co: Berlin
52. Famulok M, Winnaker E-L, Wong C-H, eds (1999) *Combinatorial Chemistry in Biology.* , Springer Verlag: Berlin
53. Merrifield B (1986) *Solid phase synthesis.* Science 232(4748): 341–347
54. Merrifield RB (1963) *Solid phase peptide synthesis. I. The synthesis of a tetrapeptide.* J Am Chem Soc 85: 2149–2154
55. Merrifield RB (1965) *Automated synthesis of peptides.* Science 150(693): 178–185
56. Hogan JC, Jr. (1996) *Directed combinatorial chemistry.* Nature 384(6604 Suppl): 17–19
57. Houghten RA, Pinilla C, Blondelle SE, Appel JR, Dooley CT, Cuervo JH (1991) *Generation and use of synthetic peptide combinatorial libraries for basic research and drug discovery.* Nature 354(6348): 84–86
58. Frank R, Döhring R (1988) *Simultaneous multiple synthesis under controlled flow conditions on cellulose paper disks as segmental solid supports.* Tetrahedron 44(19): 6031–6040
59. Frank R, Heikens W, Heisterberg-Moutsis G, Blocker H (1983) *A new general approach for the simultaneous chemical synthesis of large numbers of oligonucleotides: segmental solid supports.* Nucleic Acids Res 11(13): 4365–4377
60. Lam KS, Salmon SE, Hersh EM, Hruby VJ, Kazmierski WM, Knapp RJ (1991) *A new type of synthetic peptide library for identifying ligand–binding activity.* Nature 354(6348): 82–84
61. Furka A, Sebestyen F, Asgedom M, Dibo G (1991) *General method for rapid synthesis of multicomponent peptide mixtures.* Int J Pept Protein Res 37(6): 487–493
62. Geysen HM, Meloen RH, Barteling SJ (1984) *Use of peptide synthesis to probe viral antigens for epitopes to a resolution of a single amino acid.* Proc Natl Acad Sci USA 81(13): 3998–4002
63. Maskos U, Southern EM (1992) *Oligonucleotide hybridizations on glass supports: a novel linker for oligonucleotide synthesis and hybridization properties of oligonucleotides synthesised in situ.* Nucleic Acids Res 20(7): 1679–1684
64. Fodor SP, Read JL, Pirrung MC, Stryer L, Lu AT, Solas D (1991) *Light–directed, spatially addressable parallel chemical synthesis.* Science 251(4995): 767–773
65. Frank R (1992) *Spot synthesis: an easy techique for the positionally addressable parallel chemical synthesis on a membrane support.* Tetrahedron 48: 9217–9232
66. Frank R (2002) *High–density synthetic Peptide microarrays: emerging tools for functional genomics and proteomics.* Comb Chem High Throughput Screen 5(6): 429–440
67. Frank R (2002) *The SPOT–synthesis technique. Synthetic peptide arrays on membrane supports–principles and applications.* J Immunol Methods 267(1): 13–26
68. Geysen HM, Rodda SJ, Mason TJ (1986) *A priori delineation of a peptide which mimics a discontinuous antigenic determinant.* Mol Immunol 23(7): 709–715
69. Frank R (1995) *Simultaneous and combinatorial chemical synthesis techniques for the generation and screening of molecular diversity.* J Biotechnol 41(2–3): 259–272
70. Cantley LC, Songyang Z (1994) *Specificity in recognition of phosphopeptides by srchomology 2 domains.* J Cell Sci 18: 121–126

71. Macias M, Hyvönen M, Baraldi E, Schultz J, Sudol M, Saraste M, Oschkinat M (1996) *Structure of the WW domain of a kinase–associated protein complexed with a proline–rich peptide.* Nature 382: 646–649
72. Spink KE, Fridman SG, Weis WI (2001) *Molecular mechanisms of beta–catenin recognition by adenomatous polyposis coli revealed by the structure of an APC–beta–catenin complex.* EMBO J 20(22): 6203–6212
73. Reineke U, Kramer A, Schneider-Mergener J (1999) *Antigen sequence– and library–based mapping of linear and discontinuous protein–protein–interaction sites by spot synthesis.* Comb Chem Biol 243: 23–36
74. Darji A, Niebuhr K, Hense M, Wehland J, Chakraborty T, Weiss S (1996) *Neutralizing monoclonal antibodies against listeriolysin: mapping of epitopes involved in pore formation.* Infect Immun 64(6): 2356–2358
75. Hohne WE, et al. (1993) *Structural base of the interaction of a monoclonal antibody against p24 of HIV–1 with its peptide epitope.* Mol Immunol 30(13): 1213–1221
76. Martens W, Greiser-Wilke I, Harder TC, Dittmar K, Frank R, Orvell C, Moennig V, Liess B (1995) *Spot synthesis of overlapping peptides on paper membrane supports enables the identification of linear monoclonal antibody binding determinants on morbillivirus phosphoproteins.* Vet Microbiol 44(2–4): 289–298
77. Niebuhr K, Ebel F, Frank R, Reinhard M, Domann E, Carl UD, Walter U, Gertler FB, Wehland J, Chakraborty T (1997) *A novel proline–rich motif present in ActA of Listeria monocytogenes and cytoskeletal proteins is the ligand for the EVH1 domain, a protein module present in the Ena/VASP family.* Embo J 16(17): 5433–5444
78. Schmidt TGM, Koepke J, Frank R, Skerra A (1996) *Molecular interaction between the Strep–tag affinity peptide and its cognate target, streptavidin.* J Mol Biol 255(5): 753–766
79. Schmidt TG, Koepke J, Frank R, Skerra A (1996) *Molecular interaction between the Strep–tag affinity peptide and its cognate target, streptavidin.* J Mol Biol 255(5): 753–766
80. Voss S, Skerra A (1997) *Mutagenesis of a flexible loop in streptavidin leads to higher affinity for the Strep–tag II peptide and improved performance in recombinant protein purification.* Protein Eng 10(8): 975–982
81. Kramer A, Volkmer-Engert R, Malin R, Reineke U, Schneider-Mergener J (1993) *Simultaneous synthesis of peptide libraries on single resin and continuous cellulose membrane supports: examples for the identification of protein, metal and DNA binding peptide mixtures.* Pept Res 6(6): 314–319
82. Malin R, Steinbrecher A, Semmler W, Noll B, Johannson B, Frömmel C, Schneider-Mergener J (1995) *Identification of technetium–99m binding peptides using cellulose–bound combinatorial peptide libraries.* J Am Chem Soc 117: 11821–11822
83. Bosc C, Frank R, Denarier E, Ronjat M, Schweitzer A, Wehland J, Job D (2001) *Identification of novel bifunctional calmodulin–binding and microtubule–stabilizing motifs in STOP proteins.* J Biol Chem 276(33): 30904–30913
84. Frank R, Schneider-Mergener J (2002) SPOT–synthesis – scope and applications. In: Kocj J, Mahler M (eds) *Peptide arrays on membrane supports: a laboratory manual.* Springer Verlag, Heidelberg
85. Frank R, Hoffmann S, Kieb M, Lahmann H, Tegge W, Behm C, Gausepohl H (1996) Combinatorial synthesis on membrane supports by the SPOT technique:

imaging peptide sequence space. In: Jung G (ed) *Combinatorial peptide and non-peptide libraries – a handbook*. Verlag Chemie. pp 363–386

86. Frank R, Overwin H (1996) SPOT synthesis. *Epitope analysis with arrays of synthetic peptides prepared on cellulose membranes*. Methods Mol Biol 66: 149–169
87. Kramer A, Schneider-Mergener J (1998) *Synthesis and screening of peptide libraries on continuous cellulose membrane supports*. Methods Mol Biol 87: 25–39
88. Adler F, Türk G, Frank R, Zander N, Wu W, Volkmer-Engert R, Schneider-Mergener J, Gausepohl H (1999) A new array format for the automated parallel combinatorial synthesis by the SPOT–technique. In: Epton R (ed) Proc. International Symp. on '*Innovation and Perspectives in Solid Phase Synthesis 2000*'. Mayflower Worldwide, Kingswinford, York. pp 221
89. Pellois JP, Zhou X, Srivannavit O, Zhou T, Gulari E, Gao X (2002) *Individually addressable parallel peptide synthesis on microchips*. Nat Biotechnol 20(9): 922–926
90. Singh-Gasson S, Green RD, Yue Y, Nelson C, Blattner F, Sussman MR, Cerrina F (1999) *Maskless fabrication of light–directed oligonucleotide microarrays using a digital micromirror array*. Nat Biotechnol 17(10): 974–978
91. Gao X, LeProust E, Zhang H, Srivannavit O, Gulari E, Yu P, Nishiguchi C, Xiang Q, Zhou X (2001) *A flexible light–directed DNA chip synthesis gated by deprotection using solution photogenerated acids*. Nucleic Acids Res 29(22): 4744–4750
92. Gao X, Yu P, LeProust E, Sonigo L, Pellois JP, Zhang H (1998) *Oligonucleotide Synthesis Using Solution Photogenerated Acids*. J Am Chem Soc 120(48): 12698–12699
93. LeProust E, Pellois JP, Yu P, Zhang H, Gao X, Srivannavit O, Gulari E, Zhou X (2000) *Digital Light–Directed Synthesis. A Microarray Platform That Permits Rapid Reaction Optimization on a Combinatorial Basis*. J Comb Chem 2(4): 349–354
94. Pellois JP, Wang W, Gao X (2000) *Peptide Synthesis Based on t–Boc Chemistry and Solution Photogenerated Acids*. J Comb Chem 2(4): 355–360
95. Hughes TR, et al. (2001) *Expression profiling using microarrays fabricated by an ink–jet oligonucleotide synthesizer*. Nat Biotechnol 19(4): 342–347
96. Bischoff FR, Stadler V, Breitling F (2003) *Hochkomplexe Peptidarrays – Techniken, Anwendungen und Perspektiven*. BioSpektrum 8:654-657
97. Breitling F, Breitling F, Felgenhauer T, Fernandez S, Leibe K, Beyer M, Stadler V, Bischoff FR, Poustka A (2003) *Hochkomplexe Peptidarrays auf Computerchips*. : in press
98. Hultschig C, *Two Dimensional Screening: Towards Establishing a Novel Technique to Study Biomolecular Interactions, in Gemeinsame naturwissenschaftliche Faklutät*. 2000, Technische Universität Carolo-Wilhelmina zu Braunschweig; Germany; available under: http://www.biblio.tu–bs.de/ediss/data/20000207a/20000207a.html

15

Photoaptamer Arrays for Proteomics Applications

Drew Smith and Chad Greef

15.1 Introduction

In this chapter we describe the use of photoaptamers for protein detection in microarray format. We begin with a short review of aptamer technology in general, and a summary description of current methods for high throughput generation of photoaptamers. This section is followed by a description of making and using photoaptamer arrays for proteomics analysis.

Aptamers are nucleic acids that fold into complex shapes and have desirable properties such as ligand binding or catalysis. Aptamer technology was foreshadowed by the discovery of catalytic RNA [1,2] and was enabled by the development of efficient methods for chemical synthesis of DNA [3,4], in vitro transcription to produce RNA [5], reverse transcription of RNA to DNA [6,7] and DNA amplification by PCR [8]. These methods, combined with techniques to select interesting and useful nucleic acids, constitute the basis of SELEX, the process by which aptamers are generated [9,10].

The basic principles and practice of SELEX have been described elsewhere [11–13]. Briefly, a library of randomized sequences (typically 30–60 nt) is synthesized. Flanking the randomized region are regions of fixed sequence that serve as primer binding sites for PCR and for transcription initiation if an RNA library is to be generated. Much of the power of the SELEX process is due to the size of the starting libraries that can be generated. A library of 30 nt contains 4^{30} (10^{18}) distinct sequences; in practice, about 10^{14} sequences (1 nmol, or 25 μg) are conveniently used – a 'genome' orders of magnitude larger than any biological genome. This library can be used directly for DNA SELEX, or is transcribed for RNA SELEX. Partitioning more-active from less-active sequences is the most critical step in a SELEX experiment, and constitutes much of the art of the process. Since most SELEX experiments are aimed at obtaining ligand binders, partitioning schemes typically exploit physical differences between free and ligand-bound aptamers as their basis. Partitioning methods include filter–binding, electrophoretic and chromatographic mobility shifts, capture by immobilized targets (including cells and tissues) and variations of these techniques. Once separated, the enriched pools are recovered, reverse-transcribed (if RNA) and amplified by PCR to begin another cycle. SELEX experiments typically require 6–12 cycles to converge upon a few tens of active sequences.

Aptamers have been developed for use as therapeutic agents [14], as in vivo imaging agents [15, 16], as intra- and extra-cellular inhibitors of protein function in vivo and in tissue culture [17–22], as cell–surface labels [23, 24], as probes for target validation and drug design [25–27], as affinity purification reagents [28], and as diagnostic reagents in microwell [29–33] and microarray format [34].

The use of aptamers in the latter format is the subject of this chapter. The use of aptamers in microarray format for protein detection is a natural extension of both aptamer and microarray technologies. Like antibodies, aptamers have been discovered for a broad range of target proteins (see the Aptamer Database http://aptamer.icmb.utexas.edu/index.html), have affinities that are typically nanomolar or better, and show excellent discrimination between their targets and closely related proteins [35–40]. However, the nature of aptamers lends itself to microarray applications: nucleic acids, especially DNA, are chemically stable and resistant to degradation (except by nucleases); DNA molecules are readily synthesized by automated methods; the incorporation of modifications for array attachment is simple; and the SELEX process itself can be automated for high throughput discovery [41–43].

Our approach to microarray detection employs photoaptamers as capture agents. Photoaptamers are photoactivated crosslinking aptamers [44, 45]. The covalent complex that is generated between aptamer capture agent and target protein simplifies processing and analysis of the microarray: unbound protein can be washed away using denaturing conditions, and the captured protein can be labelled in situ for detection [46, 47]. The photocrosslinking reaction is quite specific and aids in the rejection of non-specific binding, particularly by proteins with high affinity for DNA [48]. Photocrosslinking activity is imparted by incorporating BrdU into SELEX libraries. Irradiation at 300 nm or longer wavelengths generates the 5–uridinyl radical which will react with proximal electron-rich amino acids, causing covalent complex formation [49–51].

15.2 Overview of Photoaptamer Discovery and High Throughput Production

The first photoSELEX experiments exploited electrophoretic mobility shifts to partition crosslinked from free DNA or RNA [44, 45]. Although effective, this method is time–consuming and is difficult to scale. Microbead partitioning can be adapted to a 96–well format, because suspensions of beads can be easily mixed and transferred by standard liquid handling equipment. Bead suspensions can then be converted from liquid to solid phase by filtration or magnetic partitioning. This versatility has led to the choice of microbeads for SELEX scale–up and automation at SomaLogic and in the Ellington group [41–43].

We employ three types of partitioning protocols in microbead format: protein immobilization on beads followed by photo–crosslinking of the SELEX library to the protein (bead photoSELEX); protein immobilization on beads followed by binding without crosslinking of the SELEX library to the protein (bead affinity SELEX); and photocrosslinking of the SELEX library to the target protein, followed by immobilization of the protein–DNA complexes on the beads (solution photoSELEX). These three protocols are outlined in Fig. 15.1.

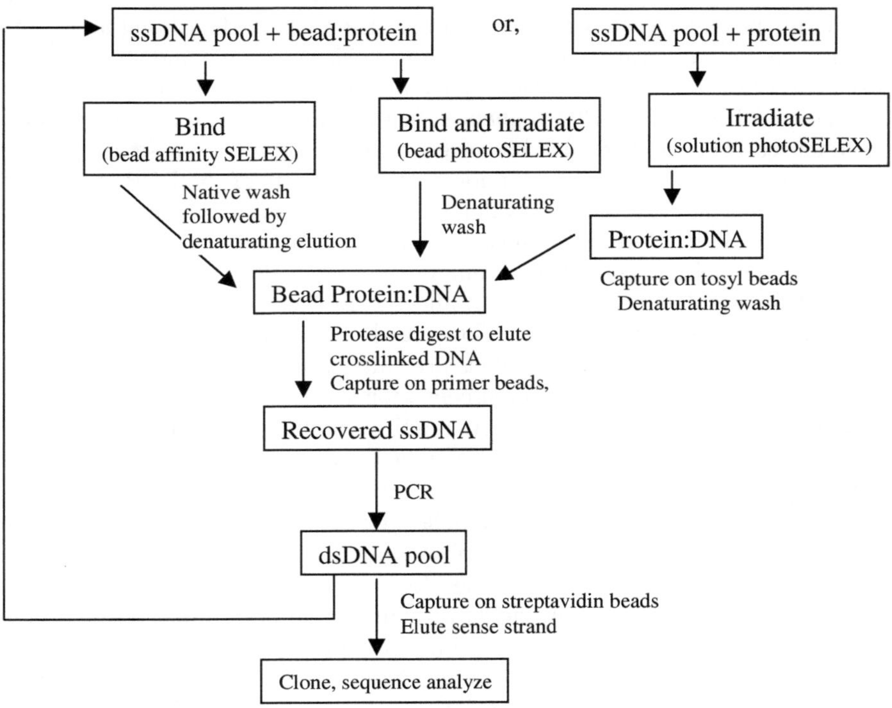

Fig. 15.1. Automated SELEX schemes. A ssDNA pool is mixed with bead-immobilized protein (the two left tracks) or with free protein (right track) and allowed to bind to target protein. For affinity selection, the beads are washed under native conditions, and sequences with binding activity are recovered by elution under denaturing conditions (bead affinity SELEX). Alternatively, the protein:DNA complexes on the beads are irradiated to form a covalent bond (bead photoSELEX). In the third alternative (solution photoSELEX) the soluble protein:DNA complexes are irradiated to form a covalent bond, and this complex is captured through the protein moiety on to tosyl-activated beads. The covalently-bound DNA is recovered by protease digestion. This DNA is then amplified and purified to start a successive round

These protocols share many common steps, and differ principally by covalent vs non-covalent binding of DNA to the target protein, and by the respective order of protein:DNA crosslinking vs protein:bead immobilization in the process. All of these processes are fully automated in 96–well format with the exception of moving plates to and from the PCR thermal cycler. The photoSELEX protocols are performed on a Cavro pipetting station modified to handle a fiber optic tool for the irradiation step. The affinity SELEX protocol can be performed on the Cavro or, more readily, on a TomTech 96–channel pipetting robot. A round of photoSELEX using these procedures requires 10–14 hours for a full 96–well plate. Affinity SELEX requires 6–8 hours per round.

Successful SELEX experiments are typically completed in 6–9 cycles; additional cycles rarely improve aptamer pool activity. Because so many SELEX experiments can be performed in parallel, the task of identifying and synthesizing active sequences generally requires much more time and effort than the selection process itself. We start by winnowing out those pools that have not converged from the $\sim 10^{14}$ sequences in the starting library to a few tens of sequences. This convergence is conveniently monitored by assessing the rate of reannealing of the double-stranded DNA products of the PCR step – a C_0t assay [52]. The PCR sample is denatured by heating to 98° in the presence of the dye SYBR Green I, which fluoresces when bound to double-stranded DNA. The DNA is cooled to a temperature that allows reannealing of fully complementary sequences ($\sim 87°$), and the gain in fluoresecence over time is monitored with a CCD camera. Converged pools typically regain full fluorescence in < 10 minutes. These pools are carried forward for activity analysis.

Photoaptamer pools can be screened for activity in solution or in microarray format. Because photoaptamers require a stable protein:DNA complex for photocrosslinking [49], active pools can be identified on the basis of their affinity for their target proteins. This is conveniently done by the nitrocellulose filter–binding assay, where radiolabelled DNA and excess protein are mixed and filtered, and the fraction of DNA bound is determined by scintillation counting or phosphorimaging of the filters [11].

The filter–binding assay yields a quantitative determination of protein–binding activity, typically expressed as a dissociation constant. A more qualitative assessment of activity can be determined in microarray format, as described below. With few exceptions, there is good agreement between the results of the two assays. The microarray assay is somewhat more stringent – pools and aptamers that have strong affinity for their target protein (< 10nM) show the best activity on microarrays, whereas lower affinity pools often (but not always) show little or no microarray activity. SELEX pools can be prepared for arraying by PCR with a 5′amino linker primer.

15.3 Using Photoaptamer Microarrays

Once pools from a SELEX experiment have been determined to have binding affinity and crosslinking activity to cognate protein they are cloned through a plasmid vector and individual sequences are determined. Alignment protocols identify sequence motifs among the different populations, as well as potential contaminants and spurious outliers. The sequences that are determined to have the greatest likelihood of representing the binding motif from the SELEX are chemically synthesized for activity screening by microarray and solution phase assays.

Chemical synthesis of photoaptamers is performed by standard phosphoramidite methods, with procedural modifications that have been optimized to maximize deprotection and recovery of full-length product. Aptamers are synthesized with appropriate attachment chemistries added as modified phosphoramidites. Product quality is confirmed with extensive HPLC, CGE, and ICR mass spec analysis. Since DNA synthesis is highly controllable the production of aptamers can be considered a very robust, manufacturable process, amenable to scale–up and quality control.

Synthetic photoaptamers are arrayed by standard contact printing methods with modifications that optimize loading, spot morphology, and aptamer activity. An important consideration is that the substrate must be chosen to minimize non-specific adsorption of DNA, as this will disrupt aptamer tertiary structure, limiting activity. We have found that commercially available microarray slides that present chemically functionalized polymer coatings allow high aptamer loading and activity. After printing, slides are processed by methods designed to render the remaining functional groups and the polymer coating inert to interaction with proteins and UPS (see below). Printed slides can be stored dry for extended periods with no loss of activity.

Photoaptamers arrays are used as discovery tools to screen pools for binding activity and individual cloned sequences for relative activity and cross–reactivity, but the ultimate goal is to evaluate protein levels in multiplex fashion from complex mixtures. In all cases the procedure to run a photoaptamer array assay is the same; protein samples are incubated over the array allowing affinity binding to occur, unbound protein is washed away, bound protein is photo–crosslinked to cognate aptamer, crosslinked protein is chemically labelled with a Universal Protein Stain (UPS), and the label is detected. Individual steps are detailed below:

Protein Binding

Protein samples are prepared in Protein Incubation Buffer (PIB) that matches as closely as possible the composition of the SELEX discovery buffer, in terms of buffer composition, salt content, and ionic strength. Carrier DNA is added, but has not been shown to be absolutely necessary. The photoaptamer arrays are prepared by equilibration in PIB for 15 minutes prior to introduction of the protein mixture. Note that conventional, protein based blocking mixtures are not necessary. Once the protein mixture is applied to the array the mixture is allowed to incubate at 30°C for at least 2 hours. For high sensitivity measurements longer incubation times may improve results. The protein incubation can be performed in either a static mode, in which the solution is allowed to interact with the array without dynamic movement, or in a flow mode, where solution is circulated over the array in either a continuous loop or a reciprocating fashion. For the static mode, simple reaction vessels are fashioned over the arrays by application of adhesive-backed wells or other similar devices, while for circulation mode more technically evolved solutions are required. Both methods yield equivalent results; the advantage of mixing is a reduction in the incubation time needed to reach maximal binding levels.

Pre-Crosslink Wash, Crosslinking, and Post-Crosslink Wash

At the end of the incubation period the protein solution is replaced with PIB, allowing removal of unbound protein while retaining cognate protein binding to aptamers through affinity interaction. The arrays are then exposed to UV irradiation, causing covalent crosslink formation between BrdU residues on the aptamers and proximal electron-rich amino acids of the cognate proteins. Optimal wavelength for the crosslink is 308 nm, which can be introduced by excimer laser excitation or broad spectrum UV which is filtered to eliminate sub–300 nm wavelengths. Optimal energy levels have been calibrated on the laser and empirically determined to be 3 J/cm^2, which gives the highest levels of specific crosslinking.

The specificity imparted to the microarray assay in the crosslinking step is a key feature of photoaptamer technology. The photoSELEX process selects those aptamers that efficiently crosslink their target protein. Because the photoactivated complex is short-lived, efficient crosslinking requires close and stable contacts between BrdU and the target amino–acid, a requirement for π–bond orbital overlap has been proposed [50]. Although polyanion–binding proteins may bind to aptamer DNA, the probability that this binding will result in productive geometry for photocrosslinking is low. We have shown that the photocrosslinking step can improve aptamer specificity by an order of magnitude or more over the specificity due to affinity interactions alone [48]. Although these measurements were made in solution, they are consistent with results obtained on microarrays, both with simple protein mixtures and with target proteins spiked in to serum [48].

Since the crosslinked aptamer–protein complex is covalently linked to the substrate it is possible to use extremely rigorous denaturing conditions to fully remove any remaining proteins from the substrate or non-cognate array features. Examples of denaturing components of washing solutions include 0.02 M NaOH, 0.1 M AcOH, 1% SDS, 2% TritonX–100, 0.5 M NaCl, 0.02 M DTT, 8 M Urea, 4 M GuHCl, 50°C, and sonication. However, since the binding interaction of aptamers to cognate proteins is extremely specific there is generally little to be gained from stringent washes at this point, but eliminating non-specifically adsorbed protein from the substrate minimizes general background signal.

Signal Generation with Universal Protein Stain (UPS)

Because the only protein molecules present on the array at this point are those that are covalently crosslinked to their cognate aptamer, a global labelling step that targets protein-specific chemical moieties is employed. Generally, direct fluorescence detection provides adequate signal/noise, but alternative methods and/or signal amplification can boost response for high sensitivity applications. A number of methods to introduce fluorescent detection molecules have been used, including lysine-specific activated ester modified dyes, thiol specific maleimide modified dyes, nitrosylation of tyrosines followed by nitrotyrosine specific Ab, and biotinylation followed by TSA detection.

Fluorescence signal from photoaptamer arrays is measured by standard microarray scanning devices, providing raw data as 16–bit TIFF images. Data processing involves fitting ROI grids to the image via standard image processing software methods, extracting mean signal intensity from features, subtracting background signal derived from no–protein controls, and evaluating resultant values by comparing to standard curves generated through dose response control experiments. An internal database processes, collates, and stores data from experiments.

Figure 15.2 shows results from a model multiplex experiment in which a series of protein mixtures were created such that each mixture contained 14 proteins at different levels, each protein was represented at some level in each mixture, and the overall protein concentration of each mixture was constant. The mixtures were each assayed on discrete arrays, and the resultant data was deconvoluted to generate multiple standard curves from one assay series. The pseudo–color image shows boxed features corresponding to the endostatin standard curve series, while the other quadruplicate groupings correspond to other proteins in the mixtures.

15.4 Discussion

Development of microarray assays capable of rapid multplexed determination of absolute and relative levels of proteins in complex mixtures will enable many new capabilities in the fields of research proteomics, drug discovery, and clinical diagnostics. Multiplexed protein analysis in the microarray format will allow researchers to explore causal relationships between relative protein levels in samples and diseased states being studied while consuming far smaller volumes of precious samples than are required for current methods.

As the number of proteins available for study increases, unique signature patterns of protein levels in diseased state samples could become apparent,

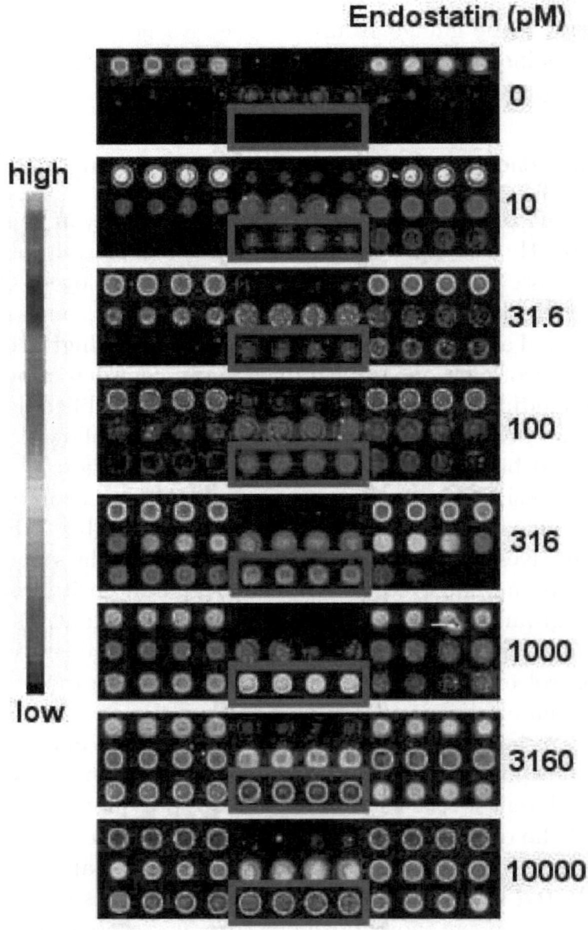

Fig. 15.2. Example of Multiplexed Standard Curve. Proteins were mixed in designed concentration combinations and assayed on photoaptamer arrays. Quadriplate photoaptamer features for Endostatin are highlighted, showing gradient in response over a range on concentrations

which will greatly expand the possibilities in early disease diagnosis, disease metabolism, and drug discovery. It is not difficult to imagine the application of diagnostic arrays in preventative medicine, in which pre-symptomatic diagnosis of many cancers and degenerative diseases will allow more effective early treatments and greatly improve probabilities of successful outcomes.

Photoaptamer microarrays are a defining methodology in the field of proteomics. The ability to select highly specific binding reagents by directed methods provides a powerful tool for protein quantitation that has not yet been available. Further, ease of manufacture and photoaptamer stability allows a wide range of applications that is unlimited by many of the constraints traditionally associated with biological reagents. The acquisition of photoaptamers is limited only by the availability of individual proteins, and as that repertoire and inventory increases the possibilities for photoaptamers will follow closely behind.

The obvious ambition for protein microarray technology is to supplant the laborious technologies now associated with proteomics: to make 2–D gel analysis and single–analyte ELISAs as obsolete for the study of protein expression as Northern blots have become for the study of mRNA expression. Microarray technology and its associated instrumentation are already cheaper, faster and more–robust than the suite of technologies associated with 2–D gel/mass spectrometry analysis. The acceptance of microarray technology for proteomic analysis now awaits the introduction of assay platforms that are as sensitive and comprehensive as the technologies we seek to replace.

The ideal microarray would combine the sensitivity of ELISA technology with the comprehensive proteome coverage of 2–D gel/MS technology. Antibody-based arrays have already shown impressive ELISA-like sensitivity in small multiplex arrays [53,54]. However, the need to identify and apply secondary labelling antibodies will soon become an important constraint on the degree of proteome coverage that can be achieved. Multiplexing with antibodies may fall well short of the coverage provided by 2–D gels. Photoaptamers start from a narrower technology base than do antibodies, but dispense with the need for a secondary reagent. As the degree of multiplexing becomes more critical in the development of protein microarray technology, the advantages of a format based on a single capture and detection reagent will become more important.

References

1. Guerrier-Takada, C., Gardiner, K. J., Marsh, T. L., Pace, N. R., and Altman, S. (1983). *The RNA Moiety Of RNase P Is The Catalytic Subunit Of The Enzyme.* Cell 35, 849–857
2. Kruger, K., Grabowski, P. J., Zaug, A. J., Sands, J., Gottschling, D. E., and Cech, T. R. (1982). *Self-splicing RNA: autoexcision and autocyclization of the ribosomal RNA in-tervening sequence of Tetrahymena.* Cell 31, 147–157
3. Beaucage, S. L., and Caruthers, M. H. (1981). *Studies on Nucleotide Chemistry V. Deoxynucleoside Phosphoramidites A New Class of Key Intermediates for Deoxypolynucleotide Synthesis.* Tetrahedron Lett 22, 1859
4. Matteucci, M. D., and Caruthers, M. H. (1981). *Studies on Nucleotide Chemistry IV. Synthesis of Deoxyoligonucleotides on a Polymer Support.* J Am Chem Soc 103, 3185
5. Studier, F. W., and Moffatt, B. A. (1986). *Use of bacteriophage T7 RNA polymerase to direct selective high–level expression of cloned genes.* J Mol Biol 189, 113–130
6. Baltimore, D. (1970). *RNA–dependent DNA polymerase in virions of RNA tumour viruses.* Nature 226, 1209–1211

7. Temin, H. M., and Mizutani, S. (1970). *RNA–dependent DNA polymerase in virions of Rous sarcoma virus.* Nature 226, 1211–1213
8. Saiki, R. K., Scharf, S., Faloona, F., Mullis, K. B., Horn, G. T., Erlich, H. A., and Arnheim, N. (1985). *Enzymatic amplification of beta–globin genomic sequences and restriction site analysis for diagnosis of sickle cell anemia.* Science 230, 1350–1354
9. Ellington, A. D., and Szostak, J. W. (1990). *In vitro selection of RNA molecules that bind specific ligands.* Nature 346, 818–822
10. Tuerk, C., and Gold, L. (1990). *Systematic evolution of ligands by exponential enrichment: RNA ligands to bacteriophage T4 DNA polymerase.* Science 249, 505–510
11. Fitzwater, T., and Polisky, B. (1996). *A SELEX primer.* Methods Enzymol 267, 275–301
12. Gold, L., Polisky, B., Uhlenbeck, O., and Yarus, M. (1995). *Diversity of oligonucleotide functions.* Annu Rev Biochem 64, 763–797
13. Tuerk, C. (1997). *Using the SELEX combinatorial chemistry process to find high affinity nucleic acid ligands to target molecules.* Methods Mol Biol 67, 219–230
14. Eyetech Study Group. (2002). *Preclinical and Phase 1a Clinical Evaluation of an Anti-Vegf Pegylated Aptamer (Eye001) for the Treatment of Exudative Age–Related Macular Degeneration.* Retina 22, 143–152
15. Charlton, J., Sennello, J., and Smith, D. (1997). *In vivo imaging of inflammation using an aptamer inhibitor of human neutrophil elastase.* Chem Biol 4, 809–816
16. Hicke, B. J., Marion, C., Chang, Y. F., Gould, T., Lynott, C. K., Parma, D., Schmidt, P. G., and Warren, S. (2001). *Tenascin–C aptamers are generated using tumor cells and purified protein.* J Biol Chem 276, 48644–48654
17. Bless, N. M., Smith, D., Charlton, J., Czermak, B. J., Schmal, H., Friedl, H. P., and Ward, P. A. (1997). *Protective effects of an aptamer inhibitor of neutrophil elastase in lung inflammatory injury.* Curr Biol 7, 877–880
18. Blind, M., Kolanus, W., and Famulok, M. (1999). *Cytoplasmic RNA modulators of an inside–out signal–transduction cascade.* Proc Natl Acad Sci USA 96, 3606–3610
19. Hicke, B. J., Watson, S. R., Koenig, A., Lynott, C. K., Bargatze, R. F., Chang, Y. F., Ringquist, S., Moon-McDermott, L., Jennings, S., Fitzwater, T., et al. (1996). *DNA aptamers block L–selectin function in vivo. Inhibition of human lymphocyte trafficking in SCID mice.* J Clin Invest 98, 2688–2692
20. Ostendorf, T., Kunter, U., Grone, H. J., Bahlmann, F., Kawachi, H., Shimizu, F., Koch, K. M., Janjic, N., and Floege, J. (2001). *Specific Antagonism of PDGF Prevents Renal Scarring in Experimental Glomerulonephritis.* J Am Soc Nephrol 12, 909–918
21. Shi, H., Hoffman, B. E., and Lis, J. T. (1999). *RNA aptamers as effective protein antagonists in a multicellular organism.* Proc Natl Acad Sci USA 96, 10033–10038
22. Vuyisich, M., and Beal, P. (2002). *Controlling protein activity with ligand–regulated RNA aptamers.* Chem Biol 9, 907
23. Davis, K. A., Lin, Y., Abrams, B., and Jayasena, S. D. (1998). *Staining of cell surface human CD4 with 2′–F–pyrimidine–containing RNA aptamers for flow cytometry.* Nucleic Acids Res 26, 3915–3924
24. Homann, M., and Goringer, H. U. (1999). *Combinatorial selection of high affinity RNA ligands to live African trypanosomes.* Nucleic Acids Res 27, 2006–2014

25. Bergan, R., Connell, Y., Fahmy, B., Kyle, E., and Neckers, L. (1994). *Aptameric inhibition of p210bcr–abl tyrosine kinase autophosphorylation by oligodeoxynucleotides of defined sequence and backbone structure.* Nucleic Acids Res 22, 2150–2154
26. Green, L. S., Bell, C., and Janjic, N. (2001). *Aptamers as reagents for high-throughput screening.* Biotechniques 30, 1094–1096, 1098, 1100 passim
27. Maurel, M. C., Biard, B., Moulinier, C., Braz, D., Nugier, J., Chaumas, I., Reboud-Ravaux, M., and Decout, J. L. (2002). *RNA–acting antibiotics: in-vitro selection of RNA aptamers for the design of new bioactive molecules less susceptible to bacterial resistance.* J Pharm Pharmacol 54, 1019–1031
28. Romig, T. S., Bell, C., and Drolet, D. W. (1999). *Aptamer affinity chromatography: combinatorial chemistry applied to protein purification.* J Chromatogr B Biomed Sci Appl 731, 275–284
29. Drolet, D. W., Moon-McDermott, L., and Romig, T. S. (1996). *An enzyme–linked oligonucleotide assay.* Nat Biotechnol 14, 1021–1025
30. Fredriksson, S., Gullberg, M., Jarvius, J., Olsson, C., Pietras, K., Gustafsdottir, S. M., Ostman, A., and Landegren, U. (2002). *Protein detection using proximity–dependent DNA ligation assays.* Nat Biotechnol 20, 473–477
31. Kato, T., Yano, K., Ikebukuro, K., and Karube, I. (2000). *Bioassay of bile acids using an enzyme–linked DNA aptamer.* Analyst 125, 1371–1373
32. Rye, P. D., and Nustad, K. (2001). *Immunomagnetic DNA aptamer assay.* Biotechniques 30, 290–292, 294–295
33. Stojanovic, M. N., and Landry, D. W. (2002). *Aptamer-Based Colorimetric Probe for Cocaine.* J Am Chem Soc 124, 9678-9679
34. Lee, M., and Walt, D. R. (2000). *A Fiber–Optic Microarray Biosensor Using Aptamers as Receptors.* Anal Biochem 282, 142–146
35. Bridonneau, P., Chang, Y. F., Buvoli, A. V., O'Connell, D., and Parma, D. (1999). *Site–directed selection of oligonucleotide antagonists by competitive elution.* Antisense Nucleic Acid Drug Dev 9, 1–11
36. Daniels, D. A., Sohal, A. K., Rees, S., and Grisshammer, R. (2002). *Generation of RNA aptamers to the G–protein–coupled receptor for neurotensin, NTS–1.* Anal Biochem 305, 214–226
37. Kensch, O., Connolly, B. A., Steinhoff, H. J., McGregor, A., Goody, R. S., and Restle, T. (2000). *HIV–1 reverse transcriptase–pseudoknot RNA aptamer interaction has a binding affinity in the low picomolar range coupled with high specificity.* J Biol Chem 275, 18271–18278
38. O'Connell, D., Koenig, A., Jennings, S., Hicke, B., Han, H. L., Fitzwater, T., Chang, Y. F., Varki, N., Parma, D., and Varki, A. (1996). *Calcium–dependent oligonucleotide antagonists specific for L– selectin.* Proc Natl Acad Sci USA 93, 5883–5887
39. Rehder, M. A., and McGown, L. B. (2001). *Open–tubular capillary electrochromatography of bovine beta–lactoglobulin variants A and B using an aptamer stationary phase.* Electrophoresis 22, 3759–3764
40. Ruckman, J., Green, L. S., Beeson, J., Waugh, S., Gillette, W. L., Henninger, D. D., Claesson-Welsh, L., and Janjic, N. (1998). *2'–Fluoropyrimidine RNA–based aptamers to the 165–amino acid form of vascular endothelial growth factor (VEGF165). Inhibition of receptor binding and VEGF–induced vascular permeability through interactions requiring the exon 7–encoded domain.* J Biol Chem 273, 20556–20567

41. Cox, J. C., and Ellington, A. D. (2001). *Automated selection of anti-protein aptamers.* Bioorg Med Chem 9, 2525–2531
42. Cox, J. C., Rajendran, M., Riedel, T., Davidson, E. A., Sooter, L. J., Bayer, T. S., Schmitz-Brown, M., and Ellington, A. D. (2002). *Automated acquisition of aptamer sequences.* Comb Chem High Throughput Screen 5, 289–299
43. Gold, L., Zichi, D., Jenison, R., and Schneider, D. WO 00/43534, Method and Apparatus for the Automated Generation of Nucleic Acid Ligands Golden, M. C., Collins, B. D., Willis, M. C., and Koch, T. H. (2000). *Diagnostic potential of PhotoSELEX–evolved ssDNA aptamers.* J Biotechnol 81, 167–178
44. Golden, M. C., Collins, B. D., Willis, M. C., and Koch, T. H. (2000). *Diagnostic potential of PhotoSELEX-evolved ssDNA aptamers.* J Biotechnol 81, 167–178
45. Jensen, K. B., Atkinson, B. L., Willis, M. C., Koch, T. H., and Gold, L. (1995). *Using in vitro selection to direct the covalent attachment of human immunodeficiency virus type 1 Rev protein to high–affinity RNA ligands.* Proc Natl Acad Sci USA 92, 12220–12224
46. Brody, E. N., and Gold, L. (2000). *Aptamers as therapeutic and diagnostic agents.* J Biotechnol 74, 5–13
47. Brody, E. N., Willis, M. C., Smith, J. D., Jayasena, S., Zichi, D., and Gold, L. (1999). *The use of aptamers in large arrays for molecular diagnostics.* Mol Diagn 4, 381–388
48. Smith, D., Collins, B. D., Heil, J., and Koch, T. H. (2003). *Sensitivity and specificity of photoaptamer probes.* Mol Cell Proteomics 2, 11–18
49. Koch, T., Smith, D., Tabacman, E., and Zichi, D. (2004). *Kinetic Analysis of Site–specific Photoaptamer–Protein Crosslinking.* J Mol Biol 336: 1159–1173
50. Meisenheimer, K. M., and Koch, T. H. (1997). *Photocrosslinking of nucleic acids to associated proteins.* Crit Rev Biochem Mol Biol 32, 101–140
51. Norris, C. L., Meisenheimer, K. M., and Koch, T. H. (1997). *Mechanistic studies relevant to bromouridine enhanced nucleoprotein photocrosslinking. Possible involvement of an excited tyrosine residue of the protein.* Photochem Photobiol 65, 201–207
52. Charlton, J., and Smith, D. (1999). *Estimation of SELEX Pool Size by Measurement of DNA Renaturation Rates.* RNA 5, 1326–1332
53. Moreno-Bondi, M. C., Alarie, J. P., and Vo-Dinh, T. (2003). *Multi–analyte analysis system using an antibody–based biochip.* Anal Bioanal Chem 375, 120–124
54. Schweitzer, B., Roberts, S., Grimwade, B., Shao, W., Wang, M., Fu, Q., Shu, Q., Laroche, I., Zhou, Z., Tchernev, V. T., et al. (2002). *Multiplexed protein profiling on microarrays by rolling–circle amplification.* Nat Biotechnol 20, 359–365

16

Biological Membrane Microarrays

Ye Fang, Anthony G. Frutos, Yulong Hong, and Joydeep Lahiri

16.1 Introduction

16.1.1 Importance of Membrane Bound Molecules

The cell membrane, in addition to providing a semipermeable barrier, is host to some of the most important molecules required for cellular function. These molecules can be classified from a molecular perspective into proteins (e.g. G protein-coupled receptors (GPCRs), receptor tyrosine kinases, ion–channels) and small molecules (e.g. glycolipids such as gangliosides and phosphatidylinositol phosphate (PIP)) [1]. Membrane–bound molecules comprise approximately 50% of all drug targets; methods to study these molecules in multiplexed, miniaturized formats are of significant interest to the pharmaceutical industry [2].

Protein profiling using protein microarrays will presumably circumvent issues associated with estimating protein abundance from mRNA levels using DNA microarrays [3–5]. There is an even more significant application for protein microarrays. Proteins are the molecules against which most drugs are designed; therefore, protein microarrays are uniquely well suited for directly determining compound binding and selectivity. In traditional drug discovery, compound libraries are tested against an identified 'target' to generate 'hits'; selectivity studies are carried out further downstream, during the progression of a 'hit' to a 'lead'. One of the primary outcomes of mRNA (or protein) profiling using DNA (or protein) microarrays will be more rapid identification of putative targets relative to conventional strategies. Therefore, technologies that enable target focused screening will become critical for keeping pace with the increased rate of target identification. Streamlining the process of drug discovery by bridging primary and secondary screening will be essential – protein microarrays, which offer selectivity information naturally, are ideally suited for meeting this challenge.

Protein arrays are difficult to fabricate because of issues related to maintaining the correctly folded conformations of proteins when immobilized.

The fabrication of membrane microarrays requires several unique considerations [6,7]. Unlike DNA or conventional protein arrays, fabricating microarrays of membranes requires the immobilization of the target and the associated lipids. Membranes on solid supports are unstable and highly susceptible to degradation when drawn through water–air interfaces [8]. This instability is undesirable as microarray based assays require immersion in different blocking and washing buffers to minimize non-specific binding. Since individual molecules are free to diffuse inside biological membranes ('the fluid mosaic model') [1], covalent immobilization of the entire supported membrane (or the embedded targets) is undesirable for the fabrication of 'biomimetic' membrane microarrays. Given these considerations, an ideal surface for membrane microarrays should seek to maximize the stability of the supported membrane while enabling lateral diffusion of individual molecules in the membrane. High stability and lateral fluidity are contradictory in nature; therefore, surfaces that balance these properties offer a practical compromise. Finally, membrane proteins contain extramembrane domains that must be correctly folded when immobilized at a surface; therefore, surfaces that offset the protein from the surface or those that are porous or deformable must be used [9].

16.1.2 Key Components of the Microarray Assay

A high quality microarray assay depends on optimization of each of the components that comprise the assay – the substrate with appropriate surface chemistry, high quality biological materials for printing, a reliable robotic printer, assay reagents, a high resolution fluorescence scanner, and finally, software for image analysis and informatics. The widespread use of DNA microarrays has resulted in the commercial availability of printers and fluorescence scanners. Due to their ready availability and ease of operation, we wanted to use these instruments for fabricating and reading membrane microarrays. However, previous work on supported membranes had emphasized the need to keep the supported membrane immersed in buffer (to prevent desorption) [8], which precluded both conventional pin–printing and scanning of slides using existing microarray scanners. Previously, membrane arrays were fabricated by immersion of patterned substrates containing lipid–binding and lipid–non binding regions in solutions of lipids. Although bioassays can be performed on such arrays by continuous flow methods, fabrication of arrays containing different immobilized membranes at different locations would require complicated fluidics [10]. Given these considerations, our research efforts were aimed at: (a) developing surface chemistries that resulted in supported membranes stable in air; (b) fabricating membrane microarrays by pin printing; (c) demonstrating the feasibility of printing membranes containing membrane proteins and ligands; and (d) developing assays for screening of compounds against membrane microarrays containing proteins or ligands.

16.1.3 Surface Chemistry

There are two general strategies for immobilization of membranes: (i) covalent or affinity-directed (e.g. streptavidin or lectin derivatized surfaces for biotin and glycosylated lipids (and proteins), respectively); and (ii) passive, non-covalent. The first approach will not be discussed in this review. There are currently two different classes of surfaces that enable the passive, non-covalent immobilization of membranes containing proteins – those presenting amphiphilic anchor molecules [11–13] and those presenting polymers that form deformable, porous surfaces (Fig. 16.1) [7, 14]. Our approach to identifying suitable surfaces involved a combination of rational surface chemistry and screening. For both approaches, we used 3 metrics to estimate the feasibility of using the surface for membrane microarrays: (i) mechanical stability of printed lipid spots as determined by the ability of printed lipid spots to resist desorption when drawn through buffer–air interfaces; (ii) long range fluidity of the supported lipids as determined by fluorescence recovery after photo-bleaching (FRAP) experiments [15] and (iii) 'functional incorporation' of membrane proteins as determined by biospecific ligand binding to membranes containing GPCRs. The choice of these metrics was based on what we felt were essential attributes for robust assays on membrane microarrays.

Raguse, Vogel and others have synthesized thiolated anchor lipids containing oligoethyleneoxide $(EG)_n$ moieties that help offset the supported membranes from the surface [11, 12]. The synthesis of these thiols is laborious and our efforts were aimed at fabricating similar surfaces using a common inter-

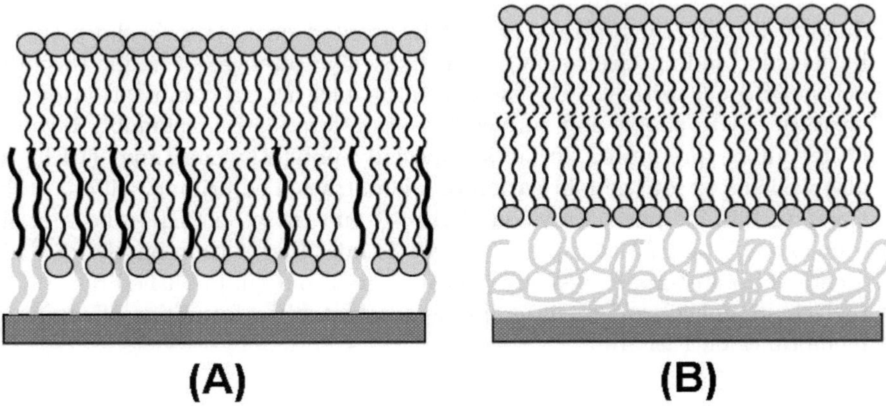

Fig. 16.1. Idealized representations of surfaces that offset supported membranes from the surface and enable the incorporation of the extramembrane domains of membrane bound proteins. **(a)** Surfaces presenting amphiphilic tethers offset the membrane by a distance determined by the length of the surface-attached hydrophilic tethers. **(b)** Surfaces that are porous and deformable can also accommodate the extramembrane domains of proteins bound to the membrane

mediate approach [16]. We made self-assembled monolayers (SAMs) of hexadecanoic acid that were activated to form interchain anhydride groups [17]; this activated surface was treated with Brij–76–amine to form the desired functionalized SAMs [13]. Arrays of supported lipids were obtained by immersion of chips containing patterns of Brij-derivatized SAMs in vesicular solutions of phosphatidycholine or by robotic pin printing of the lipids on an unpatterned Brij–presenting surface [18]. When immersed in buffer, the printed lipids stayed confined to the printed regions because of the self–limiting expansion of the lipid microspots [19]. When the lipids used were mixtures of dipalmitoylphosphatidylcholine (DPPC) and dimyristoylcholine (DMPC), the lipid microspots resisted desorption when withdrawn through air–water interfaces. Lipid microspots comprising egg–phosphatidylcholine (egg–PC) were, however, not stable on the Brij-derivatized SAMs. DPPC/DMPC lipids are in the gel phase at room temperature while egg PC is in the fluid phase; we are currently uncertain whether the phase of the lipid or issues with insertion of cis-unsaturated lipids in egg–PC causes this decreased stability. The instability of fluid phase lipid microspots on Brij was a concern and we decided to turn to a screening approach for evaluating lipid–binding surfaces.

We investigated the properties of lipids on several surfaces and found that those modified with γ–aminopropylsilane (GAPS™) had the desired properties [7, 9]. Specifically, microspots of both DPPC/DMPC and egg–PC remained stably associated with the surface even upon repeated withdrawals through buffer–air interfaces. Second, FRAP experiments revealed that supported lipids on GAPS™exhibited significant long-range lateral fluidity (approximately 50% was mobile, over the 30 minute course of the experiment). The GAPS™ surface therefore balances high mechanical stability and lateral fluidity. Finally, microarrays of GPCRs printed on GAPS™ slides showed biospecific binding (see below) to ligands. The physical basis for the interaction of lipids with GAPS™is currently unclear – a combination of electrostatic, hydrophobic and surface hydration forces are presumably involved. Other amine–presenting surfaces, especially poly(ethyleneimine) (PEI), are also well suited for the fabrication of microarrays The primary difference between membrane microarrays on GAPS™ and those on PEI is the spot size – microspots on PEI are approximately 3 times bigger than on GAPS™ for identical lipid compositions and printing conditions (unpublished results). Israelachvili and co-workers have also demonstrated the formation of supported membranes on PEI [14].

16.1.4 Pin Printing

Our first experiments for arraying membranes were carried out using a quill pin printer [13], and since the printing was successful, we have not investigated alternative printing technologies. We hypothesize that the use of alternative printers should be feasible, although there may be issues with using thermal ink jet printers that may denature proteins or cause phase transitions of the membrane.

The quill pin printer (Cartesian Technologies) is efficient and requires minimal amounts of materials for printing. A typical print run requires only a 10 μl volume of the membrane solution; each insertion of the pin reproducibly yields greater than 200 spots and at least 10 insertions of the pin into the membrane solution are possible before fresh solution needs to be added. Obtaining high-quality printing reproducibly has required a considerable amount of optimization work.

16.2 Biospecific Binding Studies Using Membrane Microarrays

Our primary objective in developing membrane microarrays is to test their use for screening compounds against membrane bound targets. To date, our research has focused on two types of membrane microarrays: (a) GPCR microarrays [7, 9]; and (b) ganglioside microarrays [20].

16.2.1 GPCR Microarrays

GPCRs are characterized by the presence of seven transmembrane helices, a glycosylated N–terminus and an intracellular C–terminus [21]. GPCRs mediate signal transduction through the binding of ligands to the extracellular side of the receptor, which leads to the activation of G proteins associated with the receptor on the intracellular side. GPCRs are extremely important pharmacological targets – 25% of the 100 top–selling drugs target GPCRs [22]. There are an estimated 400–700 GPCRs, approximately 200 of which have known ligands; GPCRs with unknown ligands, termed "orphan receptors", are also presumed to be key pharmacological targets. GPCRs can be classified into three major families: family A (rhodopsin or adrenergic receptor like family) characterized by short N–terminal tails and conserved amino acid residues within each transmembrane helix, family B (glucagons or secretin receptor like family) characterized by longer N–terminal tails and six conserved cysteine residues, and family C (metabotropic glutamate receptors) characterized by very long N–terminal tails (500–600 residues) folded as separate ligand binding domains.

GPCRs were obtained as membrane-associated suspensions in buffer from commercial vendors (Biosignal Packard or Perkin Elmer Life Sciences). Microarrays were made by printing the receptors on GAPS™ coated slides. In a typical experiment, each array was incubated with 10 μL of a solution containing labelled ligands or mixtures of the labelled ligand and unlabelled compounds for competitive binding assays. After incubation for 1 hour, the solution was carefully removed with a pipette tip attached to a vacuum pump. The slides were briefly rinsed with water, dried under a stream of nitrogen,

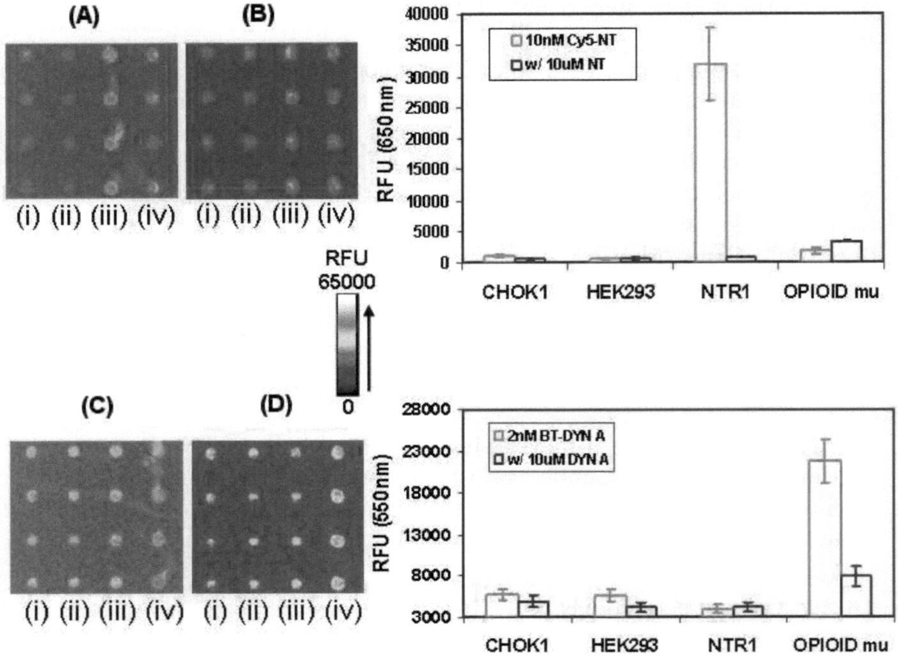

Fig. 16.2. Fluorescence images (in false color) of microarrays consisting of (from L to R) membranes from CHOK1 (i) and HEK 293 (ii) cells, and membranes containing the NTR1 (iii) and μ opioid (iv) receptors. **(a)** Image of array upon treatment with a solution containing cy5–NT (10 nM). **(b)** Image of array upon treatment with a solution containing cy5–NT (10 nM) and excess unlabelled neurotensin (10 μM). **(c)** Image of array upon treatment with BT–dyn A (2 nM). **(d)** Image of array treated with a solution containing BT–dyn A (2 nM) and excess unlabelled dynorphin (10 μM). Histograms corresponding to the amounts of binding and inhibition are shown alongside. RFU = relative fluorescence units

and imaged using a fluorescence scanner. Our experiments were designed to test: (a) the specificity of binding; (b) the selectivity of binding; and (c) the dose–dependency of binding and estimations of the binding constant.

16.2.2 Specificity of Binding

Our initial experiments were designed to test whether immobilized GPCRs retained their native specificity. Figure 16.2 shows fluorescence images of an array containing the neurotensin receptor (NTR1), the opioid receptor (μ), and membrane preparations from the cell lines used as negative controls (CHOK1 and HEK293). Figure 16.2(a, b) shows images of the array treated with solutions of fluorescently labelled neurotensin (cy5–NT) and a solution containing cy5–NT and excess unlabelled neurotensin. Neurotensin is the cognate ligand for NTR1 (Kd ∼ 1 nM). Binding of the labelled analog occurs only to microspots of NTR1; nearly complete inhibition is observed when excess neurotensin is present in the sample. These data (plotted as a histogram on the side) demonstrate that binding and inhibition are specific. Figures 16.2c and 16.2d show images of arrays that were incubated with solutions containing fluorescent dynorphin (BT–dyn), a labelled analog of dynorphin that is known to bind to the opioid receptor, and a mixture containing BT–dyn and excess unlabelled dynorphin. The highest amount of binding of BT–dyn is observed for the opioid receptor although binding, probably non-specific in nature, is observed for the other microspots. Strong inhibition is observed when unlabelled dynorphin is present in excess in the sample [23]. Taken together, these data suggest that arrayed GPCRs on GAPS^{TM} retain their native ligand specificity.

16.2.3 Selectivity of Binding

Since arrays offer selectivity information naturally, GPCR arrays of arbitrarily different structure or ligand binding specificity can be fabricated to provide

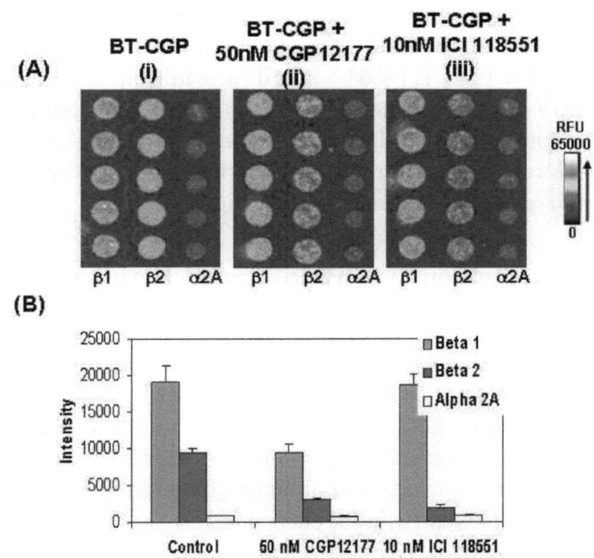

Fig. 16.3. **(a)** Fluorescence images of arrays of the β1, β2, and α2A adrenergic receptors treated with solutions containing: (i) BT–CGP12177 (5 nM); (ii) a mixture of BT–CGP12177 (5 nM) and CGP12177 (50 nM); and (iii) a mixture of BT–CGP12177 (5 nM) and ICI 118551 (10 nM). **(b)** Histograms of the data showing the non-selectivity of inhibition between the β1 and β2 receptors when CGP12177 is used as the competitive ligand and the selectivity of inhibition when ICI 118551 is used instead

information about compound design over an arbitrarily broad or narrow biological target space. While it is difficult but possible to design an inhibitor against a known GPCR, it is almost impossible to predict the pharmacological effects of that compound against other GPCRs without screening experiments. Choosing the appropriate biological target space over which to scan is equally important. For example, an antagonist chosen for being selective for the dopamine D4 receptor relative to the D2 receptor for treatment of schizophrenia was also found to be moderately potent with respect to the α1 adrenergic receptor [24]. Mutiplexed target screening is clearly essential for increasing the efficiency of discovering potent drugs without side effects – expression analysis using DNA or protein microarrays may be valuable in this regard by highlighting multiple potential targets for a given disease state and thereby enabling the design of an appropriate GPCR array.

We fabricated arrays of the adrenergic receptor (β1, β2, and α2A) to test the feasibility of using GPCR microarrays for selectivity screening. Figure 16.3a shows fluorescence images of these arrays treated with fluorescently labelled CGP12177 (BT–CGP12177), a known cognate antagonist selective for β–type adrenergic receptors. Binding occurs only to microspots corresponding to the β–type receptors. When the array is treated with a mixture containing BT–CGP12177 and unlabelled CGP12177, inhibition of binding to both the receptors is observed, which suggests that the compound has no significant selectivity between the b1 and b2 receptors. Figure 16.3c(iii) shows images of the array treated with ICI118551 – significant inhibition of binding to only the β2 receptors are observed. These data suggests that the compound is selective for the β2 receptor, in accordance with the known affinities of ICI118551 for the β1 and β2 receptors. Moreover, they demonstrate the potential of using GPCR microarrays for compound screening.

16.2.4 Dose Dependency of Binding and Estimations of the Binding Constant

A possible issue with protein microarrays is whether they can be used to provide information about the binding affinities of compounds. These estimations require measurements of small changes in the signal as a function of the compound concentration, which can be tricky to measure for an array of immobilized proteins. Despite the obvious advantages of obtaining compound affinities in a multiplexed fashion, there are few reports that demonstrate the use of protein arrays for measuring binding constants. An additional complication is that the affinity of ligands for GPCRs depends on whether the receptor is complexed to the G–protein [21]. The concern is that there may be changes in the fraction of GPCR–G protein complexes during immobilization, which can significantly impact estimations of the binding constant.

Figure 16.4 shows fluorescence images of arrays of the adrenergic receptor treated with BT–CGP12177 (Figure 16.4A) and mixtures containing BT–CGP12177 and excess unlabelled CGP12177 (Figure 16.4B). The amount of

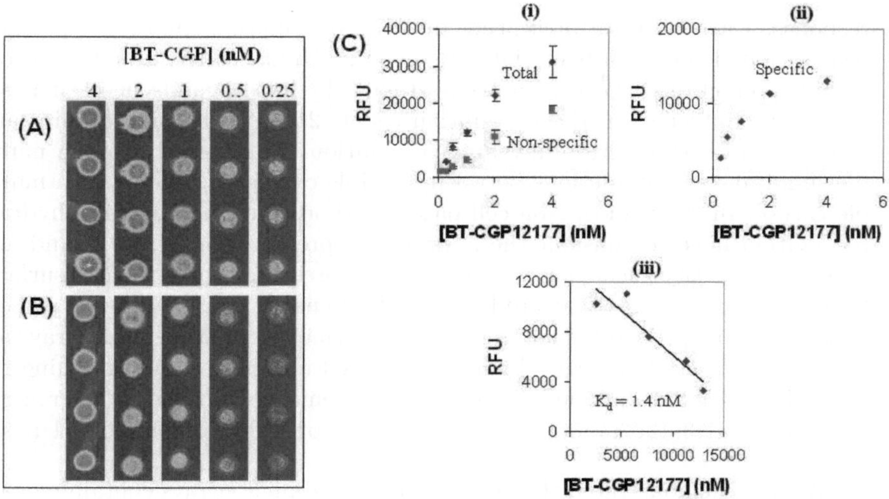

Fig. 16.4. **(a)** Profile for the binding of BT–CGP12177 to arrays of the β1 adrenergic receptor obtained by treating the array with different concentrations of the labelled ligand (0.25–4 nM). **(b)** Profile for the amount of non-specific binding of BT–CGP12177 to arrays of the β1 adrenergic receptor as a function of its concentration, estimated by the fluorescence signals observed at each concentration of BT–CGP12177 in the presence of excess CGP–12177. **(c)** (i) Plots of the amounts of total and non-specific binding as a function of the BT–CGP12177 concentration. (ii) Plot of the amount of specific binding obtained as the difference between the signals corresponding to total and non-specific binding. (iii) Scatchard analysis for binding of BT–CGP12177 to arrays of the β1 receptor

specific binding at each concentration of BTCGP12177 is determined by subtracting the fluorescence signal in the presence of excess CGP12177 (Figure 16.4c). Using Scatchard analysis, we estimate $K_d \sim 1.4$ nM. This value is similar to that obtained using other techniques, which suggests that the fraction of GPCR–G protein complexes is not significantly changed upon immobilization.

While it is possible that there are discrepancies between the affinities of compounds obtained using GPCR microarrays and conventional methods, the data obtained to date on several receptor–ligand systems has shown strong agreement. Therefore, information about compound potency can be obtained using GPCR microarrays. Importantly, the ability to make these precise measurements highlights the robustness of the GPCR microarray platform.

16.2.5 Ganglioside Microarrays

Carbohydrates appended to lipids are a key component of the cell membrane and play a role in vital processes such as cell adhesion and the immune response. Carbohydrate presenting lipids also comprise one of the primary recognition elements of bacterial pathogenesis. Unlike conventional receptor–ligand interactions, the presentation of carbohydrate ligands in itself has a significant influence on the recognition event [25, 26]. The high affinity and specificity of carbohydrate mediated recognition are achieved through multiple simultaneous interactions between multiple copies of proteins with multiple carbohydrate ligands. Non-cell based methods for studying carbohydrate recognition have to consider the appropriate presentation of the ligand and its surface density such that it mimics ligand presentation at the cell surface. Supported membranes that are laterally fluid enable this biomimetic presentation enabling processes such as ligand clustering. Membrane microarrays are well suited for studying carbohydrate mediated recognition by combining the multiplexing ability, miniaturization and convenience afforded by microarray technology with the biomimetic environment provided by supported membranes.

We have demonstrated the fabrication of lipid microarrays containing gangliosides and described their use for detecting bacterial toxins and for the screening of potential inhibitors [20]. Gangliosides are a class of carbohydrate derivatized lipids that comprise approximately 5–10% of the lipid composition of the plasma membrane of neuronal and glial cells. The interaction of the cholera and tetanus toxins with the GM1 and GT1b gangliosides, respectively, are two well-studied examples of ganglioside–toxin interactions.

Microarrays of gangliosides were made by printing sonicated dispersions of dilaurylphosphatidylcholine (DLPC) containing gangliosides (4 mol%). Figures 16.5a–f show fluorescence images of these arrays treated with solutions of toxins. When the array is treated with a solution of fluorescently labelled cholera toxin (FITC–CTx) (Fig. 16.5b) or the tetanus toxin (FITC–TTx) (Fig. 16.5c), strong fluorescence is observed from microspots containing the GM1 and GT1b gangliosides, respectively. Specific inhibition of binding of FITC–CTx to GM1 microspots is observed when the solution contains excess unlabelled cholera toxin (compare Figs. 16.5d, e, f). This inhibition is dose dependent and yields an IC50 value of $\sim$ 20 nM (data not shown).

These studies demonstrate the use of membrane microarrays for the multiplexed detection of toxins and the screening of potential inhibitors. The development of membrane microarrays for this application is especially pertinent given the recent concerns about biological warfare and the emergence of bacterial resistance to antibiotics.

16.3 Conclusions

Molecules in the membrane direct events both inside the cell and between cells, and there is hardly any aspect of cell viability that is not influenced by recognition events at the cell membrane. It is therefore not surprising

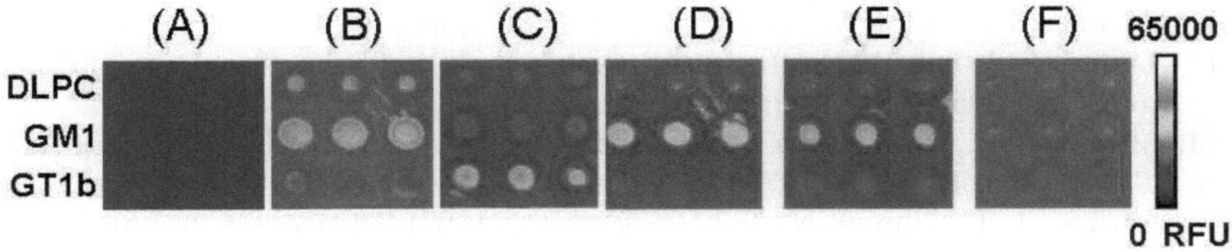

Fig. 16.5. Fluorescence images of ganglioside microarrays showing binding of labelled toxins, and estimations of inhibition using mixtures of the labelled toxins and potential unlabelled inhibitors. Each array consists of DLPC microspots (top row), DLPC and GM1 (4 mol%) (middle row), and DLPC and GT1b (bottom row). **(a)** Image of array treated with buffer only. **(b)** Image of array treated with fluorescently labelled cholera toxin (FITC–CTx) (1 nM). **(c)** Image of array treated with fluorescently labelled tetanus toxin (FITC–TTx) (2 nM). **(d)** Image of array treated with a mixture containing FITC–CTx (1 nM) and unlabelled tetanus toxin (100 nM). **(e)** Image of array treated with a mixture containing FITC–CTx (1 nM) and unlabelled bungarotoxin (100 nM). **(f)** Image of array treated with a mixture containing FITC–CTx (1 nM) and unlabelled cholera toxin (100 nM)

that membrane bound molecules constitute nearly half of current drug targets. GPCR and ganglioside microarrays are but two examples of membrane microarrays; based on our current learnings, fabricating other types of membrane microarrays (e.g. microarrays of receptor tyrosine kinases, ion channels, etc) should be feasible. Since the user has control of the membrane composition, membrane arrays of any arbitrary composition can be fabricated, which may enable, beyond compound screening, studies of fundamental aspects of biomolecular recognition at surfaces.

References

1. Lodish, H., Baltimore, D., Berk, A., Zipursky, S. L., Matsudaria, P. and Darnell, J. (1997) Molecular Cell Biology (Scientific American Books, Inc., Oxford)
2. Subrahmanyam, S., Piletsky, S. A. and Turner, A. P. F. (2002) Anal. Chem. 74, 3942–3951
3. MacBeath, G. and Schreiber, S. L. (2000) Science 289, 1760–1761
4. Zhu, H., Bilgin, M., Bangham, R., Hall, D., Casamayor, A., Bertone, P., Lan, N., Jansen, R., Bidlingmaier, S., Houfek, T., Mitchell, T., Miller, P., Dean, R. A., Gerstein, M. and Snyder, M. (2001) Science 293, 2101–2105
5. Zhu, H. and Snyder, M. (2001) Curr. Opin. Chem. Biol. 5, 40–45
6. Mitchell, P. (2002) Nature Biotecnol. 20, 225–229
7. Fang, Y., Frutos, A. G. and Lahiri, J. (2002) J. Am. Chem. Soc. 124, 2394–2395
8. Cremer, P. S. and Boxer, S. G. (1999) J. Phys. Chem. B 103, 2554–2559
9. Fang, Y., Frutos, A. G. and Lahiri, J. (2002) ChemBioChem 3, 987–991
10. Cremer, P. S. and Yang, T. (1999) J. Am. Chem. Soc. 121, 8130–8131
11. Lang, H., Duschl, C. and Vogel, H. (1994) Langmuir 10, 197–210

12. Raguse, B., Braach–Maksvytis, V., Cornell, B. A., King, L. G., Osman, P. D. J., Pace, R. J. and Wieczorek, L. (1998) Langmuir 14, 648–659
13. Lahiri, J., Kalal, P., Frutos, A. G., Jonas, S. J. and Schaeffler, R. (2000) Langmuir 16, 7805–7810
14. Majewski, J., Wong, J. Y., Park, C. K., Seitz, M., Israelachvili, J. and Smith, G. S. (1998) Biophys. J. 75, 2363–2367
15. Groves, J. T., Ulman, N. and Boxer, S. G. (1997) Science 275, 651–653
16. Lahiri, J., Isaacs, L., Tien, J. and Whitesides, G. M. (1999) Anal. Chem. 71, 777–790
17. Yan, L., Marzolin, C., Terfort, A. and Whitesides, G. M. (1997) Langmuir 13, 6704–6712
18. Lahiri, J., Jonas, S. J., Frutos, A. G., Kalal, P. and Fang, Y. (2001) Biomed. Microdevices 3, 157–164
19. Hovis, J. and Boxer, S. G. (2000) Langmuir 16, 894–897
20. Fang, Y., Frutos, A. G. and Lahiri, J. Langmuir, (2003) Langmuir 19, 1500-1505
21. Haga, T. and Berstein, G. (1999) in CRC Methods in Signal Transduction Series (CRC Press, Boca Raton, FL)
22. Klabunde, T. and Hessler, G. (2002) ChemBioChem 3, 928–944
23. Bell, K. M. and Traynor, J. R. (1998) Can. J. Physiol. Pharmacol. 76, 325–333
24. Hendrix, J. A., Shimshock, S. J., Shutske, G. M., Tomer IV, J. D., Kapples, K. J., Palermo, M. G., Corbett, T. J., Vargas, H. M., Kafka, S., Brooks, K. M., Laws-Ricker, L., Leee, D. K. H., Lannoy, I. d., Bordeleau, M., Rizkalla, G., Owolabi, J. and Kamboj, R. K. (2002) ChemBioChem 3, 999–1009
25. Mammen, M., Seok-Ki, C. and Whitesides, G. M. (1998) Angew. Chem. Int. Ed. 37, 2755–2794
26. Liang, R., Loebach, J., Horan, N., Ge, M., Thompson, C., Yan, L. and Kahne, D. (1997) Proc. Natl. Acad. Sci. U. S. A. 94, 10554–10559

Part IV

Cell & Tissue Microarrays

17

Use of Reporter Systems for Reverse Transfection Cell Arrays

Brian L. Webb

17.1 Introduction

The ability to transfer exogenous recombinant genes into cultured mammalian cells has revolutionized the study of gene function and gene regulation [1]. Originally, the ability of viruses to transmit their genetic material across the plasma membrane of target cells was exploited as the means to shuttle desired genes into cells. Due to the highly efficient nature of viral infection, highjacking the viral genome with a desired recombinant gene of interest results in expression of the desired protein in nearly all target cells [2]. However, the multi–step process required to develop recombinant viruses as well as biosafety issues led to the development of more convenient means of gene transfer. A variety of DNA transfection methods were the result. One method involves the use of diethylaminoethyl (DEAE)–dextran, a positively charged dextran molecule that interacts with the negatively charged phosphate backbone of DNA. DNA–DEAE dextran complexes can adsorb onto the cell surface and can be taken up by endocytosis, leading to the in vivo expression of the target gene [3]. Another method involves mixing calcium chloride, DNA, and phosphate buffer to produce small, insoluble particles of calcium phosphate containing entrapped DNA [4–6]. These DNA–calcium phosphate complexes settle onto adherent cultured cells and are taken up by phagocytosis. Perhaps the easiest and thus most popular transfection method to date involves using cationic lipid reagents [7–9]. Cationic lipids, such as Lipofectamine, form unilamellar vesicles in an aqueous environment [10]. Positively charged cationic lipid vesicles bind to negatively charged DNA, forming liposome–DNA complexes. These complexes can be taken up by mammalian cells by endocytosis. Thus, conventional transfections are performed by mixing DNA with a transfection reagent to form DNA complexes and then adding these complexes onto target cells attached to a growth support surface. Optimization of lipid compositions have yielded lipid reagents with low toxicity and high transfection efficiencies in a wide range of eukaryotic cells.

Expression of exogenous genes using DNA transfection has enabled the study of gene function in vivo. For example, the function of unknown genes can be discovered by examining the effect of their overexpression in transfected cells using a variety of cell-based assays. This approach has led to the identification of many novel drug targets. However, the incredible speed of gene cloning and sequencing brought about by the genomic revolution has outpaced conventional gene discovery approaches in the pharmaceutical industry. One potential answer to this challenge is reverse transfection, a high throughput gene expression method for examining the function of hundreds to thousands of genes in parallel.

The notion of performing surface-mediated transfection was first described by Paulson et al. [11]. As contact between DNA and the target cells is a requirement for successful transfection, Paulson and coworkers suggested immobilizing the DNA particles onto a cell growth surface prior to attaching the target cells. Subsequent addition and attachment of target cells to the DNA-loaded surface can lead to higher probability of cell–DNA contact, potentially leading to higher transfection efficiencies. More recently, the appeal of performing DNA transfections off a solid surface for gene therapy applications has lead to numerous reports of surface-mediated transfection, on such surfaces as biodegradable polymers and modified silica nanoparticles [12,13]. Two groups recognized the potential of merging surface-mediated transfection technology with DNA microarray technology. Genova Pharmaceuticals filed a patent application on a method of simultaneously screening large numbers of genes using surface-mediated transfection of arrayed libraries of cDNAs [14]. Immobilization of individual cDNA clones in unique locations on a surface was achieved using hybridization to arrayed oligo linkers. Simultaneous transfection of the arrayed cDNA library would thus generate patches of transfected cells which could be screened for any desired gene function using cell-based or biochemical assays.

A more straightforward immobilization approach was described by Zhaudin and Sabatini, who coined the phrase "reverse transfection" to describe surface-mediated transfection of cDNAs spotted in an array format on a cell growth surface by a conventional arrayer [15]. Following treatment with a transfection reagent, the surface is overlayed with adherent cells, which become transfected in patches with the various cDNAs. The term "reverse" was used because the order of addition of the target cells and DNA to the surface is reversed compared to conventional transfection techniques. Although a uniform lawn of mammalian cells is cultured on the array surface, only those cells that adhere to the spots of arrayed DNA become transfected, producing localized patches of transfected cells each expressing a unique protein. As with the Genova method, the array of transfected cell clusters produced by reverse transfection can be used for high throughput analysis of gene function.

Transfection cell arrays can be viewed as specialized protein microarrays, with several key advantages. First, the proteins to be studied can be expressed and characterized in their native cellular environment as opposed to being isolated proteins immobilized on a surface. Second, proteins which may be difficult to purify, such as membrane-associated proteins, can be studied using transfection arrays. And third, the shelf life of transfection arrays following fabrication is very long, since they are essentially immobilized DNA spots until subsequent addition of mammalian cells, compared to the uncertain stability of immobilized purified proteins.

Zhiauddin and Sabatini have published a detailed protocol for producing transfection arrays, which is available on the internet (http://staffa.wi.mit.edu/sabatini_public/reverse_transfection.htm). One limitation of this reverse transfection technology as described by Zhiauddin and Sabatini is the need for extensive post-transfection processing of the array to detect protein activity, including fixing and permeabilizing the cells and multiple antibody incubation steps. At Corning we have investigated the use of reporter constructs co-transfected along with other genes of interest as a convenient means to monitor and screen gene function on reverse transfection microarrays. Reporter systems are commonly used for conventional transfections as a means to monitor the activity of transfected proteins. We have demonstrated the usefulness of reporter systems for assessing the activity of putative signaling proteins produced by reverse transfection. Thus, the focus of this chapter will be a description of how to use reporter constructs for reverse transfection microarray assays. The reverse transfection protocol we use is essentially that described by Zhiauddin and Sabatini except that it was modified to include the co-transfection of a reporter plasmid.

17.2 Reporter Systems for Reverse Transfection

Signal transduction is essential for cellular proliferation, differentiation, and regulation of key cellular activities inside the cell. It is the process by which extracellular signals are transmitted through the membrane via receptors into the nucleus to trigger transcriptional responses. Enhancer elements within promoters are the convergent points for the majority of signal transduction pathways. AP–1, CRE, SRE, NF–kB and SRF binding elements are examples of enhancer elements contained within promoters that are responsive to various signaling pathways [16]. Incorporation of these elements into reporter systems represents a simple and rapid means for assessment of the in vivo activation of these pathways. A host of reporters linked to enhancer elements have been developed, including luciferase, secreted alkaline phosphatase, chloramphenicol acetyltransferase (CAT), β–galactosidase, and green fluorescent protein (GFP) [17]. Assays are performed by co-transfection of a reporter with a gene of interest into a target cell line. Activation of the reporter indicates involvement of the gene of interest in that particular signaling pathway. Thus, the activity of unknown genes can be screened conveniently using reporter activation as a read-out.

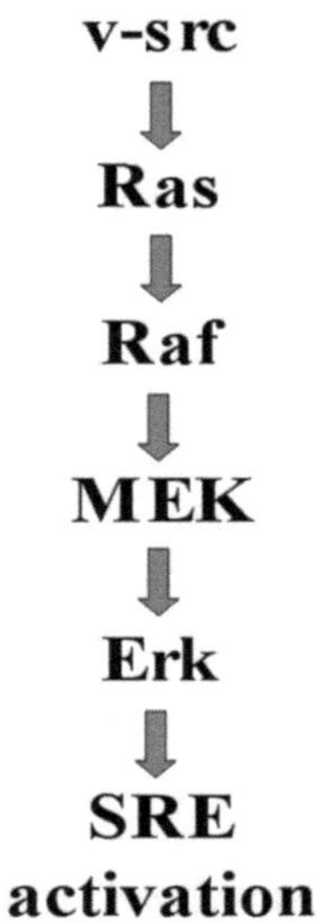

Fig. 17.1. MAP kinase signaling pathway. Activation of the MAP kinase signaling pathway by the oncogene v–src leads to transcriptional activation of genes containing the SRE enhancer element in their promoters

We have used GFP linked to the SRE enhancer element as a model system for studying MAP kinase signaling on cell transfection arrays. MAP kinases are rapidly phophorylated and activated in response to various extracellular stimuli, such as certain growth factors [18]. Activation of the MAP kinase Erk by an upstream signaling cascade ultimately leads to transcriptional activation of promoters containing an SRE enhancer element, as shown in Fig. 17.1. We developed an SRE reporter linked to the GFP protein to demonstrate the usefulness of reporter systems to monitor the activity of MAP kinase signaling proteins produced by reverse transfection. As GFP–SRE reporter systems are not commercially available, we cloned the GFP gene into the pSRE–Luc vector, swapping the GFP gene for the luciferase gene. The resulting pSRE–GFP plasmid produces GFP protein in response to SRE activation.

The activated mutants of three different genes involved in the MAP kinase signaling pathway known to activate the SRE (v–src, RasV12, and Raf–CAAX) were used to test this reporter system. Conventional co-transfection experiments performed in HEK293 cells indicated specific activation of the pSRE–GFP reporter by all three of the activated signaling genes (Fig. 17.2). Activation was assessed by GFP protein expression using fluorescence microscopy. Very little GFP signal was seen in control cells co-transfected with a control vector and the pSRE–GFP reporter, indicating low basal SRE activation in these cells. Strong GFP expression was induced by all of the three activated genes, demonstrating the utility of the GFP reporter for monitoring SRE activation.

Fabrication of the reverse transfection arrays was performed essentially according to the protocol of Zhiauddin and Sabatini [15]. Briefly, plasmid DNAs at the indicated concentrations were mixed with gelatin (final concen-

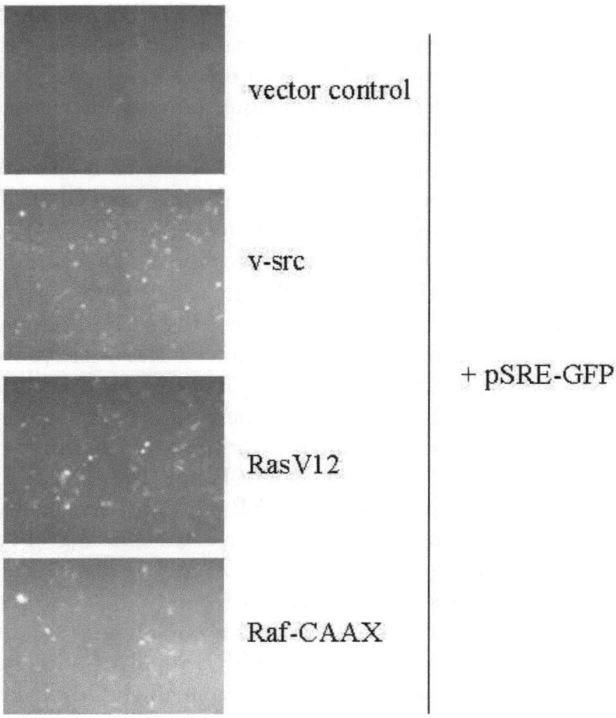

Fig. 17.2. Activation of pSRE–GFP reporter by mutationally-activated MAP kinase pathway signaling proteins. The pSRE–GFP reporter construct was generated by linking the green fluorescence gene to the SRE enhancer element. Conventional co-transfection experiments were performed using pSRE–GFP along with DNAs encoding for three activated mutant signaling proteins, v–src, RasV12, and Raf–CAAX in HEK293 cells. Following 48 hours, GFP-producing cells were visualized using fluorescence microscopy

tration of 0.2%). DNA/gelatin solutions were printed in an array format on Corning GAPS^{TM} slides using a Cartesian PixSys 5500 printer. The printed slides were dried in a vacuum dessicator for two hours. Effectine transfection reagent for each slide was prepared by mixing 150 µl EC Buffer, 16 µl Enhancer, and 25 µl Effectine transfection reagent in a 1.5 ml micro–centrifuge tube. This solution was added to a CoverWell Incubation Chambers (Grace BioLabs catalog #PC200) and the slide was pressed down onto the CoverWell Chamber, sealing the transfection reagent between the slide and the chamber. Incubation of the array with the transfection reagent between 15–20 minutes is optimal. Following the incubation, the CoverWell was peeled off the slide, excess reagent removed from the slide, and the slide was placed in a Quad-Perm cell culture device. During the Effectine incubation, HEK293 cells were prepared as follows. HEK293 cells grown in T75 flasks were trypsinized, resuspended in Iscove's DMEM media containing 10% FBS, 50 units/ml penicillin and 50 μg/ml streptomycin, and counted using a Coulter Counter. Then, 5–7.5 $\times 10^6$ HEK293 cells resuspended in 10 mL media were carefully added to each well of the QuadriPerm chamber containing an Effectine-treated slide and incubated at 37°C. Typically patches of transfected cells can first be detected after 16–24 hours and are assayed after 48–72 hours. To validate the reverse transfection protocol, a plasmid encoding the GFP gene under the strong constitutively–expressing CMV promoter (pQBI25–fPA) was printed on a GAPS^{TM} slide in an array as described above. This plasmid mixed with gelatin (0.2%) was printed using 3 different sized microarray pins to determine the optimal pin size for printing reverse transfection arrays. After incubation (36–48 hours) to allow for expression of the GFP protein, the slides were scanned on a GenePix4000B scanner. As shown in Fig. 17.3, the number of transfected cells within each 'patch' increased as the pin size increased, with the most uniform patches having the greatest number of successfully transfected cells occurring with the CMP10B pin (Fig. 17.3a). Using this pin, patches of 30–50 cells expressing the GFP protein were consistently visible within 48 hours, indicating successful reverse transfection (Fig. 17.3b shows a high magnification image of one CMP10B patch). Therefore, the CMP10B pin was used for all subsequent experiments.

Reverse transfection arrays with plasmids encoding v–src and Raf–CAAX were produced first in the absence of the SRE reporter and conventional immunofluorescence techniques were used to confirm the MAP kinase signaling activity of these two mutationally activated proteins. Multiple replicate spots of each of these two DNAs were printed on the array. Following 48 hours incubation to allow expression of the arrayed genes, the levels of phosphotyrosine and phosphorylated Erk within the transfected cells were assayed using conventional immunofluorescence techniques. To do this, the media was removed from the cells, the cells were washed 2× with PBS, fixed for 10 minutes with 4% formaldehyde, washed 3× with PBS, and permeabilized for 5 minutes with 0.2% Triton X–100. The fixed cells were then blocked with PBS/10% goat serum for 30 minutes to reduce non-specific antibody binding. To evaluate the phosphotyrosine levels in the transfected cells, one set of slides was incubated with a phosphotyrosine specific antibody followed by a Cy3-labelled goat anti-mouse secondary antibody. As shown in Fig. 17.4a, the cells in the patches transfected with v–src displayed significantly elevated levels of phosphotyrosine, consistent with the overexpression of the tyrosine kinase v–src. A higher magnification fluorescent microscope image of one v–src transfected cell patch is shown in Fig. 17.4a. Neither the cells transfected with a control vector nor those transfected with the Raf–CAAX construct displayed phospho–tyrosine antibody staining above background levels. This indicates that expression of v–src by reverse transfection produces functional v–src protein with tyrosine kinase activity.

To determine if the MAP kinase pathway was activated by these overexpressed signaling proteins (Fig. 17.1), the levels of activated, phosphory-

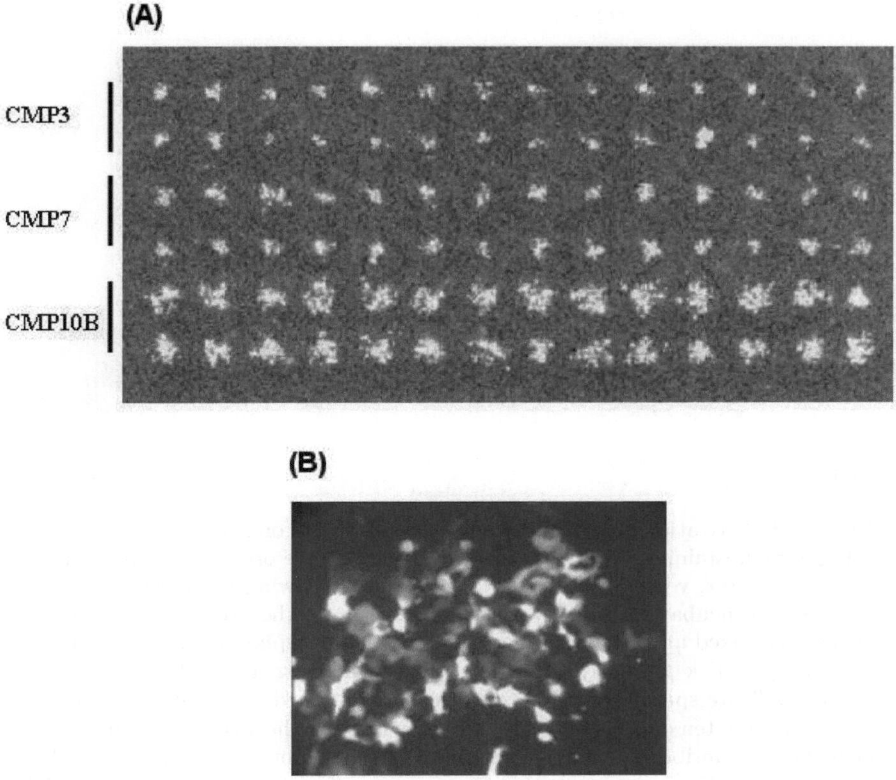

Fig. 17.3. Comparison of pin sizes for producing reverse transfection arrays. **(a)** A plasmid encoding for GFP under the strong constitutive CMV promoter, pQBI25-fPA (0.05 μg/μl), was mixed with gelatin (0.2%) and printed on a Corning GAPS^{TM} slide using either a CMP3, CMP7, or CMP10B pin. Two rows of 14 duplicate spots were printed using each pin. The slide was treated with Effectine reagent followed by the addition of HEK293 cells, as described in the text. The cells were fixed and imaged on a GenePix 4000B scanner after 48 hours. **(b)** A representative 'cluster' of cells expressing GFP protein printed with the CMP10B pin is shown at higher magnification (40×)

lated Erk were assayed using a phospho–Erk antibody (detected using a Cy3-labelled goat anti-rabbit secondary antibody). Patches of cells transfected with v–src displayed high levels of phospho–Erk staining, indicating significant activation of the MAP kinase pathway by v–src (Fig. 17.4b). Again, a higher magnification image of one v–src transfected cell patch stained with anti-phospho–Erk antibody is shown in Fig. 17.4b. The cell patches transfected with Raf–CAAX also showed elevated levels of phospo–Erk staining compared to the vector control cells, though the extent was much less than seen with v–src (Fig. 17.4a). Thus, activation of the MAP kinase pathway by

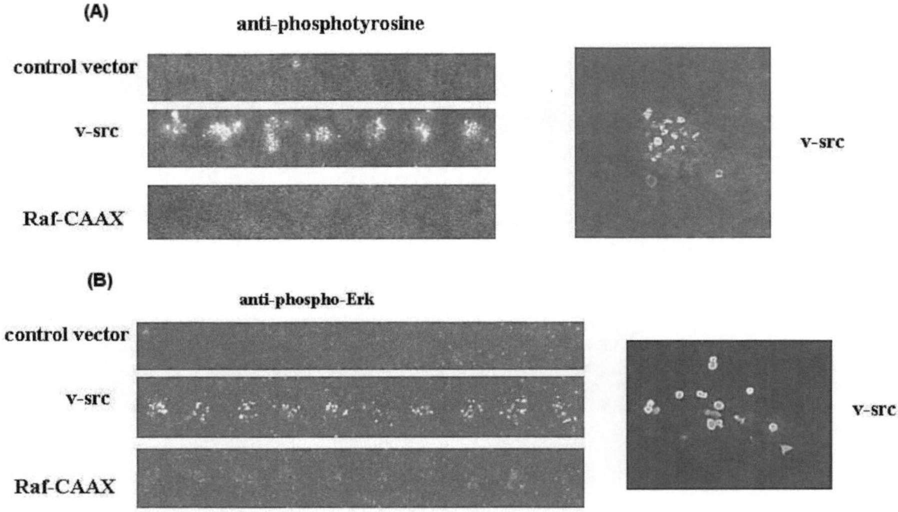

Fig. 17.4. Activation of MAP kinase pathway detected on reverse transfection array using immunostaining. Reverse transfection arrays were produced by printing either a control vector, v–src, or Raf–CAAX plasmids. Following the reverse transfection process and incubation to allow for the expression of the proteins, the cells on the arrays were fixed and stained either with **(a)** anti-phosphotyrosine antibody or **(b)** anti-phospho–Erk antibody. Slides were scanned using a GenePix4000B scanner. Seven duplicate spots are shown from the array stained with anti-phosphotyrosine antibody and ten duplicate spots are shown from the array stained with anti-phospho–Erk antibody. A higher magnification image of a representative cell patch transfected with v–src taken with a fluorescence microscope is shown to the right of each panel

both v–src and Raf–CAAX can be detected on a reverse transfection array using conventional immunostaining.

This system was then used to demonstrate the convenience of using co-transfections of the pSRE–GFP reporter on reverse transfection cell arrays. Reverse transfection arrays were printed with a mixture of Raf–CAAX and pSRE–GFP DNA. The ratio of reporter construct to gene–of–interest construct used for conventional reporter transfection experiments is typically 1:10, ensuring that each cell transfected with a reporter construct also receives the second gene construct. To illustrate the optimal ratio for reporter transfection arrays, a titration experiment was performed using various amounts of the reporter construct pSRE–GFP and the construct encoding for Raf–CAAX (Fig. 17.5). For establishing the position of each cell cluster within the array, a row of 10 duplicate spots of constitutively expressed CMV–promoter driven GFP vector (pQBI25–fPA) was printed at the top and the bottom of the array. In between these border rows were printed spots of mixtures of pcDNA3–Raf–CAAX vector and pSRE–GFP reporter vector at the indicated

(A)

pSRE-GFP (µg/µL)	pCDNA3-Raf-CAAX (µg/µL)
0 (vector)	0.1
0.01	0
0.01	0.01
0.01	0.025
0.01	0.05
0.01	0.10
0.025	0
0.025	0.01
0.025	0.025
0.025	0.05
0.025	0.10
0.05	0
0.05	0.01
0.05	0.025
0.05	0.05
0.05	0..1
0.10	0
0.10	0.01
0.10	0.025
0.10	0.05
0.10	0.1

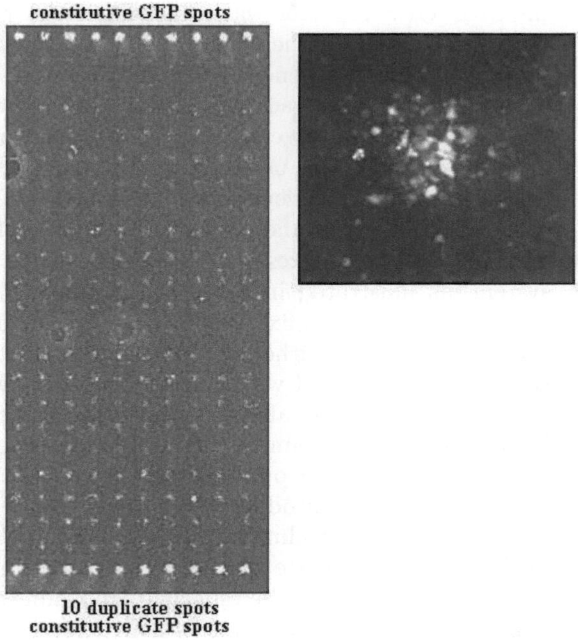

Fig. 17.5. SRE activation detected using a SRE–GFP reporter co-transfected with Raf–CAAX on a reverse transfection array. **(a)** To determine the optimal concentration of pSRE–GFP reporter and the pcDNA3–Raf–CAAX DNA, a titration experiment was performed using various amounts of each, as indicated. For establishing the position of each cell cluster within the array, a row of ten duplicate spots of constitutively ex-pressed CMV–promoter driven GFP vector (pCMV–GFP) was printed at the top and the bottom of the array. Following reverse transfection, the resulting array was imaged without fixing the cells on a GenePix4000B scanner. Activation of the SRE by Raf–CAAX was clearly detected by the production of the GFP protein in transfected cells. **(b)** A higher magnification image of one patch of cells co-transfected with 0.10 µg/µL pSRE–GFP and 0.025 µg/µL pcDNA–Raf–CAAX obtained using a fluorescence microscope is shown

concentrations. Following reverse transfection, the constitutively expressed GFP vector border spots produced cell clusters of GFP–expressing cells. The patches of cells co-transfected with pSRE–GFP and the control pcDNA3 vector displayed very little GFP fluorescent signal, indicating low background SRE activation. As seen in the conventional transfections, the cell patches co-transfected with pSRE–GFP and pcDNA3–Raf–CAAX showed elevated levels of GFP fluorescence compared to the control vector spots. Not surprising, the intensity of the GFP signal in cells transfected by the SRE-regulated GFP construct was lower than that observed in the border cell patches transfected with pCMV–GFP, where the GFP expression is driven by the strong CMV promoter. In addition, the absolute number of cells co-transfected with the SRE–GFP reporter was somewhat lower than the number of cells that were transtected by the single pCMV–GFP plasmid. Nonetheless, the GFP signal generated by gene-specific activation of the SRE promoter was easily detectable above the background signal using both laser scanning (Fig. 17.5a) and fluorescence microscopy (Fig. 17.5b). The highest reporter signal was seen in the co-transfections using a relatively high pSRE–GFP reporter concentration compared to conventional transfections (Fig. 17.5b). Thus, the optimal range of pSRE–GFP and co-transfected gene–of–interest is 0.025–0.10 μg/μL and 0.01–0.05 μg/μL, respectively. A key advantage of using the GFP reporter system, as illustrated in Fig. 17.5, is that SRE activation can be assessed and quantitated in unfixed, unprocessed cells. The reverse transfection array shown was imaged without fixing the cells. Instead, media was removed and the array was covered with a coverslip and imaged immediately. A substantial time savings was afforded using this reporter method compared to the immunofluorescent staining method described in Fig. 17.4.

Thus, this chapter outlines the use of reporter constructs to monitor the activity of proteins produced by reverse transfection. We have demonstrated the utility of this technique using a model MAP kinase system. The simplicity and convenience of this reporter co-transfection method for reverse transfection arrays will be especially appealing for high throughput screening applications where post-transfection processing would be cumbersome and prohibitive. This method could be extended to larger reverse transfection arrays used for screening genes of unknown function simply by including the reporter construct in the gelatin printing solution. In addition, the development of other reporter systems that are more quantitative than GFP and are still suitable for array applications would make reverse transfection reporter systems even more attractive.

17.3 Reagents and Protocols

- Gelatin, Type B: 225 Bloom (Sigma #G–9391)
- GAPS™ slides (Corning #2549)
- CMP3, CMP7, and CMP10B Micro Spotting Pins (Telechem International, Inc.)
- PixSys 5500 Robotic Arrayer (Cartesian Technologies, Model AD20A5)
- CoverWell Incubation Chambers (Grace BioLabs #PC200)
- QuadriPerm chambers (Sigma)
- Effectine reagent (Qiagen #301425)
- pQBI25–fPA encoding for GFP (Qbiogene)
- pcDNA3–v–src, pcDNA3–Raf–CAAX were kindly provided by Dr. Steve Martin.
- pcDNA3–HA–KRasV12 was kindly provided by Dr. Steve Taylor.

- pSRE–Luc (Stratagene)
- Antibodies used for immunofluoresence (Phospho–Tyrosine Monoclonal Antibody (P–Tyr–100) #9411 and Phospho–p44/42 MAP Kinase (Thr202/ Tyr204) Antibody #9101) were from Cell Signaling.

Preparation of gelatin solutions and transfection array slides were performed according to the published protocol of Zhiauddin and Sabatini (http:// staffa.wi.mit.edu/sabatini_public/reverse_transfection.htm).

We used a PixSys5500 Robotic Arrayer with Telechem's ArrayIt CMP10B pins to print the DNA/gelatin solutions. The size of the printed DNA spots using this pin was approximately 250 μm and the spots were printed 600 μm apart.

References

1. Ravid, K. and Freshney, R.I., ed. (1998) *DNA Transfer to Cultured Cells*, Wiley-Liss, New York
2. Ausubel, F.M. et al. eds. (1991) *Current protocols in molecular biology*, New York, Wiley Interscience
3. Vaheri, A. and Pagano, J.S. (1965). Virology 27, 434
4. Graham, F.L., and Ven der Eb, A.J. (1973) Virology 52, 456–467
5. Chen, C., and Okayama, H. (1987) Mol. Cell. Biol. 7, 2745
6. Wigler, M., Silverstein, S., Lee, L.S., Pellicer, A., Cheng, Y.C., and Axel, R. (1977) Cell 11, 223
7. Felgner, P.L. et al. (1987). Proc. Natl. Acad. Sci. USA 84, 7413
8. Lasic, D.D., ed.(1997) *Lipsosomes in Gene Therapy*, CRC Press, New York
9. Felgner, P.L., Gadek, T.R., Holm, M., Roman, R., Chan H.W., Wenz, M., Northrop, J.P., Ringold, G.M., and Danielsen M. (1987) Proc Natl Acad Sci USA 84, 7413–7417
10. Hawley-Nelson, P., Ciccarone, V., Gebeyehu, G., Jessee, J., and Felgner, P.L. (1993) Focus 15, 73–78
11. Paulson, B. O., Clarke, M. F., anc Chuck, S.Y. United States Patent 5, 811, 274
12. Zheng, J., Manuel, W. S., and Hornsby, P. J. Biotehnol. Prog. (2000) 16, 254–257
13. Kneuer, C., Sameti, M., Haltner, E.G., Schiestel, T., Schirra, H., Schmidt, H., and Lehr, C. Int. J. Pharm. (2000) 196, 257–261
14. Cen, H., and Sun, S., WO 99/55886
15. Ziauddin, J. and Sabatini, D.M. Nature (2001) 411, 107–110
16. Cochran B.H. (1993) NIDA Res Monogr 125, 3–24
17. Alam, J., and Cook, J.L. (1990) Anal Biochem 188, 245–254
18. Lange-Carter, C.A., Pleiman, C.M., Gardner, A.M., Blumer, K.J., and Johnson, G.L. (1993) Science 260, 315–9

18

Whole Cell Microarrays

Ravi Kapur

18.1 Introduction

The post-genomic revolution is changing the face of drug discovery into a cell centric focus. It is predicted that cell-based screening in biopharmaceuticals will increase from 30% to 50% of all screening activities by 2005. The mapping of the genome has created a significant challenge of validating gene targets for specific disease states. Functional genomics within living cells is seen as a solution. Industrialization of cell biology will follow the path of industrialization of molecular biology; development of tools and techniques to gather and manage data with high throughput. The market drivers of gene sequencing, faster and cheaper, will also be drivers for extraction of the knowledge of the cellome. Additionally, the emerging marketplace for point–of–care diagnostics (POCD) presently focused on DNA and protein analysis will rapidly evolve into cell-based point–of–care diagnostics. It is projected that the growth rate of cell-based POCD will eventually exceed the growth rate of adoption of cell-based screening in biopharmaceuticals. In the recent past, cell-based assays have been assessed for utility as functional assays for detection, classification and identification of chemical and biological agents considered to be environmental pollutants or toxicants. As detection elements, living cells may play a critical role in early detection of change in the cellular milieu affected by chemical or biological threat agents.

The use of whole cells to screen and diagnose drugs, target disorders, or environmental toxicants is presently rate-limited by the throughput, cost and meaningful interpretation of the intracellular pathways modulated by such agents. The tools and techniques responsible for revolutionizing the genomic era will similarly come into play for cell-based screening: hardware for high throughput data generation, and software for data management, information extraction and knowledge generation towards diagnosis.

18.2 The Need

Functional cell-based assays serve as an early biological filter in various stages of the drug discovery process. They can serve the role of assays to tease out the validity of gene targets implicated in disease state in addition to testing the drug–responsiveness of said targets; in secondary screening to screen and rank–list the in-vitro safety and efficacy of lead compounds; for early toxicity profiling of lead compounds; and for early adsorption, distribution, metabolism and excretion (ADME) profiling across cells from multiple tissue types.

Similarly cells captured from patients with pathological states can be probed for surface markers or intracellular chromosomal abnormalities to detect and diagnose the target disorder whether it be viral infection or fetal/maternal genetic disorders.

The use of a panel of cell types such as mucosal, endothelial, immune and neurological can be used to profile the cellular signature in response to known toxicants of chemical and biological origin for eventual use in detection and classification of unknown chemical/biological samples.

The ultimate success of cell-based assays as functional tools for screening, detection, and diagnosis requires building of a knowledge base of cellular responses across multiple cell types and multiple chemical/biological molecules. The ability to generate this cellular knowledge base to enable in the future either a priori prediction of cellular activity or minimization of empirical experiments requires generation of a massive quantity of cellular information; the shotgun approach to cell biology. The ability to generate, manage and extract information from massive amounts of data in a cost-effective way from live whole cell-based experiments is the cornerstone of the knowledge base of the cellome. Tools to enable massively parallel number of experiments will be required to decipher the cell much like the automation approach to decipher its predecessor, DNA.

18.3 The Solution

18.3.1 High Density Microplates

Automation of processes is the cornerstone of enabling high throughput yield, while miniaturization positively impacts both throughput and cost. The adoption of 96 well microplates, designed for enzyme linked immunosorbent assays, for culturing cells for use in screening was an attempt to increase throughput of data by parallelization of experiments. The continued drive for higher throughput at lower cost is leading to the migration of cell-based assays onto 384 well plates, and it is projected that 50% of cell-based assays will have migrated to the 384 well plate format over the next 4 years.

Though there have been attempts to migrate cellular assays onto even higher density microplate formats, such as 1536 and 3456 well reaction plates, the success has been variable and constrained. The physical geometry of the high density micro wells impedes homogenous distribution of cells due to surface tension forces pulling the liquid to the edges and walls of the cylindrical or rectangular wells. Additionally, the low volume of each well, 1 μl–3 μl, necessitates a very tight control on evaporation-mediated compromise in cellular viability. This limits the practical utility of these high density cellular assay platforms to a few robust cell types for short incubation experiments.

In addition to the constraints of surface tension artifacts, higher density microplate platforms are likely to have intrinsic engineering issues related to optical flatness resulting in sphericity and astigmatism. Additionally, the interstitial material between wells can contribute to light piping between wells. This problem is compounded when scanning multiple wells in one scan and limits the throughput of readout.

18.3.2 Microarrays

For ultra-high density cellular platforms to be successful, there will need to be a departure from the large area footprint of traditional high density microplates. New planar platforms such as glass slides or plastic substrates with small footprints engineered and optimized for cell adhesion and optical microscopy, coupled to fluid delivery platforms will provide the solution for high throughput and low cost cell screening. The microarrays of cells on said planar substrates will reduce cost by reduction in consumption of cells, reagents and compounds. Increased throughput of screening will result from increased density of the cellular islands on a small macroscopic footprint permitting imaging of all cellular domains in one optical pass. The addressability provided by distinct pre-defined geometric localization of the cells, will further enable rapid high resolution readout of cellular domains positive for target activity. The planar substrates engineered for optical microscopy (optically flat, thin, and with low autofluorescence) will further enhance the throughput and quality of collected data.

Two functional classes of cellular microarrays can be envisioned to meet the needs of biopharma and biotech: 1) Single cell type high density arrays of one cell type for high throughput screening of multiple compounds, and 2) Multiple distinct cellular populations on a single chip screened across a single compound. The former serves the high throughput screening efforts, while the latter supports assay development, target validation and ADME–Tox.

18.3.3 Single Cell Type High Density Microarrays

Arraying a single cell type in distinct domains on a planar substrate followed by addressing each cellular domain with a distinct compound can enable high throughput screening of multiple compounds. The cell domain size can be controlled to accommodate the required number of cells, and the interstitial space between domains can be adjusted depending on the modality of delivery of compounds and reagents to the cellular domains.

Microarray Fabrication can be achieved by selective deposition of cell-adhesive and cell-repulsive chemicals onto glass or plastic substrates. The cell adhesive chemistry can be deposited selectively via a stencil or mask using solution or vapor phase deposition. The cell repulsive chemistry can be backfilled in bulk. A cell adhesive molecule includes compounds that introduce charge or are polar, contain sulfur or amines, and are capable of binding cells or other cell binding molecules such as proteins, peptides and synthetic ligands for cell surface receptors. Cell repulsive molecules include hydrophobic organosilanes or hydrophilic molecules such as polyethylene glycol that repel protein adsorption. Surfaces with cell-repulsive and cell-adhesive chemistries when incubated with cells, will post-wash result in retention of cells on the adhesive regions.

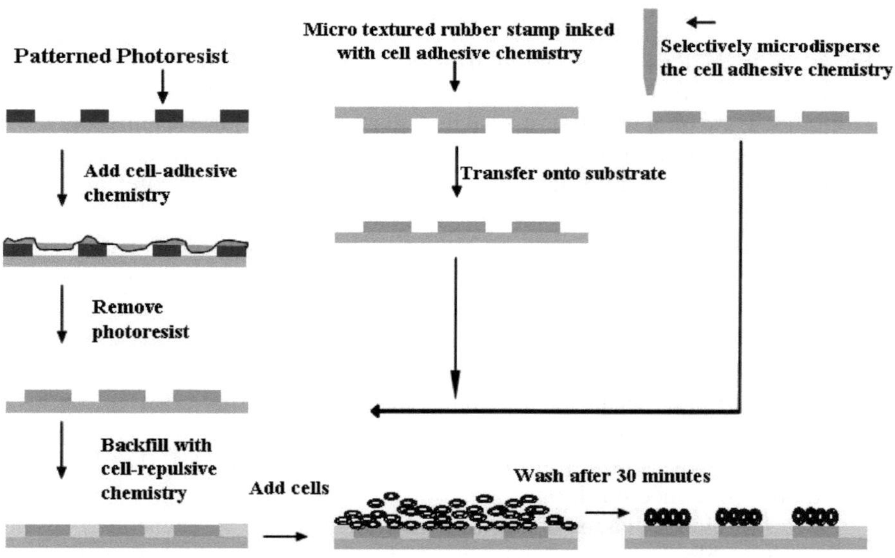

Fig. 18.1. Schematic process of fabricating cellular arrays

There are many published methods for fabricating chemically modified substrates for formation of cellular microarrays, as reviewed in Chaps. 2, 3, 16 and 17 of this book and in [1–7]. The choice of thiols, organosilanes, cell adhesive peptides/proteins or other chemistries is dictated by access to technology, ease–of–use, desired pattern fidelity (ratio of number of cells in desired domains versus cells in interstitial regions), and desired time of retention of cells in domains (using chemistry as the barrier between 2 cellular domains is a time limited process; the chemical barrier degrades in its efficiency to resist protein adsorption and cell adhesion over time). The choice of micro–stamping, photoresist masking or micro–dispensing of the cell-adhesive chemistry is dictated by access to the technology, desired throughput and reliability, and desired density of cellular domains. Figure 18.1 is a schematic depiction of the various approaches to creating chemically selective surfaces to enable formation of microarrays of living cells.

An additional emerging way of creating microarrays is to selectively micro–dispense the cells mixed with protein rich medium directly onto a highly hydrophobic and naturally cell repulsive substrate. A candidate material is poly(cyclic) olefin that appears to have fairly high resistance to breakdown of pattern fidelity of the microarrayed cells. Figure 18.2 shows an example of microarrayed cells on 1020R (polycyclic olefin available from Zeon Chemicals) fabricated by selective micro–dispensing. There is no cell-repulsive chemistry backfilled in the interstitial space. The cells are directly dispensed in fibronectin enriched medium onto spots of approximately 500–750 µm diameter.

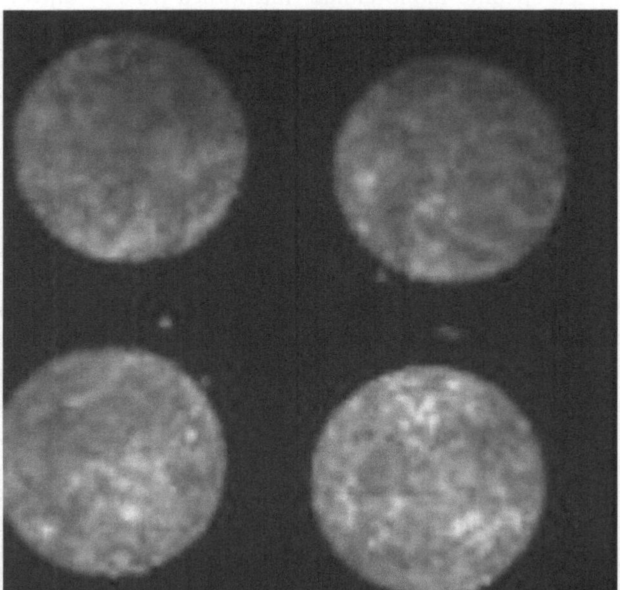

Fig. 18.2. Micro-dispensed cells on polycyclic olefin after 48 hours in culture

18.3.4 Multiple Cellular Population Microarrays

For functional genomics, there is a need for high throughput analysis of gene function within living cells. Ziauddin and co-workers [8], using microarrays of full-length cDNA in expression vectors, demonstrated a recent innovation in high throughput functional genomics. Plating of living cells onto the cDNA arrayed glass slides resulted in uptake and expression of specific proteins in spatially distinct groups of cells residing on a common substrate. These 200 spatially distinct cell clusters, each expressing a unique intracellular or cell membrane protein, can be used to screen for the effect of a single drug across 200 protein targets in one experiment. Additionally, the effect of genes on cellular phenotype can be addressed with this model (see also Chap. 17).

Multiple tissue specific cell types can additionally be arrayed on a glass/ plastic substrate to serve in applications such as ADME–Tox (adsorption, distribution, metabolism, excretion and toxicology). The ability to measure the effect of a single drug across multiple tissue specific cells enables an understanding of its side–effects away from intended targets and generation of a toxicology profile across tissue types. Such arrays can be fabricated by microarraying cell-specific mono-clonal antibodies (mAB) onto a glass/ plastic substrate followed by incubation of cell–types with antigens specific to the arrayed antibodies. Eurogentec in collaboration with GenomicDevices & Diagnostics has developed a method of antibody based cell capture on chips which can be followed by a PCR or RT–PCR analysis [9]. The specificity of the antigen–antibody reaction will determine the efficiency of sorting of the cells and associated noise and cross–contamination within the array. This technique works well for sorting blood cells and is aided by the commercially available high purity antibodies for blood cell specific antigens. Incubating the mAB arrayed substrate with one cell type, followed by a wash, and incubation with a second cell type decreases the non-specific adsorption mediated cross–contamination as compared to incubating a mixture of all cell types on the substrate. Commercially available mAB arrays from Beckton Dickenson or home-brewed arrays (with control on spot size, type of antibody and array density) can be used to generate microarrays of multiple cell types on a common substrate.

A third approach to generating multi-cellular arrays with a wide bandwidth of cell types is the use of microarrays of cell differentiating factors to induce on-chip differentiation of totipotent/pluripotent cells into tissue specific cells [10]. In this approach, stem cell differentiating factors are microarrayed on a glass/plastic substrate using commercial off–the–shelf automated liquid handling tools. The interstitial region between domains is chemically modified to prevent cell–adhesion. Totipotent or pluripotent cells are incubated with the substrate and bind to the domains containing distinct cell differentiating factors. Interaction of the cells with the underlying differentiating molecules results in each domain having a cellular phenotype and genotype corresponding to its differentiated state.

Figure 18.3 is a schematic depiction of the various approaches to creating microarrays of multiple cell types or single cell-type expressing distinct proteins in discrete clusters.

18.4 Challenges and Opportunities for Cellular Micrroarrays

18.4.1 Challenges

While it is easy to draw on the development and adoption of DNA microarray technology as a baseline guide for development and adoption of cellular microarrays, the distinction between the two technologies lies in the complexity of the biological entity being miniaturized. Cells–on–a–chip is not 'lab–on–a–chip' it is 'life–on–a–chip'. The extreme sensitivity of cells to pH, temperature, humidity, nutrients, and waste products exponentially increases the challenge associated with creating stable and reproducible arrays. The differential adhesivity of cells to surfaces and their change in functional response on adhesion to artificial substrates further compounds the complexity of using cellular microarrays for screening or diagnostics. Unlike DNA microarrays that can be stabilized for extended shelf life, cellular microarrays have a functional finite life in culture (24–72 hours) further reducing their flexibility of use. The

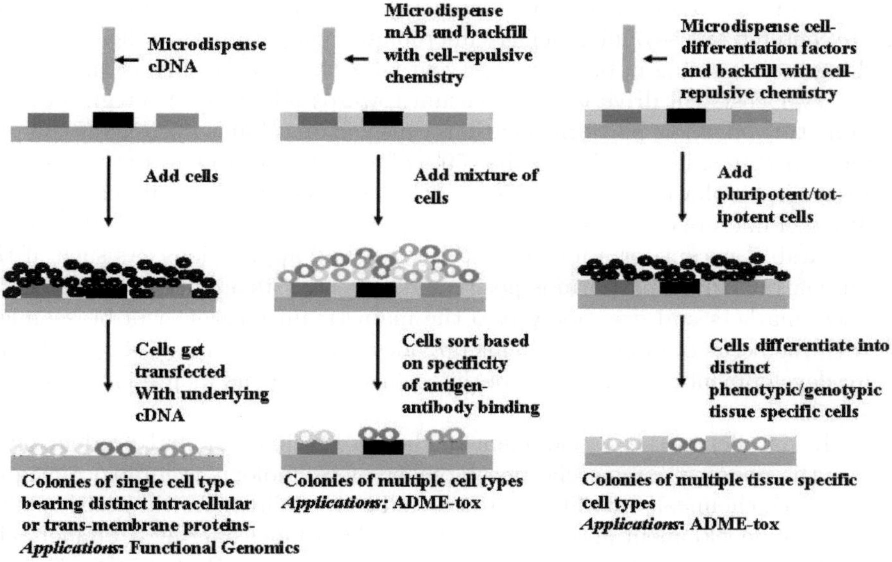

Fig. 18.3. Schematic depiction of process of reverse transfection, Ziauddin et al. **(a)**; monoclonal antibody mediated cell sorting **(b)**; and cell–differentiation mediated multi-cellular microarrays **(c)**

density of cellular microarrays will be limited by the large biological variance in cell populations. The large baseline variance of functionality of cells in culture, more profound in primary cells, places sharp statistical limits on the minimum number of cells required to make an accurate determination of change in functionality in response to a compound. Theoretically, use of single cells for screening/diagnosis is feasible for highly controlled model cell systems exhibiting very low variance in baseline response. Practically, for real world cellular lines and primary cell types, a minimum of 100 cells is required to make a statistically relevant detection. This limits the absolute obtainable density for cellular microarrays. For open systems requiring exposure of the planar cellular microarrays to a liquid dispensing device for spatially controlled treatment of the cellular domains with distinct compounds, sterility and evaporation will require careful management. These requirements will add to the technical challenge and cost of developing the technology for adoption by mainstream markets. Lastly, to bring cellular microarrays to practice as a tool for high throughput screening and point–of–care diagnostics will require the development and standardization of hardware, software, biological reagents, cell lines, and processes.

18.4.2 Opportunities

The rapidly growing cell-based screening market (compound annual growth rate at 3%) in biopharma is the single most important determinant for the successful adoption of cellular microarrays. The present screening platforms are centered on use of high density microplates compatible with the liquid handling tools residing in biopharma. Eventually, the drive for higher throughput at lower cost will drive the momentum towards adoption of integrated, and miniaturized whole platform solutions centered on cellular microarrays on planar substrates. It is projected that 50% of all assays will migrate to cell-based assays in biopharma by 2005. Most of this conversion will be driven in 96 and 384 well microplates. The use of 1536 well microplates for cell-based assays is unsuitable, except for a few niche cell types and applications. As such, if the microarray driven platform is positioned correctly, its adoption into the early stage markets and eventually into the mainstream markets will be seamless with the needs of biopharma. This provides a 4–5 year window of opportunity for development and validation of the technology beyond its present prototype stage.

In parallel to the development of the core technology and product offering, the ongoing commercial development of technologies centered on liquid handling, chemically modified surfaces and cell stabilization will positively impact the development and utility of the whole product offering. Commercially available liquid handling tools (such as from Cartesian, Packard, Picoliter) to array cells on commercially available chemically microarrayed substrates [11] will hasten the development and standardization of tools and techniques to serve the core technology development. The ongoing development of technologies for cell preservation and stabilization by means as varied as cryopreservation, freeze–drying or room temperature drying will dramatically impact the utility and flexibility of the whole product offering by enabling extended shelf–life of the consumable microarrayed substrates.

The ultimate success of cellular microarrays will be driven by the ability of the technology to deliver on the promise of faster, cheaper, smaller and better to enable industrialization of cell biology.

References

1. Singhvi, *Engineering cell shape and function*, Science, Vol. 264, pp 696
2. Thomas, *Surfaces designed to control the projected area and shape of individual cells*, Journal of Biomechanical Engineering, Vol. 121, pp 40, 1994
3. Bhatia, *Controlling cell interactions by micropatterning in co-cultures: Hepatocytes and 3T3 fibroblasts*, J. Biomed. Mater. Res., Vol. 34, pp 189, 1997
4. Kapur, *Cellular and cytoskeleton morphology and strength of adhesion of cells on self–assembled monolayers of organosilanes*, Exp. Cell Res., Vol. 244, pp 275, 1998
5. Matsuda, *Development of micropatterning technology for cultured cells*, ASAIO Trans, Vol. 36, pp 559, 1990
6. Matsuda, *Development of surface photochemical modification method for micropatterning of cultured cells*, J. Biomed. Mater. Res., Vol. 29, pp 749, 1995
7. Mrksich, *Using microcontact printing to pattern the attachment of mammalian cells to self–assembled monolayers of alkanethiolates on transparent films of gold and silver*, Exp. Cell Res., Vol. 235, pp 305, 1997
8. Ziauddin, *Microarrays of cells expressing defined cDNAs*, Nature, Vol. 411, pp 107, 2001
9. Eurogentec Inc., http://www.eurogentec.be
10. Kapur, International Patent Publication Number WO 00/60356, October 12, 2000
11. Creative Scientific Methods– http://www.cre8ive–sci.com Erie Scientific Co., Schott Glass Co

19

Tissue Microarrays for Miniaturized High-Throughput Molecular Profiling of Tumors

Ronald Simon, Martina Mirlacher, and Guido Sauter

19.1 Introduction

High throughput expression screening methods, like cDNA microarrays which allow the simultaneous expression analysis of tens of thousands of genes in one experiment, have fundamentally changed the way potentially significant genes are discovered. More recently, modern proteomics tools have been employed to survey the expression of hundreds or thousands of genes at the protein level [1]. Such methods are now extensively used in both academic and industrial research. As a result, hundreds or thousands of ESTs, genes or gene products with a potential role in non-neoplastic or neoplastic diseases have been discovered.

Many of these findings may eventually lead to clinically useful applications. For example, disease specific overexpression of a gene can be exploited in a diagnostic test. In the best case, a gene being overexpressed or functionally altered in a particular disease could serve as a therapeutic target. To further investigate the potential utility of a newly detected gene alteration, it is important to collect profound information on the epidemiology of the candidate gene expression in a multitude of diseased and non-diseased tissues. New technology is also facilitating high throughput analysis of multiple different tissues. For example, this can be achieved by multi–tissue Northern blots, protein arrays, or high throughput real time PCR facilities [2–5]. However, all these methods share the disadvantage that disintegrated tissues are used and that the cell types expressing a gene of interest cannot be identified. This is problematic because candidate genes can be expressed in multiple different tissue compartments. In-situ technologies such as immunohistochemistry (IHC), RNA in-situ hybridization (RNA–ISH) or fluorescence in situ hybridization (FISH) are therefore optimal for molecular epidemiology studies. However, such large-scale in-situ tissue analyses were cumbersome and slow when traditional methods of molecular pathology were used. Moreover, cutting of traditional tissue sections for in-situ analysis would rapidly exhaust valuable tissue resources since not more than 200 sections can typically be made from one tissue block. To overcome these shortcomings we have recently developed a tissue microarray (TMA) technique [6]. In this method up to 1,000 different tissue samples can be combined on one microscope glass slide and then be simultaneously analyzed by in-situ analysis methods.

19.2 The TMA Technology

The availability of a large collection of well-characterized tissues – optimally with attached clinical data – is the most important prerequisite to benefit from the TMA technology. Accordingly, most of the work related to the manufacturing of TMAs is similar to classical molecular pathology studies and includes collecting potentially relevant tissues, reviewing all the corresponding slides, and selecting blocks for subsequent arraying. Depending on the degree of organization of a tissue archive and its related databases, the time needed for this part of the project varies greatly.

The tissue arraying process itself is simple. The key components of the commercially available tissue microarraying devices are two needles with a slightly different diameter. With the smaller needle (outer diameter 0.6 mm), holes are punched into empty 'recipient' paraffin blocks. Subsequently, a slightly larger needle (inner diameter 0.6 mm) is utilized to transfer tissue cylinders from preexisting 'donor' paraffin blocks into these pre-made holes at specific coordinates. Regular microtomes can then be used to cut tissue microarray sections. An adhesive coated slide system (Instrumedics, Hackensack, New Jersey) facilitates the cutting. TMA sections can be used for all types of in situ analyses including immunohistochemistry (IHC), fluorescence in situ hybridization (FISH) or RNA in situ hybridization. Figure 19.1 shows an overview of an H&E stained TMA section as well as examples of IHC and ISH results.

19.3 The Representativity Issue

The question of whether or not a small sample measuring 0.6 mm in diameter can be representative of an entire, potentially heterogeneous tumor has been a major concern in the early period of using TMAs [7–11]. At least 20 studies have compared IHC findings on TMAs and their corresponding traditional 'large' sections [7,9,10,12–28], with the vast majority of them revealing a high level of concordance of results [7,9,10,12,13,15,17,18,20,21,23–28]. In several of these studies, multiple samples were taken from the donor blocks in order to determine how many samples are needed to obtain results on TMAs that are sufficiently concordant to those observed in large section analyses. In general, these studies found that two or three samples provided more representative information than a single sample [7, 9, 12, 13, 24] and that adding more than

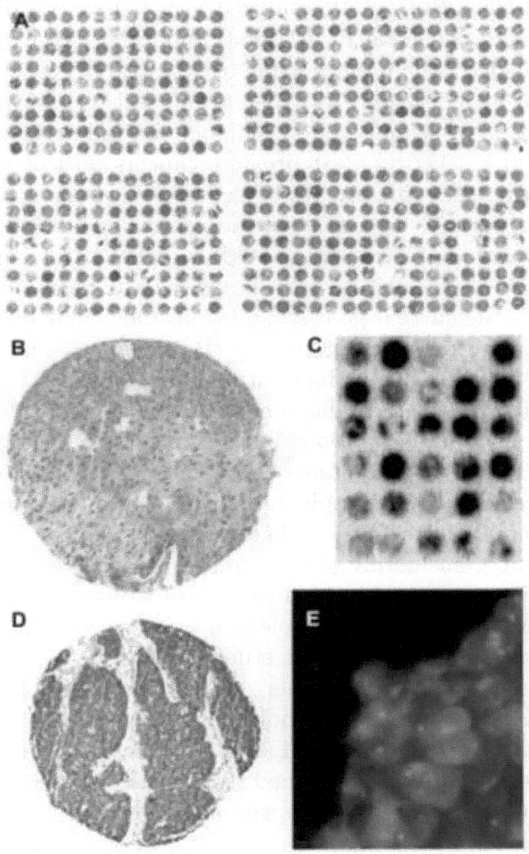

Fig. 19.1. Examples for TMAs. H&E stained bladder cancer tissue micro array section **(a)**, and a magnification of one H&E stained tissue spot **(b)**. **(c)** Autoradiography of RNA in-situ hybridization against Vimentin mRNA on a small TMA. The black staining intensity level indicates the Vimentin expression level. **(d)** Immunohistochemical detection of the Egfr protein. The panel E shows a FISH analysis of the Topoisomerase 2 alpha (TOP2A) gene. Blue staining indicates cell nuclei. Each nucleus contains 2 green (centromere 17) signals and multiple red (TOP2A) signals, indication TOP2A gene amplification

four or five samples would not lead to a massive improvement of the concordance level [7,24]. Camp et al. studied expression of ER, PR, and Her2 in 2–10 tissue cores obtained from the same donor blocks in a set of 38 invasive breast carcinomas. They found that analysis of 2 cores was sufficient to obtain identical results as compared to the corresponding whole tissue sections in 95% of cases. 99% concordance was reached if 4 cores were analyzed, and analysis of additional cores did not result in a significant further increase of concordance [7]. Similarly, Hoos et al. analyzed 1–3 tissue cores from 59 fibroblastic tumors with heterogeneous Ki–67, p53, and pRB expression. Analysis of 3 tissue cores yielded concordance rates of 91% (pRB), 96% (Ki–67), and 98% (p53) respectively, compared to whole tissue sections [9]. Recently, Rubin et al. determined the optimal sample number for immunohistochemical Ki–67 measurement in 1–10 cores of 88 prostate cancers. In this study, 3 cores were required to optimally represent Ki–67 expression with respect to the standard tumor slide, whereas 3–4 cores gave the optimal predictive value for clinical outcomes. More than 4 cores did not add significant information [24].

However, all these studies were based on the assumption that classical large sections – the current gold standard for molecular tumor tissue analysis – is representative of an entire tumor. It is very possible that this notion is not always true. In the optimal case, a 'large' section will contain tumor tissue measuring 3 × 2 cm in diameter. Given a section thickness of 3 μm the examined tumor volume is about 0.0018 cm^3. This volume represents only 1/19,000 of a tumor with a diameter of 4 cm or 1/150,000 of a tumor with a diameter of 8 cm. A TMA sample measuring 0.6 mm in diameter represents a tumor volume of 0.00000108 cm^3 that is 1/1,600 of a 3 × 2 cm tumor area on a 'large' section. Considering these numbers, the representativity problem is about 1,000 times greater between the entire tumor and a traditional 'large' section than between a TMA sample and a 'large' section.

These calculations suggest that studies investigating the utility of molecular analysis methods should rather address the question of whether or not established associations between molecular features and tumor phenotype or clinical outcome can be found. In fact, all studies that we are aware of using TMAs to reproduce firmly established associations between molecular features and tumor phenotype or prognosis revealed the expected significant results. For example, expected associations with clinical outcome were found in TMA studies for the KI67 labelling index in urinary bladder cancer [10], soft tissue sarcoma [29], and in Hurthle cell carcinoma [30], for vimentin expression in kidney cancer [20], and for expression of estrogen and progesterone receptor proteins [26] or HER–2 alterations in breast cancer patients [31]. The associations with prognosis that were obtained in a TMA analysis are shown for HER2 overexpression and HER2 amplifications in a set of 553 breast cancers in Fig. 19.2. Another study confirmed the known frequencies of amplification for Cyclin–D1, c–myc and HER2 in various cancer types [32]. A multitude of studies found associations between gene amplification or protein overexpression and tumor phenotype, e.g. cyclin E [33], FGFR1, RAF1 [34], MDM2 or CDK4 [35] amplification or MAGE–A4 expression [36] and stage and grade in bladder cancer, CK7 and CK20 expression and grade in colorectal carcinoma [37], IGFBP2 expression and hormone–refractory state [38], EIF3S3 amplification and stage [39], aneusomy and grade [40] or E–cadherin expression and tumor size [41] in prostate cancer, aneusomy and tumor type in brain tumors [42], particular expression profiles and histological subtypes in breast cancer [43] and synovial sarcoma [44], or SHP1 expression and tumor development in lymphomas [45]. In addition, it has been demonstrated that TMAs can be utilized for comprehensive analyses of amplicon architecture [35, 46]. Overall, these data clearly show that relevant data can be obtained in TMA studies. This is especially true if the TMAs used are large enough to provide sufficient power for statistical analyses.

19.4 TMA Applications

More than 100 publications reviewing or using the TMA approach had been published at the end of 2002. Obviously there is a large variety of possible TMA applications. Virtually all research involving in-situ tissue analysis can be done in a TMA format. Most published studies have utilized TMAs in cancer research. TMAs that were applied in these projects can be divided into 5 different categories: prevalence TMAs, normal tissue TMAs, progression TMAs, prognostic TMAs, and TMAs composed of experimental tissues. Prevalence TMAs contain tumor samples without clinico–pathological data attached. Despite this limitation, they are highly useful to determine the prevalence of a given alteration in tumor entities of interest. Remarkably, tumor entities that can be successfully analyzed on prevalence TMAs include Hodgkin's lymphoma [14, 17, 27]. This could not necessarily be expected since these tumors predominantly consist of reactive inflammatory cells with only few dispersed neoplastic Hodgkin or Reed Sternberg cells. Prevalence TMAs can contain tissue samples from various different tumor entities. The largest 'multitumor' TMA manufactured in our laboratory contained 4,788 different samples from 130 different tumor types [47]. This TMA is currently utilized for the analysis of multiple different markers on the DNA and protein level. In one study the frequency of 17q23 amplifications, which is linked to poor prognosis in breast cancer, was analyzed using FISH. The multitumor TMA analysis revealed that 17q23 amplification can occur in 18 additional tumor categories besides breast cancer, including tumors of the adrenal gland, lung, ovary, skin, soft tissue, stomach, thyroid gland, urinary bladder, and uterus [47].

Normal TMAs are especially important if candidate genes are evaluated for their potential utility as diagnostic reagents or therapeutic targets. For such applications, it is important to see whether candidate genes are also expressed in normal tissues. In case of potential therapeutic targets it would be most important to know whether vital organs like brain, heart, kidney, liver or bone marrow cells expressed a candidate gene.

Progression TMAs contain samples of different stages of one particular tumor type [6, 48–50]. For example, an ideal prostate cancer progression TMA would contain samples of either normal prostate, benign prostatic hyperplasia (BPH), prostatic intraepithelial neoplasia (PIN), incidental carcinomas (stage pT1), organ confined carcinomas (pT2), or carcinomas with extraprostatic growth (pT3–4), as well as metastases and recurrences after androgen withdrawal treatment. TMAs are also suited to study progression within tu-

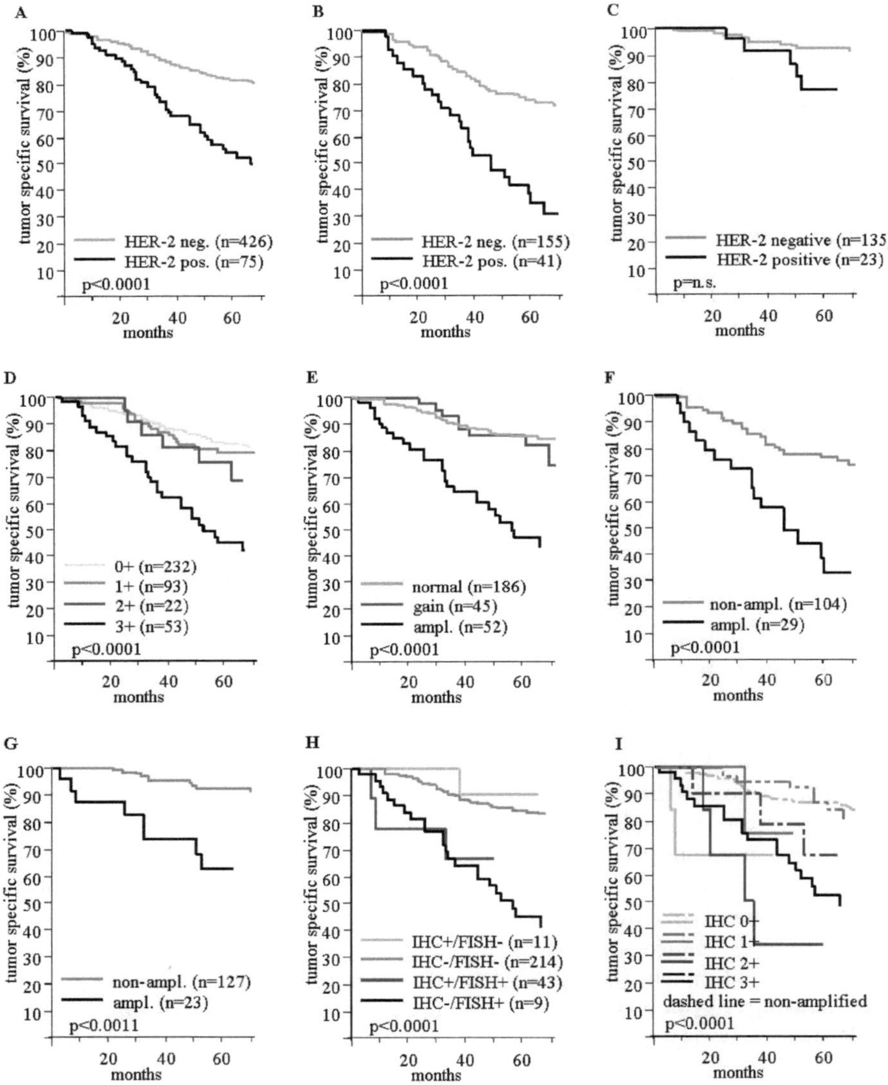

Fig. 19.2. HER–2 protein overexpression / gene amplification and tumor specific survival in patients with ductal breast cancer. The curves show the associations of HER–2 protein overexpression with poor prognosis in all patients **(a)**, and in the subgroups of nodal positive **(b)**, and nodal negative tumors **(c)**. The influence of the HercepTest score on prognosis is shown in **(d)**. The relationship of HER–2 gene amplification with prognosis is shown for all patients **(e)**, nodal positive **(f)**, and nodal negative tumors **(g)**. For (e) a HER–2 gain is defined as a HER–2/centromere 17 ration ratio of > 1 and < 3. The prognostic impact of combined FISH and IHC results is shown in **(h)** and **(i)**

mors. TMAs can easily include large numbers of pairs of primary tumors and their non-invasive precursor lesions, metastases, or recurrences after specific treatment. In our laboratory we have constructed a TMA composed of tissues from 196 nodal positive breast carcinomas. From each tumor, one sample was taken from the primary tumor and from each of three different metastases. Together with samples from 196 nodal negative breast carcinomas this 'breast cancer metastasis TMA' contains almost 1000 tissue samples. In a recent study, we used this array to demonstrate a high concordance in the HER2 amplification/overexpression between primary tumors and their nodal metastases [51].

Prognosis TMAs contain samples from tumors with available clinical follow–up data. Molecular features were analyzed for their prognostic significance in bladder [33, 35, 52], breast [15, 26, 31, 53–55], prostate [56–58], brain [25,59,60], liver [61], kidney [20], and colorectal tumors [62–64], Hodgkin's lymphoma [14], and malignant melanoma [65]. Although all recent prognosis TMAs comprised tissues from retrospective studies from heterogeneously treated patients, these TMAs proved to be highly useful. For example, significant associations were found between 17q23 amplifications [31] or Cox2 expression [55] and breast cancer prognosis, between Top2A expression and prognosis in glioblastoma [59], between MYC and AIB1 expression and prognosis in hepatocellular carcinoma [61], and between IGFBP2 and prostate cancer prognosis [38]. Future prognosis TMAs will increasingly contain homogeneously treated tumors as clinical trial groups are implementing the making of TMAs from patients included in clinical trials as part of their protocols.

TMAs can also be made from experimental tissues like cell lines [35,66] or xenografts. Cell line TMAs are especially useful for selections of optimal cell lines for subsequent functional analyses. For example, it is possible to screen hundreds of arrayed cell lines for amplification of a gene of interest. Amplified cell lines can then be ordered and, for example, utilized for testing potentially inhibiting drug candidates.

Obviously the use of TMAs is not limited to cancer research. TMAs have also been used in quality control. For example, TMAs can be used to compare the results of IHC analysis between different laboratories [67,68]. It has also been suggested to place small TMAs containing a variety of normal tissues on slides that are used for diagnostic IHC thus providing optimal negative and positive controls [69].

19.5 Future Directions

TMA technology has become a widely accepted standard technology. Several attempts are under way to further improve and automate the technology. Prototype versions of automated tissue arrayers have now become commercially available. When they are operational, good quality TMAs can be produced.

However, automated tissue arrayers will not noticeably improve the availability of TMAs since the assembling of a TMA is only a minor part of the entire TMA making process. Much more promising is the possibility of automated TMA analysis. Since one technician can manually stain more than 200,000 tissue samples per week, the reading of the TMA slides has become the major bottleneck in the system. In principle, TMAs are optimally suited for automated IHC analysis. The most critical step for automation of IHC analysis is the selection of the area to be analyzed. This selection has already taken place in TMAs. It is expected, that systems will soon become available that will automatically scan TMA slides and measure the intensity of staining for each individual TMA spot. In one of our studies we compared manual versus automated analysis of p21 staining on a colon cancer TMA, and we were able to identify a similar association with prognosis using our home made TMA analysis software to that detected after manual analysis (Marcel Ramseier, Simon Hänggi, personal communication). In another study using a commercial system we found a 92.1% concordance in the interpretation of the Her2 status between manual and automated scoring [53]. However, Her2 is an easy to measure protein. Her2 is hardly expressed in non-neoplastic tissues, overexpression in tumors is usually at a high level, and excellent IHC staining kits are available. Automated measurement will be much more difficult for many other gene products, especially if expression occurs in multiple different cell types or cellular compartments or in case of significant background staining. Once automated imaging with or without image analysis can be performed, it is possible to link these data to other databases containing molecular, pathological or clinical data. For example, Manley et al. constructed an Internet based database comprised of interrelated data from 336 prostate cancer patients transferred into 19 TMA blocks with 5451 TMA biopsy cores. Automatically acquired digital images of the TMA spots were successfully analyzed over the Internet for several immunohistochemical biomarkers including E–cadherin, prostate-specific antigen, p27 (Kip1), and Ki–67 labelling index, and attached clinico–pathological data were used for subsequent statistical analyses [70]. This study shows nicely how TMA data with clinical and pathology information linked to an Internet database can assist collaborative multi–institutional studies.

19.6 Protocol

Manufacturing TMAs is a four–step process including sample collection, preparation of recipient blocks, construction of TMA blocks, and sectioning. The required materials and recommended laboratory procedures are briefly described below.

19.6.1 Sample Collection

- Exactly define the TMA to be made. Include normal tissues.

- Collect all slides of these tissues from the archive.
- One pathologist must review all sections from all candidate specimens to select the optimal slide. Tissue areas suited for subsequent punching should be marked.
- Collect the tissue blocks that correspond to the selected slides. These blocks and their corresponding marked slides must be matched and sorted in the order of appearance on the TMA.

19.6.2 Preparing Recipient Blocks

- Melt paraffin at 60°C, filtrate and pour it into a stainless steel mold. In contrast to normal paraffin blocks, tissue microarray blocks are cut at room temperature. Therefore, a special type of paraffin ('Peel–A–Way' paraffin; Polysciences Inc., PA, USA) is recommended with a melting temperature between 53 and 55°C.
- Place a slotted plastic embedding cassette (as used in every histology lab) on the top of the warm paraffin.
- Cool paraffin block down for 2 hours at room temperature and for 2 additional hours at 4°C. Large recipient blocks (for example $30 \times 45 \times 10$ mm) are easier to handle than the smaller blocks.

19.6.3 TMA Block Constuction

Only if all this preparatory work has been done can a tissue–arraying device be employed. At least two different tissue–arraying systems are now commercially available. Several groups have introduced inexpensive modifications to the existing commercially available manual non-automated arrayers, which markedly improve performance and facilitate arraying of frozen tissue. The TMA manufacturing process consists of five steps that are repeated for each sample placed on the TMA:

- punching a hole into an empty (recipient) paraffin block
- removing and discarding the wax cylinder from the needle used for recipient block punching
- removing a cylindrical sample from a donor paraffin block
- placing the cylindrical tissue sample in the pre-made hole in the recipient block
- proceeding to the new coordinates for the next tissue sample

Exact positioning of the tip of the tissue cylinder at the level of the recipient block surface is crucial for the quality and the yield of the TMA block. Placing the tissue too deeply into the recipient block results in empty spots in the first sections taken from the TMA block. Positioning the tissue cylinder not deep enough causes empty spots in the last sections taken from this TMA. As soon as all tissue elements are filled into the recipient block, the block is heated at 40°C for 10 minutes. Protruding tissue cylinders are then gently pressed deeper into the warmed TMA block using a glass slide.

19.6.4 TMA Block Sectioning

Regular sections can be taken from TMA blocks using standard microtomes. However, the more samples a TMA block contains, the more difficult regular cutting becomes. As a consequence, the number of slides of inadequate quality increases with the size of the TMA. In turn, fewer sections from the TMA block can effectively be analyzed. Using a tape sectioning kit (Instrumedics Inc., NY, USA) facilitates cutting and leads to highly regular non-distorted sections (ideal for automated analysis). The use of the tape sectioning system is described below:

- Place an adhesive tape on the TMA block in the microtome immediately before cutting.
- Cut a section (usually 5 µm). The tissue slice is now adhering to the tape.
- Place the tissue slice on a special 'glued' slide
- Expose the slide (tissue on the bottom) to UV light for 35 seconds (This leads to polymerization of the glue on the slide and on the tape).
- Dip the slide into TPC solution (Instrumedics) at room temperature for 5–10 seconds.
- Gently remove the tape from the glass slide leaving the tissue on the slide.
- Air dry slides at room temperature.

References

1. Schweitzer, B. and Kingsmore, S. F. *Measuring proteins on microarrays.* Curr Opin Biotechnol, 13: 14–19, 2002
2. Belin, D. *The use of RNA probes for the analysis of gene expression. Northern blot hybridization and ribonuclease protection assay.* Methods Mol Biol, 86: 87–102, 1998
3. Bichsel, V. E., Liotta, L. A., and Petricoin, E. F., *3rd Cancer proteomics: from biomarker discovery to signal pathway profiling.* Cancer J, 7: 69–78, 2001
4. Kallioniemi, O. P. *Biochip technologies in cancer research.* Ann Med, 33: 142–147, 2001
5. Walker, N. J. *Real–time and quantitative PCR: applications to mechanism–based toxicology.* J Biochem Mol Toxicol, 15: 121–127, 2001
6. Kononen, J., Bubendorf, L., Kallioniemi, A., Bärlund, M., Schraml, P., Leighton, S., Torhorst, J., Mihatsch, M., Sauter, G., and Kallioniemi, O. *Tissue microarrays for high–throughput molecular profiling of hundreds of specimens.* Nat Med, 4: 844–847, 1998
7. Camp, R. L., Charette, L. A., and Rimm, D. L. *Validation of tissue microarray technology in breast carcinoma.* Lab Invest, 80: 1943–1949, 2000
8. Gancberg, D., Di Leo, A., Rouas, G., Jarvinen, T., Verhest, A., Isola, J., Piccart, M. J., and Larsimont, D. *Reliability of the tissue microarray based FISH for evaluation of the HER–2 oncogene in breast carcinoma.* J Clin Pathol, 55: 315–317, 2002

9. Hoos, A., Urist, M. J., Stojadinovic, A., Mastorides, S., Dudas, M. E., Leung, D. H., Kuo, D., Brennan, M. F., Lewis, J. J., and Cordon-Cardo, C. *Validation of tissue microarrays for immunohistochemical profiling of cancer specimens using the example of human fibroblastic tumors.* Am J Pathol, 158: 1245–1251, 2001
10. Nocito, A., Bubendorf, L., Maria Tinner, E., Suess, K., Wagner, U., Forster, T., Kononen, J., Fijan, A., Bruderer, J., Schmid, U., Ackermann, D., Maurer, R., Alund, G., Knonagel, H., Rist, M., Anabitarte, M., Hering, F., Hardmeier, T., Schoenenberger, A. J., Flury, R., Jager, P., Luc Fehr, J., Schraml, P., Moch, H., Mihatsch, M. J., Gasser, T., and Sauter, G. *Microarrays of bladder cancer tissue are highly representative of proliferation index and histological grade.* J Pathol, 194: 349–357, 2001
11. Rimm, D. L., Camp, R. L., Charette, L. A., Costa, J., Olsen, D. A., and Reiss, M. *Tissue microarray: a new technology for amplification of tissue resources.* Cancer J, 7: 24–31, 2001
12. Engellau, J., Akerman, M., Anderson, H., Domanski, H. A., Rambech, E., Alvegard, T. A., and Nilbert, M. *Tissue microarray technique in soft tissue sarcoma: immunohisto–chemical Ki–67 expression in malignant fibrous histiocytoma.* Appl Immunohistochem Mol Morphol, 9: 358–363, 2001
13. Fernebro, E., Dictor, M., Bendahl, P. O., Ferno, M., and Nilbert, M. *Evaluation of the tissue microarray technique for immunohistochemical analysis in rectal cancer.* Arch Pathol Lab Med, 126: 702–705, 2002
14. Garcia, J. F., Camacho, F. I., Morente, M., Fraga, M., Montalban, C., Alavaro, T., Bellas, C., Castano, A., Diez, A., Flores, T., Martin, C., Martinez, M. A., Mazorra, F., Menarguez, J., Mestre, M. J., Mollejo, M., Saez, A. I., Sanchez, L., and Piris, M. A. *Hodgkin's and Reed–Sternberg cells harbor alterations in the major tumor suppressor pathways and cell–cycle checkpoints: analyses using tissue–microarrays.* Blood, 12: 12, 2002
15. Ginestier, C., Charafe-Jauffret, E., Bertucci, F., Eisinger, F., Geneix, J., Bechlian, D., Conte, N., Adelaide, J., Toiron, Y., Nguyen, C., Viens, P., Mozziconacci, M. J., Houlgatte, R., Birnbaum, D., and Jacquemier, J. *Distinct and complementary information provided by use of tissue and DNA microarrays in the study of breast tumor markers.* Am J Pathol, 161: 1223–1233, 2002
16. Gulmann, C., Butler, D., Kay, E., Grace, A., and Leader, M. *Biopsy of a biopsy: validation of immunoprofiling in gastric cancer biopsy tissue microarrays.* Histopathology, 42: 70–76, 2003
17. Hedvat, C. V., Hegde, A., Chaganti, R. S., Chen, B., Qin, J., Filippa, D. A., Nimer, S. D., and Teruya-Feldstein, J. *Application of tissue microarray technology to the study of non-Hodgkin's and Hodgkin's lymphoma.* Hum Pathol, 33: 968–974, 2002
18. Hendriks, Y., Franken, P., Dierssen, J. W., De Leeuw, W., Wijnen, J., Dreef, E., Tops, C., Breuning, M., Brocker-Vriends, A., Vasen, H., Fodde, R., and Morreau, H. *Conventional and tissue microarray immunohistochemical expression analysis of mismatch repair in hereditary colorectal tumors.* Am J Pathol, 162: 469–477, 2003
19. Merseburger, A. S., Kuczyk, M. A., Serth, J., Bokemeyer, C., Young, D. Y., Sun, L., Connelly, R. R., McLeod, D. G., Mostofi, F. K., Srivastava, S. K., Stenzl, A., Moul, J. W., and Sesterhenn, I. A. *Limitations of tissue microarrays in the evaluation of focal alterations of bcl–2 and p53 in whole mount derived prostate tissues.* Oncol Rep, 10: 223–228, 2003

20. Moch, H., Schraml, P., Bubendorf, L., Mirlacher, M., Kononen, J., Gasser, T., Mihatsch, M. J., Kallioniemi, O. P., and Sauter, G. *High–throughput tissue microarray analysis to evaluate genes uncovered by cDNA microarray screening in renal cell carcinoma.* Am J Pathol, 154: 981–986, 1999
21. Mucci, N. R., Akdas, G., Manely, S., and Rubin, M. A. *Neuroendocrine expression in metastatic prostate cancer: evaluation of high throughput tissue microarrays to detect heterogeneous protein expression.* Hum Pathol, 31: 406–414, 2000
22. Natkunam, Y., Warnke, R. A., Montgomery, K., Falini, B., and van De Rijn, M. *Analysis of mum1/irf4 protein expression using tissue microarrays and immunohistochemistry.* Mod Pathol, 14: 686–694, 2001
23. Rassidakis, G. Z., Jones, D., Thomaides, A., Sen, F., Lai, R., Cabanillas, F., McDonnell, T. J., and Medeiros, L. J. *Apoptotic rate in peripheral T–cell lymphomas. A study using a tissue microarray with validation on full tissue sections.* Am J Clin Pathol, 118: 328–334, 2002
24. Rubin, M. A., Dunn, R., Strawderman, M., and Pienta, K. J. *Tissue microarray sampling strategy for prostate cancer bio-marker analysis.* Am J Surg Pathol, 26: 312–319, 2002
25. Sallinen, S. L., Sallinen, P. K., Haapasalo, H. K., Helin, H. J., Helen, P. T., Schraml, P., Kallioniemi, O. P., and Kononen, J. *Identification of differentially expressed genes in human gliomas by DNA microarray and tissue chip techniques.* Cancer Res, 60: 6617–6622, 2000
26. Torhorst, J., Bucher, C., Kononen, J., Haas, P., Zuber, M., Kochli, O. R., Mross, F., Dieterich, H., Moch, H., Mihatsch, M., Kallioniemi, O. P., and Sauter, G. *Tissue microarrays for rapid linking of molecular changes to clinical endpoints.* Am J Pathol, 159: 2249–2256, 2001
27. Tzankov, A., Zimpfer, A., Lugli, A., Krugmann, J., Went, P., Schraml, P., Maurer, R., Ascani, S., Pileri, S., Geley, S., and Dirnhofer, S. *High–throughput tissue microarray analysis of G1–cyclin alterations in classical Hodgkin's lymphoma indicates overexpression of cyclin E1.* J Pathol, 199: 201–207, 2003
28. Yosepovich, A. and Kopolovic, J. *Tissue microarray technology–a new and powerful tool for the molecular profiling of tumors.* Harefuah, 141: 1039–1041, 1090, 2002
29. Hoos, A., Stojadinovic, A., Mastorides, S., Urist, M. J., Polsky, D., Di Como, C. J., Brennan, M. F., and Cordon-Cardo, C. *High Ki–67 proliferative index predicts disease specific survival in patients with high–risk soft tissue sarcomas.* Cancer, 92: 869–874, 2001
30. Hoos, A., Stojadinovic, A., Singh, B., Dudas, M. E., Leung, D. H., Shaha, A. R., Shah, J. P., Brennan, M. F., Cordon-Cardo, C., and Ghossein, R. *Clinical significance of molecular expression profiles of Hurthle cell tumors of the thyroid gland analyzed via tissue microarrays.* Am J Pathol, 160: 175–183, 2002
31. Barlund, M., Forozan, F., Kononen, J., Bubendorf, L., Chen, Y., Bittner, M. L., Torhorst, J., Haas, P., Bucher, C., Sauter, G., Kallioniemi, O. P., and Kallioniemi, A. *Detecting activation of ribosomal protein S6 kinase by complementary DNA and tissue microarray analysis.* J Natl Cancer Inst, 92: 1252–1259, 2000
32. Schraml, P., Kononen, J., Bubendorf, L., Moch, H., Bissig, H., Nocito, A., Mihatsch, M., Kallioniemi, O., and Sauter, G. *Tissue microarrays for gene amplification surveys in many different tumor types.* Clin Cancer Res, 5: 1966–1975, 1999

33. Richter, J., Wagner, U., Kononen, J., Fijan, A., Bruderer, J., Schmid, U., Ackermann, D., Maurer, R., Alund, G., Knonagel, H., Rist, M., Wilber, K., Anabitarte, M., Hering, F., Hardmeier, T., Schonenberger, A., Flury, R., Jager, P., Fehr, J. L., Schraml, P., Moch, H., Mihatsch, M. J., Gasser, T., Kallioniemi, O. P., and Sauter, G. *High–throughput tissue microarray analysis of cyclin E gene amplification and over–expression in urinary bladder cancer.* Am J Pathol, 157: 787–794, 2000
34. Simon, R., Richter, J., Wagner, U., Fijan, A., Bruderer, J., Schmid, U., Ackermann, D., Maurer, R., Alund, G., Knonagel, H., Rist, M., Wilber, K., Anabitarte, M., Hering, F., Hardmeier, T., Schonenberger, A., Flury, R., Jager, P., Fehr, J. L., Schraml, P., Moch, H., Mihatsch, M. J., Gasser, T., and Sauter, G. *High–throughput tissue microarray analysis of 3p25 (RAF1) and 8p12 (FGFR1) copy number alterations in urinary bladder cancer.* Cancer Res, 61: 4514–4519, 2001
35. Simon, R., Struckmann, K., Schraml, P., Wagner, U., Forster, T., Moch, H., Fijan, A., Bruderer, J., Wilber, K., Mihatsch, M. J., Gasser, T., and Sauter, G. *Amplification pattern of 12q13–q15 genes (MDM2, CDK4, GLI) in urinary bladder cancer.* Oncogene, 21: 2476–2483, 2002
36. Kocher, T., Zheng, M., Bolli, M., Simon, R., Forster, T., Schultz-Thater, E., Remmel, E., Noppen, C., Schmid, U., Ackermann, D., Mihatsch, M. J., Gasser, T., Heberer, M., Sauter, G., and Spagnoli, G. C. *Prognostic relevance of MAGE–A4 tumor antigen expression in transitional cell carcinoma of the urinary bladder: a tissue microarray study.* Int J Cancer, 100: 702–705, 2002
37. Park, S. Y., Kim, H. S., Hong, E. K., and Kim, W. H. *Expression of cytokeratins 7 and 20 in primary carcinomas of the stomach and colorectum and their value in the differential diagnosis of metastatic carcinomas to the ovary.* Hum Pathol, 33: 1078–1085, 2002
38. Bubendorf, L., Kolmer, M., Kononen, J., Koivisto, P., Mousses, S., Chen, Y., Mahlamaki, E., Schraml, P., Moch, H., Willi, N., Elkahloun, A. G., Pretlow, T. G., Gasser, T. C., Mihatsch, M. J., Sauter, G., and Kallioniemi, O. P. *Hormone therapy failure in human prostate cancer: analysis by complementary DNA and tissue microarrays.* J Natl Cancer Inst, 91: 1758–1764, 1999
39. Saramaki, O., Willi, N., Bratt, O., Gasser, T. C., Koivisto, P., Nupponen, N. N., Bubendorf, L., and Visakorpi, T. *Amplification of EIF3S3 gene is associated with advanced stage in prostate cancer.* Am J Pathol, 159: 2089–2094, 2001
40. Skacel, M., Ormsby, A. H., Pettay, J. D., Tsiftsakis, E. K., Liou, L. S., Klein, E. A., Levin, H. S., Zippe, C. D., and Tubbs, R. R. *Aneusomy of chromosomes 7, 8, and 17 and amplification of HER–2/neu and epidermal growth factor receptor in Gleason score 7 prostate carcinoma: a differential fluorescent in situ hybridization study of Gleason pattern 3 and 4 using tissue microarray.* Hum Pathol, 32: 1392–1397, 2001
41. Rubin, M. A., Mucci, N. R., Figurski, J., Fecko, A., Pienta, K. J., and Day, M. L. *E–cadherin expression in prostate cancer: a broad survey using high–density tissue microarray technology.* Hum Pathol, 32: 690–697, 2001
42. Fuller, C. E., Wang, H., Zhang, W., Fuller, G. N., and Perry, A. *High–throughput molecular profiling of high–grade astro–cytomas: the utility of fluorescence in situ hybridization on tissue microarrays (TMA–FISH).* J Neuropathol Exp Neurol, 61: 1078–1084, 2002
43. Korsching, E., Packeisen, J., Agelopoulos, K., Eisenacher, M., Voss, R., Isola, J., van Diest, P. J., Brandt, B., Boecker, W., and Buerger, H. *Cytogenetic al-*

terations and cytokeratin expression patterns in breast cancer: integrating a new model of breast differentiation into cytogenetic pathways of breast carcinogenesis. Lab Invest, 82: 1525–1533, 2002

44. Allander, S. V., Illei, P. B., Chen, Y., Antonescu, C. R., Bittner, M., Ladanyi, M., and Meltzer, P. S. *Expression profiling of synovial sarcoma by cDNA microarrays: association of ERBB2, IGFBP2, and ELF3 with epithelial differentiation.* Am J Pathol, 161: 1587–1595, 2002
45. Oka, T., Yoshino, T., Hayashi, K., Ohara, N., Nakanishi, T., Yamaai, Y., Hiraki, A., Sogawa, C. A., Kondo, E., Teramoto, N., Takahashi, K., Tsuchiyama, J., and Akagi, T. *Reduction of hematopoietic cell–specific tyrosine phosphatase SHP–1 gene expression in natural killer cell lymphoma and various types of lymphomas/leukemias : combination analysis with cDNA expression array and tissue microarray.* Am J Pathol, 159: 1495–1505, 2001
46. Monni, O., Barlund, M., Mousses, S., Kononen, J., Sauter, G., Heiskanen, M., Paavola, P., Avela, K., Chen, Y., Bittner, M. L., and Kallioniemi, A. *Comprehensive copy number and gene expression profiling of the 17q23 amplicon in human breast cancer.* Proc Natl Acad Sci U S A, 98: 5711–5716, 2001
47. Andersen, C. L., Monni, O., Wagner, U., Kononen, J., Barlund, M., Bucher, C., Haas, P., Nocito, A., Bissig, H., Sauter, G., and Kallioniemi, A. *High–throughput copy number analysis of 17q23 in 3520 tissue specimens by fluorescence in situ hybridization to tissue microarrays.* Am J Pathol, 161: 73–79, 2002
48. Bubendorf, L., Kolmer, M., Kononen, J., Koivisto, P., Mousses, S., Chen1, Y., Mahlamäki, E., Schraml, P., Moch, H., Willi, N., Elkahlhoun, A., Pretlow, T., Gasser, T., Mihatsch, M., Sauter, G., and Kallioniemi, O. *Molecular mechanisms of hormone therapy failure in human prostate cancer analyzed by a combination of cDNA and tissue microarrays.* J Natl Cancer Inst, 91: 1758–1764, 1999
49. Bubendorf, L., Kononen, J., Koivisto, P., Schraml, P., Moch, H., Gasser, T., Willi, N., Mihatsch, M., Sauter, G., and Kallioniemi, O. *Survey of gene amplifications during prostate cancer progression by high–throughput fluorescence in situ hybridization on tissue microarrays.* Cancer Res, 59: 803–806, 1999
50. Richter, J., Wagner, U., Kononen, J., Fijan, A., Bruderer, J., Schmid, U., Ackermann, D., Maurer, R., Alund, G., Knönagel, H., Rist, M., Wilber, K., Anabitarte, M., Hering, F., Hardmeier, T., Schönenberger, A., Flury, R., Jager, P., Fehr, J. L., Schraml, P., Moch, H., Mihatsch, M. J., Gasser, T., Kallioniemi, O. P., and Sauter, G. *High–throughput tissue microarray analysis of cyclin E gene amplification and over–expression in urinary bladder cancer.* Am J Pathol, 157: 787–794, 2000
51. Simon, R., Nocito, A., Hubscher, T., Bucher, C., Torhorst, J., Schraml, P., Bubendorf, L., Mihatsch, M. M., Moch, H., Wilber, K., Schotzau, A., Kononen, J., and Sauter, G. *Patterns of her–2/neu amplification and overexpression in primary and metastatic breast cancer.* J Natl Cancer Inst, 93: 1141–1146, 2001
52. Rao, J., Seligson, D., Visapaa, H., Horvath, S., Eeva, M., Michel, K., Pantuck, A., Belldegrun, A., and Palotie, A. *Tissue microarray analysis of cytoskeletal actin–associated bio-markers gelsolin and E–cadherin in urothelial carcinoma.* Cancer, 95: 1247–1257, 2002
53. Bucher, C., Torhorst, J., Kononen, J., Haas, P., Askaa, J., Godtfredsen, S. E., Bauer, K. D., Seelig, S., Kallioniemi, O., and Sauter, G. *Automated, High–Throughput Tissue Microarray Analysis for Assessing the Significance of HER–2 In-volvement in Breast Cancer.* In: Proceedings of the ASCO annual meeting, Abstr. #2388, New Orleans, LA, 2000

54. Poremba, C., Heine, B., Diallo, R., Heinecke, A., Wai, D., Schaefer, K. L., Braun, Y., Schuck, A., Lanvers, C., Bank-falvi, A., Kneif, S., Torhorst, J., Zuber, M., Kochli, O. R., Mross, F., Dieterich, H., Sauter, G., Stein, H., Fogt, F., and Boecker, W. *Telomerase as a prognostic marker in breast cancer: high–throughput tissue microarray analysis of hTERT and hTR.* J Pathol, 198: 181–189, 2002
55. Ristimaki, A., Sivula, A., Lundin, J., Lundin, M., Salminen, T., Haglund, C., Joensuu, H., and Isola, J. *Prognostic significance of elevated cyclooxygenase–2 expression in breast cancer.* Cancer Res, 62: 632–635, 2002
56. Bubendorf, L., Kononen, J., Koivisto, P., Schraml, P., Moch, H., Gasser, T. C., Willi, N., Mihatsch, M. J., Sauter, G., and Kallioniemi, O. P. *Survey of gene amplifications during prostate cancer progression by high–throughout fluorescence in situ hybridization on tissue microarrays.* Cancer Res, 59: 803–806, 1999
57. Dhanasekaran, S. M., Barrette, T. R., Ghosh, D., Shah, R., Varambally, S., Kurachi, K., Pienta, K. J., Rubin, M. A., and Chinnaiyan, A. M. *Delineation of prognostic biomarkers in prostate cancer.* Nature, 412: 822–826, 2001
58. Mousses, S., Bubendorf, L., Wagner, U., Hostetter, G., Kononen, J., Cornelison, R., Goldberger, N., Elkahloun, A. G., Willi, N., Koivisto, P., Ferhle, W., Raffeld, M., Sauter, G., and Kallioniemi, O. P. *Clinical validation of candidate genes associated with prostate cancer progression in the CWR22 model system using tissue microarrays.* Cancer Res, 62: 1256–1260, 2002
59. Miettinen, H. E., Jarvinen, T. A., Kellner, U., Kauraniemi, P., Parwaresch, R., Rantala, I., Kalimo, H., Paljarvi, L., Isola, J., and Haapasalo, H. *High topoisomerase IIalpha expression associates with high proliferation rate and and poor prognosis in oligodendrogliomas.* Neuropathol Appl Neurobiol, 26: 504–512, 2000
60. Miettinen, H. E., Paunu, N., Rantala, I., Kalimo, H., Paljarvi, L., Helin, H., and Haapasalo, H. *Cell cycle regulators (p21, p53, pRb) in oligodendrocytic tumors: a study by novel tumor microarray technique.* J Neurooncol, 55: 29–37, 2001
61. Wang, Y., Wu, M. C., Sham, J. S., Zhang, W., Wu, W. Q., and Guan, X. Y. *Prognostic significance of c–myc and AIB1 amplification in hepatocellular carcinoma. A broad survey using high–throughput tissue microarray.* Cancer, 95: 2346–2352, 2002
62. Chung, G. G., Provost, E., Kielhorn, E. P., Charette, L. A., Smith, B. L., and Rimm, D. L. *Tissue microarray analysis of beta–catenin in colorectal cancer shows nuclear phospho–beta–catenin is associated with a better prognosis.* Clin Cancer Res, 7: 4013–4020, 2001
63. Hoos, A., Nissan, A., Stojadinovic, A., Shia, J., Hedvat, C. V., Leung, D. H., Paty, P. B., Klimstra, D., Cordon-Cardo, C., and Wong, W. D. *Tissue Microarray Molecular Profiling of Early, Node–negative Adenocarcinoma of the Rectum: A Comprehensive Analysis.* Clin Cancer Res, 8: 3841–3849, 2002
64. Otsuka, M., Kato, M., Yoshikawa, T., Chen, H., Brown, E. J., Masuho, Y., Omata, M., and Seki, N. *Differential expression of the L–plastin gene in human colorectal cancer progression and metastasis.* Biochem Biophys Res Commun, 289: 876–881, 2001
65. Kielhorn, E., Provost, E., Olsen, D., D'Aquila, T. G., Smith, B. L., Camp, R. L., and Rimm, D. L. *Tissue microarray–based analysis shows phospho–beta–catenin expression in malignant melanoma is associated with poor outcome.* Int J Cancer, 103: 652–656, 2003
66. Hoos, A. and Cordon–Cardo, C. *Tissue microarray profiling of cancer specimens and cell lines: opportunities and limitations.* Lab Invest, 81: 1331–1338, 2001

67. Mengel, M., von Wasielewski, R., Wiese, B., Rudiger, T., Muller-Hermelink, H. K., and Kreipe, H. *Inter–laboratory and inter–observer reproducibility of immunohistochemical assessment of the Ki–67 labelling index in a large multi–centre trial.* J Pathol, 198: 292–299, 2002
68. von Wasielewski, R., Mengel, M., Wiese, B., Rudiger, T., Muller–Hermelink, H. K., and Kreipe, H. *Tissue array technology for testing interlaboratory and interobserver reproducibility of immunohistochemical estrogen receptor analysis in a large multicenter trial.* Am J Clin Pathol, 118: 675–682, 2002
69. Packeisen, J., Buerger, H., Krech, R., and Boecker, W. *Tissue microarrays: a new approach for quality control in immunohistochemistry.* J Clin Pathol, 55: 613–615, 2002
70. Manley, S., Mucci, N. R., De Marzo, A. M., and Rubin, M. A. *Relational database structure to manage high–density tissue microarray data and images for pathology studies focusing on clinical outcome: the prostate specialized program of research excellence model.* Am J Pathol, 159: 837–843, 2001

20

Application of Microarray Technologies for Translational Genomics

Spyro Mousses, Natasha Caplen, Mark Basik, Anne Kallioniemi, and Olli Kallioniemi

20.1 Introduction

There has been an exponential growth in the rate at which the human genome is being decoded to decipher its genetic information. New enabling technologies have been developed to accelerate throughput in both structural and functional genomics, rapidly expanding our capacity to extract data from the genome. The human genome project reported the near completion of the first draft of the 3 billion base pair human genome and a catalogue of more than 34 thousand human genes [1–5]. The promise of this milestone in scientific achievement is that it will lead to a better understanding of biological processes, and facilitate medical breakthroughs by the discovery of new disease-related genes.

Besides the new sequencing technologies that have led to the rapid completion of the genome sequence, the need to apply these discoveries has given birth to innovative high throughput technologies, which have made it possible to interrogate the expression and sequence variation of thousands of genes in parallel. The most popular and powerful example is the DNA microarray [6–9], which can be used to simultaneously quantify the expression of thousands of genes, thereby producing insight into the expressed 'transcriptome'. Thousands of studies have used DNA microarrays for genome scale analysis of gene expression or sequence variation and have generated long lists of candidate genes associated with various disease states [10–13]. Based solely on the microarray data however, the utility of these candidate genes in clinical diagnostics and therapeutics can only be hypothesized. Since traditional functional and clinical validation of candidate genes is carried out one gene at a time, it is becoming increasingly apparent that these studies are generating hypotheses at a rate that far exceeds the rate for testing these hypotheses with current approaches. Indeed, a major bottleneck is present in the translation of genomic information into medical advances. High throughput hypothesis testing platforms therefore need to be developed and applied before the full potential of the genomic revolution can truly be realized. In this chapter, two new ‘translational genomics’ technologies will be described: tissue microarrays and live cell microarrays. These novel technologies can enable high throughput hypothesis testing so as to rapidly translate genomic data into scientific knowledge and medical discoveries.

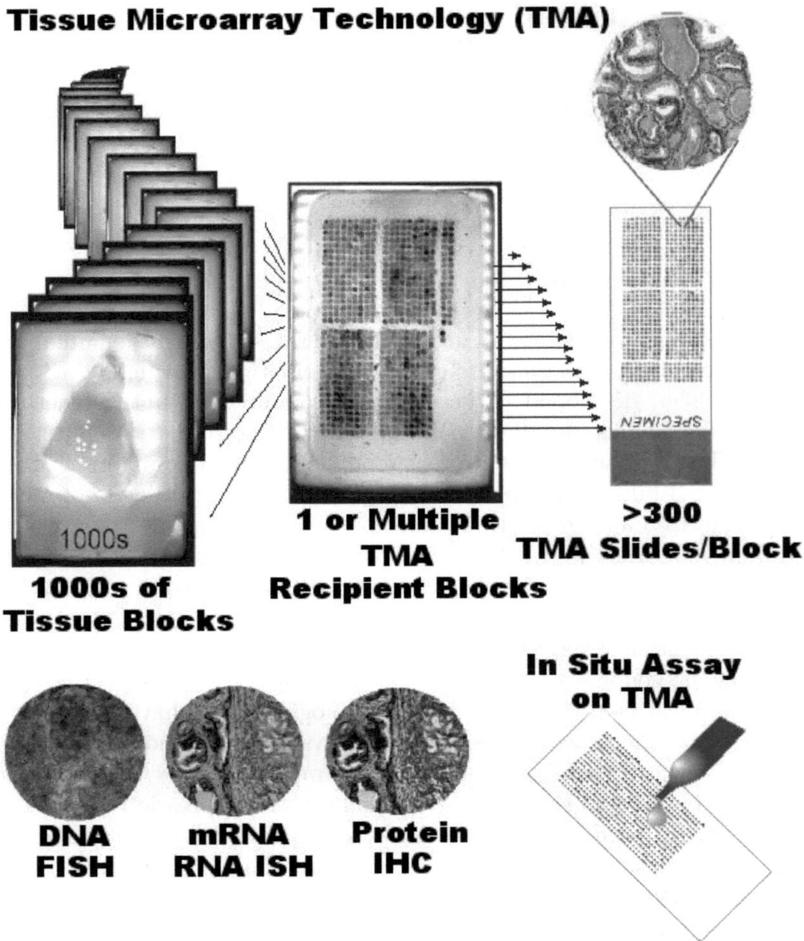

Fig. 20.1. Tissue microarray technology: Thousands of paraffin-embedded fixed tissue blocks are selected and core biopsies taken to be arrayed onto a recipient paraffin block. The recipient block is then sectioned over 300 times and the sections placed onto microscope slides. Each slides has the same tissues in the same coordinates as the recipient block. These slides can then be used for in-situ assays including FISH, RNA in-situ hybridization and protein immunostaining (modified from [14] - Hum Mol Genet 2001, 10:657–662)

20.2 High Throughput Clinical Target Validation Using Tissue Microarrays

The actual clinical relevance and prevalence of molecular alterations discovered by DNA microarrays must be evaluated in order to justify further preclinical and clinical testing of these candidate gene targets. The evaluation of each of these gene alterations one by one is very time–consuming. It requires access to, collection, preparation and examination of large resources of clinical material usually found in pathology departments. Even if the use of sophisticated data mining methods allows one to narrow down the list to twenty or even ten candidate gene targets, their full clinical validation remains a daunting if not impossible task for most genomic labs.

One solution to this clinical validation challenge is to assemble clinical samples on a miniaturized scale on a microarray platform that facilitates parallel analysis. The need to invent new ways to validate multiple molecular alterations in our laboratory led to the development of tissue microarray (TMA) technology [15], (also see Chap. 19 of this book). This technology permits high throughput in situ analysis of specific molecular targets in hundreds or thousands of tissue specimens at once. TMAs are miniaturized collections of arrayed tissue spots on a microscope glass slide that provide a template for highly parallel organization of molecular targets. These arrayed tissue samples can then be interrogated either at the DNA, RNA or protein level (Fig. 20.1).

The use of TMAs allows the discovery of relationships between the presence of molecular alterations and tissue, cell and subcellular morphology as well as with clinical correlates such as patient outcome, which are associated with the specimens. TMAs are thus ideally suited for large-scale translational studies of candidate molecular targets [14].

In practice, the construction of TMAs is relatively simple: successive cylindrical core biopsies are punched from selected areas on paraffin embedded fixed tissue blocks, such as those found in any pathology department. These core biopsies are inserted in an arrayed manner into a recipient paraffin block, which is pre-punched to accept placement of these biopsies. (Fig. 20.2). Detailed technical information on the construction of the TMAs was recently reviewed by Kononen et al. [16]. The most time–consuming and laborious step is often the selection and collection of paraffin blocks of samples to be arrayed on a TMA. The next step is the selection of the exact area of morphological interest on a regular H&E stained section cut from each of the chosen blocks. Over 1000 individual tissue biopsies can then be arrayed onto the recipient block, which can be sectioned with a regular microtome for up to 300 thin sections, depending on the depth of the biopsies. Each of these sections has the identical configuration of tissue spots (rows and columns) found on the recipient block. These sections are placed on glass slides, which can be used immediately or stored for months or years. TMA slides can be applied for analyses of DNA, RNA and protein targets using various techniques, such as fluorescence in situ hybridization (FISH), mRNA in situ hybridization, or

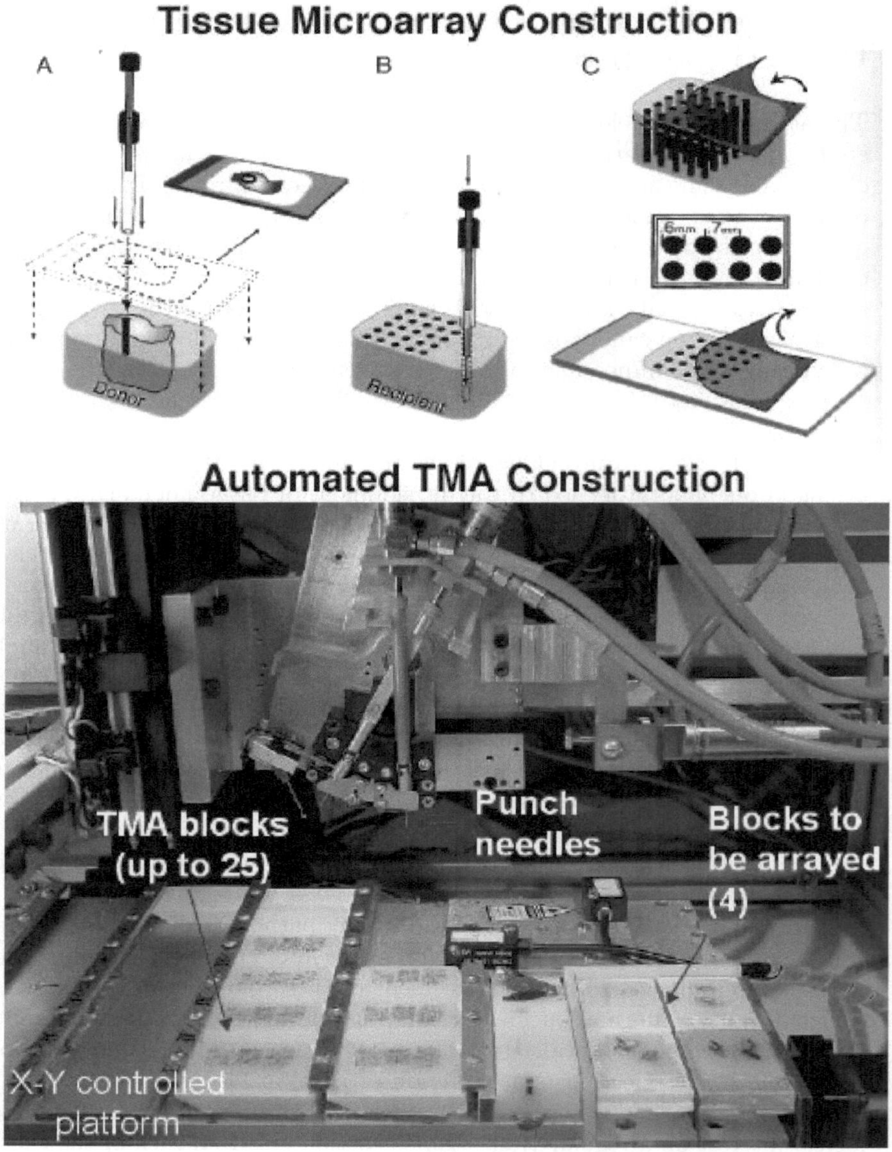

Fig. 20.2. Tissue microarray construction and automated TMA construction: TMAs are constructed by identifying the site of interest on the donor recipient block **(a)**, placing the biopsy into the recipient block in an organized way **(b)**, and sectioning the block using the tape transfer method **(c)** (Instrumedics Inc., New Jersey). This process can be automated as in the prototype model displayed in which multiple blocks can be simultaneously biopsed and cores inserted into multiple recipient TMA blocks, under computer control

immunohistochemistry (IHC). In fact, it is possible to interrogate with all 3 methods a virtually identical cohort of tissue samples using a series of successively sectioned TMA slides. Moreover, by using small (0.6 mm) diameter biopsies, TMA technology prevents the loss of precious archival material. In fact, because of this small size of the biopsies, it is feasible to take several biopsies from each donor paraffin block in order to construct replicate TMA blocks in one sitting without destroying the original block. For example, construction of 10 replicate TMA blocks from a starting material of 1000 tissues would enable one to produce up to 3000 TMA slides. This would only remove ten 0.6 mm cylindrical cores of each of the 1000 tissue blocks. These 3000 TMA slides can each be used with a different probe or assay to analyze up to 3000 different genes of interest, in 1000 specimens per assay. This produces a total capacity of up to 3 million individual spot measurements from precious clinical tissue material. TMAs therefore make it possible to perform large-scale clinical studies on a single microscope slide.

Since clinical epidemiology studies require large case numbers, TMAs are ideal for the efficient use of the large tissue resources available in pathology laboratory archives. If matching clinical data such as survival and treatment response exists for these specimens, rapid extraction of clinicopathological correlates in over 1000 of these specimens can be performed in a single TMA experiment. Since TMA slides are usually created as multiple sets containing the same clinical specimens populations, data from multiple genes can be analyzed across that population to determine patterns of involvement amongst related genes and gene products. For instance, all of the members of a signaling pathway can be studied on successive TMA slides. Another TMA example is that of creating a 'progression TMA' in which multiple tissue samples of different stages of a disease can be arrayed on one TMA, so as to permit rapid determination of the onset of a molecular event in relationship to the stages of disease progression. It is thus clear that the throughput and uniformity of TMAs can be used for a variety of creative applications to produce data of a scale, quality, and nature that is unique to this platform.

20.3 Examples of Studies Integrating DNA and Tissue Microarray Technologies for the Rapid Clinical Translation of Genomic Discoveries

Tissue microarrays can be used for the high throughput analysis of a variety of specimens including different tissue and organ types from various disease and normal states. TMAs have also been constructed from cell lines and from tissues from various model organisms. However, most studies reporting the use of TMAs have focused on their application in the study of human disease, especially cancer. Given the current proliferation of lists of candidate genes generated by DNA microarrays, TMAs have already been used to validate and prioritize molecular targets in a variety of ways, including some already mentioned:

1. Clinical validation in patient tissue samples of results obtained from the analysis of cell lines or rodent disease model systems in vivo.
2. Integration of information about the same molecular target at the DNA, RNA and protein level.
3. Extension of results obtained from the analysis of a limited number of tissue samples by cDNA microarrays to an epidemiologically representative cohort by TMAs.
4. Assessment of the prevalence of molecular alterations at various stages of tumor progression.
5. Correlation of molecular data with clinicopathological and patient outcome variables.
6. Determination of the cellular and subcellular distribution of the targets.

Many studies illustrating each of these prospects have already been published. Some examples follow:

Example A) In a study using cDNA microarrays, Bärlund et al. [17] reported that the ribosomal protein S6 kinase gene is one of several markedly overexpressed and amplified genes in breast cancer cell lines. TMAs containing over 600 clinical breast cancers confirmed that this gene is amplified and highly expressed at the protein level in 10–15% of primary breast tumors. Moreover, concomitant overexpression and amplification of the S6 kinase gene was found to be a significant poor prognostic indicator in breast cancer.

Example B) Moch et al. [18] used cDNA microarrays to identify transcripts that were differentially expressed between a renal carcinoma cell line and normal kidney tissue. One of these genes, vimentin, was further evaluated for protein expression using a TMA containing 532 renal cell carcinoma samples. They reported clear differences in vimentin protein expression among different histological subtypes of renal cell carcinomas as well as an association between vimentin expression and poor prognosis in patients with renal cancer.

Example C) Sugita et al [19] performed microarray analysis on 4 lung cancer cell lines and generated a list of 20 highly expressed genes. Using a TMA containing 187 non-small cell lung cancers, they found that the overexpression of one of these, the MAGE–A gene, was more specific for a histological subtype of these cancers, squamous cell carcinoma of the lung. Thus this gene may become a marker for this histological subtype of lung cancer.

Example D) Global gene expression in primary human gliomas was compared to the gene expression profile of normal brains by Sallinen et al. [20] using cDNA microarrays. A set of differentially expressed genes was produced, which included the IGFBP2 gene. As in prostate cancer, overexpression of IGFBP2 was found to be associated with tumor progression. Immunohistochemical analysis of IGFBP2 expression levels in 418 brain tumors in a TMA confirmed the cDNA microarray results and also revealed that the IGFBP2 overexpression was associated with poor patient–survival.

Example E) Ginestier et al [21] compared mRNA expression levels on cDNA microarrays with protein expression on TMAs for 15 molecules with a proven or suspected role in breast cancer in 55 breast tumors. A good correlation was found only in 5 of these, thus underlining the necessity for confirming cDNA microarray findings. A TMA of 600 breast tumors was used to identify a prognostic value for one of the molecules, MUC1. On the other hand, RNA levels and not protein expression had a prognostic value for the THBS1 gene. This study highlights the need to combine these microarray technologies in order to obtain clinically useful and valid information.

Example F) Mousses et. al. [22] used cDNA microarrays to identify genetic alterations occurring in human prostate cancer xenografts during the progression of hormone sensitive tumors to hormone refractory tumors. Three key genes were found to be involved in the resistance to the growth suppressive effects of hormone therapy in these xenografts. S100P mRNA expression was increased in xenografts, while CRYM and LMO4 mRNA expression were decreased. To clinically validate these results, a prostate cancer progression microarray was probed with antibodies against each of the three gene products. S100P protein expression was directly correlated with stage of disease, while levels of CRYM and LMO4 proteins were both lower in a significant number of advanced hormone refractory tumors compared to a population of primary tumors, thereby validating in the clinical context the trends observed in the xenografts.

Example G) Using cDNA microarrays Dhanasekaran et al. [23] studied alterations in gene expression in different stages of prostate cancer. Several genes with significant expression changes between different groups of tumors were identified. Two of these genes, hepsin and pim–1, were selected for further study using TMAs. A positive correlation between expression of these two genes and measures of clinical outcome was observed.

These studies are but some of the many examples which illustrate the power of the TMA technology for rapid translation of cDNA microarray results into clinically meaningful information. An analysis of hundreds of tumor samples was performed within the short period of a few weeks, a task that would otherwise have taken years to accomplish using traditional techniques. We predict that this powerful research approach will be increasingly applied in the future, as more and more investigators seek the validation and prioritization of their early cDNA microarray leads. Ongoing development of improved tissue–arraying instrumentation including automated (robotic) TMA construction, automated digital image acquisition, storage, analysis and standardization will facilitate further expansion of the technology.

20.4 High Throughput Characterization of Gene Function Using Live Cell Microarrays

Alterations in gene or protein expression levels tell us very little about the biological function of the gene, its potential clinical impact or suitability as a drug target. Besides clinical validation, it is also necessary to 'functionally' validate target genes identified by microarray screening, i.e. to verify whether the observed molecular alterations are responsible for phenotypic or functional changes in the target tissue. Functional validation is traditionally performed in molecular- and cell-based assays on a gene–by–gene basis. This is the second major bottleneck in translational genomics. A variety of tailor-made assay formats often have to be specifically designed for each candidate target. For example, investigators may screen for the phenotypic effects of gene overexpression by knocking down gene expression with anti-sense molecules. Protein interactions may be elucidated using the yeast two–hybrid strategy [24]. Specific biochemical assays such as assays for enzymatic activity may have to be developed for some targets in order to search for small molecule inhibitors [25] from compound libraries. Such high throughput screening has usually been carried out in a microtiter plate format for each gene target, but the plethora of targets arising from genomics and proteomics surveys will require parallel approaches to rapidly investigate their function.

A recent innovation in high throughput functional characterization was the application of a well–less microarray platform in place of a traditional microtiter plate platform. Ziauddin and Sabatini [26] demonstrated how parallel transfection of hundreds of genes can be carried out in a microarray format using a technique they termed 'reverse transfection'. Plasmid expression vectors containing full-length cDNAs were complexed with a lipid transfection reagent and then printed at a high density on a glass slide. The slide is placed in a cell culture plate in which viable cells are grown. These cells will eventually cover the plasmid microarray with a lawn of adherent cells. Cells which are growing on top of the DNA spots are transfected, while other cells are not, resulting in expression of specific proteins in spatially distinct groups of cells (Fig. 20.3). The phenotypic effects of this 'reverse transfection' of hundreds of genes can be detected using specific cell-based bioassays. (see also Chaps. 17 and 18 of this book).

Ziauddin and Sabatini [26] showed that this cell-based array system using cDNAs as transgenes can identify drug–target interactions and evaluate phenotypic changes resulting from the expression of specific proteins in the cells.

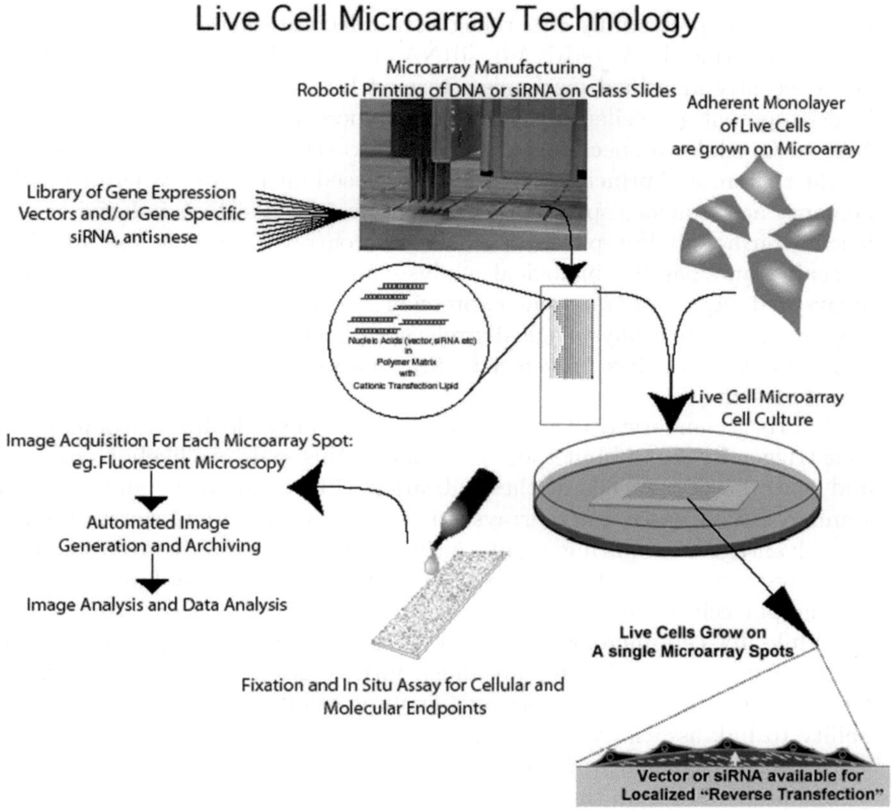

Fig. 20.3. Live cell microarray technology. Live cell arrays are created starting with a library of vectors or siRNAs which are printed onto glass slides within a polymer matrix containing cationic lipid transfection agent. Slides are placed into a tissue culture dish onto which live cells are placed to grow as a monolayer on the slides. The cells growing over the spotted vectors or siRNAs are transfected and are assayed for cellular and molecular endpoints of interest. Finally image acquisition by fluorescent microscopy is followed by image analysis and archiving

The power of this technology lies in the parallel nature and miniaturization of gene transfer into live cells for analysis of the molecular and the phenotypic effects that the expression of specific transgenes have. This method depends on the availability of libraries of full-length genes in expression vectors. It is likely that this limitation will be removed as various applied genomics programs, such as the FLEX database at the Harvard Institute of Proteomics (http://134.174.168.120/YFlex/wall) are completed.

An exciting future potential alternative of live cell expression microarray based technology lies in the specific silencing of genes in a sequence-specific manner. The concept is to use a live cell microarray like platform for the inhibition of gene expression by either single stranded antisense oligonucleotides, or small interfering RNAs (siRNAs). siRNA are RNA duplexes [27–29] that trigger a recently identified mechanism termed RNA interference (RNAi) which leads to potent gene silencing. Many researchers are now routinely using siRNAs to knockdown specific genes in order to study their function. We have conducted proof of principle live cell RNAi based microarray experiments that demonstrate sequence specific and spatially confined siRNA induced gene silencing on a well–less platform. RNAi microarrays are ideal for functional screening and parallel biological analysis and may have an advantage over arrays making use of transgene expression as over–expression of a given gene may not generate a physiologically relevant phenotype whereas the inhibition of gene expression has proven a highly successfully method for delineating gene function.

Efforts are underway to generate human genome wide libraries of molecules that trigger RNAi [30] but these reagents on this scale are likely to be costly and plate based analysis of these libraries will be expensive and time consuming. RNAi based microarrays on a miniaturized platform would have the advantage of requiring significantly less material than conventional well based systems and can be easily adapted for a broad range of functional, high throughput cell-based assays.

While live cell microarray technology using either overexpression or inhibition of gene expression require much further development, their potential for enabling genomic scale functional analysis could significantly speed up our ability to link associative gene expression data with a functional effect. One of the biggest challenges for either type of live cell array will be extracting quantitative data from the cells on the microarray spots. Traditional scanners do not provide the resolution required to extract single cell level information and it may be necessary to apply automated high content screening based instrumentation. Fortunately, the development of imaging systems for tissue microarray analysis can be directly applied to imaging of cell microarrays treated with various stains and assays. For example, a fluorescent microscopy system fitted with automated stage control for high throughput fluorescent image acquisition of DNA FISH of tissue microarrays (Fig. 20.4) can easily be adapted and utilized for capturing images from fluorescent endpoints on cells sitting on cell transfection microarrays. Similarly, data management systems developed for tissue microarray images and image analysis can be directly modified and adapted for the needs of cell transfection microarrays.

20.5 Conclusions

High throughput genomic and proteomic screening technologies have led to a massive increase in the rate of data generation, greatly exceeding the rate at which biological significance and clinical relevance can be determined. The consequence of the new discovery technologies is that the validation of tar-

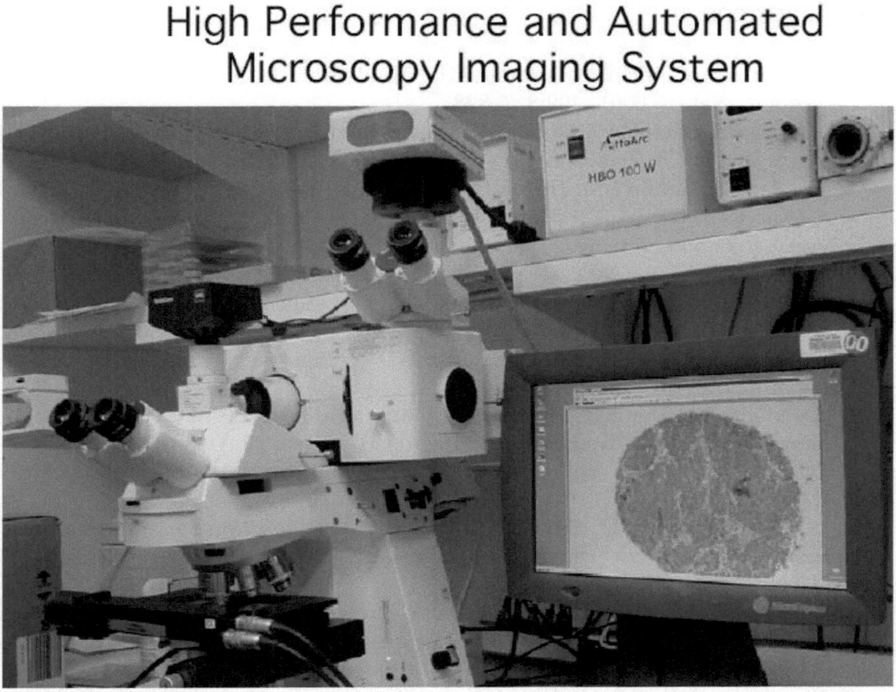

Fig. 20.4. High performance and automated microscopy imaging system

gets has become the rate–limiting step in translating genomic and proteomic information to clinical and therapeutic applications. This limitation has hindered the promise of new biological insight and medical discoveries resulting from the completion of the Human Genome Project. We have presented a two–stage microarray based validation strategy, which can follow the analysis of gene expression patterns with cDNA microarrays: a clinical validation using tissue microarrays for the analysis of the clinical significance of alterations in candidate gene targets, and a functional validation using cell-based arrays for high throughput knockdown of gene targets. Although these solid phase platforms differ in many ways, DNA, tissue, and live cell transfection microarrays have some common unifying themes, including high throughput, miniaturization, and a the parallel nature of data generation. These different microarray based approaches can be integrated into translational genomics systems to greatly increase the flow of information from the genome to the bed–side.

References

1. Birney E, Bateman A, Clamp ME, Hubbard TJ. *Mining the draft human genome.* Nature 2001, 409:827–828
2. Collins FS, Patrinos A, Jordan E, Chakravarti A, Gesteland R, Walters L. *New goals for the US Human Genome Project.* 1998–2003. Science 1998, 282:682–689
3. Lander ES, Linton LM, Birren B, Nusbaum C, Zody MC, Baldwin J, Devon K, Dewar K, Doyle M, FitzHugh W et al.. *Initial sequencing and analysis of the human genome.* Nature 2001, 409:860–921
4. Venter JC, Adams MD, Myers EW, Li PW, Mural RJ, Sutton GG, Smith HO, Yandell M, Evans CA, Holt RA et al.. *The sequence of the human genome.* Science 2001, 291:1304–1351
5. Wolfsberg TG, McEntyre J, Schuler GD. *Guide to the draft human genome.* Nature 2001, 409:824–826
6. Lockhart DJ, Dong H, Byrne MC, Follettie MT, Gallo MV, Chee MS, Mittmann M, Wang C, Kobayashi M, Horton H, Brown EL. *Expression monitoring by hybridization to high–density oligonucleotide arrays.* Nat Biotechnol 1996, 14:1675–1680
7. Schena M, Shalon D, Davis RW, Brown PO. *Quantitative monitoring of gene expression patterns with a complementary DNA microarray.* Science 1995, 270:467–470
8. Schena M, Shalon D, Heller R, Chai A, Brown PO, Davis RW. *Parallel human genome analysis: microarray–based expression monitoring of 1000 genes.* Proc Natl Acad Sci USA 1996, 93:10614–10619
9. Shoemaker DD, Schadt EE, Armour CD, He YD, Garrett-Engele P, McDonagh PD, Loerch PM, Leonardson A, Lum PY, Cavet G et al.. *Experimental annotation of the human genome using microarray technology.* Nature 2001, 409:922–927
10. Bittner M, Meltzer P, Chen Y, Jiang Y, Seftor E, Hendrix M, Radmacher M, Simon R, Yakhini Z, Ben-Dor A et al.. *Molecular classification of cutaneous malignant melanoma by gene expression profiling.* Nature 2000, 406:536–540
11. DeRisi J, Penland L, Brown PO, Bittner ML, Meltzer PS, Ray M, Chen Y, Su YA, Trent JM. *Use of a cDNA microarray to analyse gene expression patterns in human cancer.* Nat Genet 1996, 14:457–460
12. Golub TR, Slonim DK, Tamayo P, Huard C, Gaasenbeek M, Mesirov JP, Coller H, Loh ML, Downing JR, Caligiuri MA et al.. *Molecular classification of cancer: class discovery and class prediction by gene expression monitoring.* Science 1999, 286:531–537
13. Perou CM, Sorlie T, Eisen MB, van de Rijn M, Jeffrey SS, Rees CA, Pollack JR, Ross DT, Johnsen H, Akslen LA et al.. *Molecular portraits of human breast tumours.* Nature 2000, 406:747–752
14. Kallioniemi OP, Wagner U, Kononen J, Sauter G. *Tissue microarray technology for high–throughput molecular profiling of cancer.* Hum Mol Genet 2001, 10:657–662
15. Kononen J, Bubendorf L, Kallioniemi A, Bärlund M, Schraml P, Leighton S, Torhorst J, Mihatsch MJ, Sauter G, Kallioniemi OP. *Tissue microarrays for high–throughput molecular profiling of tumor specimens.* Nat Med 1998, 4:844–847

16. Kononen J, Hostetter G, Sauter G, Kallioniemi OP. *Construction of tissue microarrays.* In A Companion to Molecular Cloning, vol 4, edn 3. Edited by Bowtell D, Sambrook J. Cold Spring Harbor: Cold Spring Harbor Laboratory Press, 2001: in press
17. Bärlund M, Forozan F, Kononen J, Bubendorf L, Chen Y, Bittner ML, Torhorst J, Haas P, Bucher C, Sauter G et al.. *Detecting activation of ribosomal protein S6 kinase by complementary DNA and tissue microarray analysis.* J Natl Cancer Inst 2000, 92:1252–1259
18. Moch H, Schraml P, Bubendorf L, Mirlacher M, Kononen J, Gasser T, Mihatsch MJ, Kallioniemi OP, Sauter G. *High–throughput tissue microarray analysis to evaluate genes uncovered by cDNA microarray screening in renal cell carcinoma.* Am J Pathol 1999, 154:981–986
19. Sugita M, Geraci M, Gao B, Powell RL, Hirsch FR, Johnson G, Lapadat R, Gabrielson E, Bremnes R, Bunn PA and Franklin WA. *Combined use of oligonucleotide and tissue microarrays identifies cancer/testis antigens as bio-markers in lung carcinoma.* Cancer Res 2002;62(14): 3971–9
20. Sallinen SL, Sallinen PK, Haapasalo HK, Helin HJ, Helen PT, Schraml P, Kallioniemi OP, Kononen J. *Identification of differentially expressed genes in human gliomas by DNA microarray and tissue chip techniques.* Cancer Res 2000, 60:6617–6622
21. Ginestier C, Charafe-Jauffret E, Bertucci F, Eisinger F, Geneix J, Bechlian D, Conte N, Adelaide J, Toiron Y, Nguyen C, Viens P, Mozziconacci MJ, Houlgatte R, Birnbaum D, Jacquemier J. *Distinct and complementary information provided by use of tissue and DNA microarrays in the study of breast tumor markers.* Am J Path 2002;161(4):1223–33
22. Mousses, S., Bubendorf, L., Wagner, U., Hostetter, G., Kononen, J., Cornelison, R., Goldberger, N., Elkahloun, A.G., Willi, N., Koivisto, P., Ferhle, W., Raffeld, M., Sauter, G., Kallioniemi, OP. *Clinical validation of candidate genes associated with prostate cancer progression in the CWR22 model system using tissue microarrays.* Cancer Research 2002; 62(5): 1256–60
23. Dhanasekaran SM, Barrette TR, Ghosh D, Shah R, Varambally S, Kurachi K, Pienta KJ, Rubin MA, Chinnaiyan AM. *Delineation of prognostic bio-markers in prostate cancer.* Nature 2001, 412:822–826
24. Toby GG, Golemis EA. *Using the yeast interaction trap and other two–hybrid-based approaches to study protein–protein interactions.* Methods 2001, 24:201–217
25. Vidal M, Endoh H. *Prospects for drug screening using the reverse two–hybrid system.* Trends Biotechnol 1999, 17:374–381
26. Ziauddin J, Sabatini DM. *Microarrays of cells expressing defined cDNAs.* Nature 2001, 411:107–110.1
27. Caplen NJ, Parrish S, Imani F, Fire A, Morgan RA. *Specific inhibition of gene expression by small double–stranded RNAs in invertebrate and vertebrate systems.* Proc Natl Acad Sci USA 2001, 98:9742–9747
28. Elbashir SM, Lendeckel W, Tuschl T. *RNA interference is mediated by 21– and 22–nucleotide RNAs.* Genes Dev 2001, 15:188–200
29. Elbashir SM, Harborth J, Lendeckel W, Yalcin A, Weber K, Tuschl T. *Duplexes of 21–nucleotide RNAs mediate RNA interference in cultured mammalian cells.* Nature 2001, 411:494–498
30. Hannon GJ. *RNA interference.* Nature 2002;418:244–51

31. Khan J, Wei JS, Ringner M, Saal LH, Ladanyi M, Westermann F, Berthold F, Schwab M, Antonescu CR, Peterson C, Meltzer PS. *Classification and diagnostic prediction of cancers using gene expression profiling and artificial neural networks.* Nat Med 2001, 7:673–679

Index

MIX
Papier aus verantwortungsvollen Quellen
Paper from responsible sources
FSC® C105338

If you have any concerns about our products,
you can contact us on
ProductSafety@springernature.com

In case Publisher is established outside the EU,
the EU authorized representative is:
Springer Nature Customer Service Center GmbH
Europaplatz 3, 69115 Heidelberg, Germany

Printed by Libri Plureos GmbH
in Hamburg, Germany